T0270955

Systematic Design of Analog CMOS Circuits

Using Pre-Computed Lookup Tables

Discover a fresh approach to efficient and insight-driven analog integrated circuit design in nanoscale-CMOS with this hands-on guide.

- Expert authors present a sizing methodology that employs SPICE-generated lookup tables, enabling close agreement between hand analysis and simulation.
- Illustrates the exploration of analog circuit tradeoffs using the g_m/I_D ratio as a central variable in script-based design flows, captured in downloadable Matlab code.
- Includes over forty detailed worked examples, including the design of low-noise and low-distortion gain stages, and operational transconductance amplifiers.

Whether you are a professional analog circuit designer, a researcher, or a graduate student, this book will provide you with the theoretical know-how and practical tools you need to acquire a systematic and re-use oriented design style for analog integrated circuits in modern CMOS.

Paul G. A. Jespers is a Professor Emeritus of the Université Catholique de Louvain and a Life Fellow of the IEEE.

Boris Murmann is a Professor of Electrical Engineering at Stanford University, and a Fellow of the IEEE.

"Analog design generates insight, but requires expertise. To build up such expertise, analytic models are used to create design procedures. Indeed, analytic models easily allow device sizing from specifications. They lack accuracy, however. The models of present-day nanometer MOS transistors have become rather complicated. On the other hand SPICE simulations do provide the required accuracy but don't generate as much insight. The use of SPICE-generated lookup tables, as described in this book, provides an excellent compromise. The accuracy is derived from SPICE and the design procedure itself is made through MATLAB employing parameters like g_m/I_D. As a result a considerable amount of intuition can be built up. Such design procedure is highly recommended to whoever wants to gain insight by doing analog design, without losing the accuracy of real SPICE simulations."

Willy Sansen, KU Leuven

"With the advent of sub-micron MOS transistors more than two decades ago, traditional design based on the square-law model is no longer adequate. Alternatives such as 'tweaking' with SPICE or relying on more sophisticated device models do not provide the circuit insight necessary for optimized design or are too mathematically complex.

The design methodology presented in this book overcomes these shortcomings. A focus on fundamental design parameters – dynamic range, bandwidth, power dissipation – naturally leads to optimized solutions, while relying on transistor data extracted with the simulator ensures agreement between design and verification. Comprehensive design examples of common blocks such as OTAs show how to readily apply these concepts in practice.

This book fixes what has been broken with analog design for more than twenty years. I recommend it to experts and novices alike."

Bernhard Boser, University of California, Berkeley

"The authors present a clever solution to capture the precision of the best MOSFET models, current or future, in a comprehensive and efficient design flow compatible with nanometric CMOS processes. In this book, you will also enjoy a wealth of invaluable information to deepen your analog design skills."

Yves Leduc, Polytech Nice Sophia

Systematic Design of Analog CMOS Circuits

Using Pre-Computed Lookup Tables

PAUL G. A. JESPERS
Université Catholique de Louvain, Belgium

BORIS MURMANN
Stanford University

CAMBRIDGE
UNIVERSITY PRESS

CAMBRIDGE
UNIVERSITY PRESS

University Printing House, Cambridge CB2 8BS, United Kingdom

One Liberty Plaza, 20th Floor, New York, NY 10006, USA

477 Williamstown Road, Port Melbourne, VIC 3207, Australia

314-321, 3rd Floor, Plot 3, Splendor Forum, Jasola District Centre, New Delhi - 110025, India

103 Penang Road, #05-06/07, Visioncrest Commercial, Singapore 238467

Cambridge University Press is part of the University of Cambridge.

It furthers the University's mission by disseminating knowledge in the pursuit of
education, learning and research at the highest international levels of excellence.

www.cambridge.org
Information on this title: www.cambridge.org/9781107192256
DOI: 10.1017/9781108125840

First published 2017

A catalogue record for this publication is available from the British Library

ISBN 978-1-107-19225-6 Hardback

Additional resources for this title are available at www.cambridge.org/Jespers

Contents

Symbols and Acronyms

A_v	Small-signal voltage gain
A_{v0}	Low frequency small-signal voltage gain
A_{intr}	Intrinsic gain
A_{VT}	Pelgrom coefficient for threshold voltage mismatch
A_β	Pelgrom coefficient for current factor mismatch
ACM	Advanced Compact Model
CLM	Channel Length Modulation
CSM	Charge Sheet Model
C	Capacitor value
C_{ox}	Oxide capacitance per unit area
C_{gb}	Gate-to-bulk capacitance
C_{gd}	Gate-to-drain capacitance
C_{gs}	Gate-to-source capacitance
C_j	Junction capacitance
C_C	Compensation capacitance
CMOS	Complementary Metal Oxide Semiconductor
C_{self}	Self-loading capacitance of an amplifier
D	Diffusion constant
DIBL	Drain-Induced Barrier Lowering
EKV	Enz, Krumenacher and Vittoz compact model
FO	Fan-out (ratio between load and input capacitances of a circuit)
f	Frequency in Hz
f_c	Cutoff frequency (-3dB frequency)
f_T	Transit frequency
f_u	Unity gain frequency (where $|A_v| = 1$)
g_{ds}	Output conductance
g_m	Gate transconductance
g_{mk}	k^{th} derivative of I_D with respect to V_{GS}
g_{mb}	Bulk transconductance
g_{ms}	Source transconductance
HD$_2$, HD$_3$	Fractional harmonic distortion of order 2, 3, ...
i	Normalized drain current
IGS	Intrinsic Gain Stage

I_D	DC drain current
I_S	Specific current
I_{Ssq}	Square specific current ($W = L$)
I_{Su}	Unary specific current ($W = 1$ μm)
J_D	Drain current density (I_D/W)
L	Gate length
N	Impurity concentration
n	Subthreshold slope factor
q	Normalized mobile charge density
q_S, q_D	Normalized mobile charge density at the source and drain
Q_i	Mobile charge density
RHP	Right Half Plane
S_{VT0}	Threshold voltage sensitivity factor with respect to V_{DS}
S_{IS}	Specific current sensitivity factor with respect to V_{DS}
U_T	Thermal voltage kT/q
V_X	DC voltage component at node x
v_x	AC voltage component at node x
v_X	Total voltage at node x, $v_X = V_X + v_x$
V_{EA}	Early voltage
V_I	DC component of input voltage
v_i	AC component of input voltage
v_I	Total input voltage $v_I = V_I + v_i$
v_{id}	Differential input voltage, AC component
V_S, V_G, V_D	Source, gate and drain voltage with respect to bulk (DC)
V_{GS}, V_{DS}	Gate and drain voltage with respect to the source (DC)
v_{gs}, v_{ds}	Incremental gate and drain voltage with respect to the source
$v_{gs,pk}, v_{ds,pk}$	Incremental gate and drain voltage amplitude (sinusoid)
V_P	Pinch-off voltage with respect to the bulk
V_{Dsat}	Drain saturation voltage
v_{sat}	Saturation velocity of mobile carriers
V_T	Threshold voltage
V_{OV}	Gate overdrive voltage, $V_{GS} - V_T$
W	Transistor width
WI, MI, SI	Weak, moderate and strong-inversion
β	Current factor[1] ($\mu C_{ox} W/L$)
γ	Backgate effect parameter
γ_n, γ_p	Thermal noise factor for n-channel and p-channel devices[2]
μ	Mobility
μ_o	Low-field mobility
ρ	Normalized transconductance efficiency

---

[1] The symbol β is also used to denote the feedback factor in amplifier circuits. The distinction is usually clear from the context.

[2] The distinction from the backgate effect parameter γ is usually clear from the context.

ψ_S Surface potential
ω Angular frequency ($2\pi f$)
ω_c Angular cutoff frequency ($2\pi f_c$)
ω_T Angular transit frequency ($2\pi f_T$)
ω_{Ti} Angular transit frequency considering only C_{gs} (instead of C_{gg})

1 Introduction

1.1 Motivation

Since the 1960s, the square-law model for complementary metal-oxide-semiconductor (CMOS) transistors has been used extensively to analyze and design analog and digital integrated circuits. An advantage of the square-law equations is that they are easy to derive from basic solid-state physics, algebraically simple and yet useful for gaining insight into basic CMOS circuit behavior. As a result, the square-law model remains useful as a "warm-up tool" for students in circuit design, and it is featured in all popular analog integrated circuit textbooks (examples include [1], [2]).

On the other hand, it is well known that the square-law MOS model is plagued by several limitations, especially when it comes to short-channel transistors:

- Modern MOSFETs are impaired by numerous mobility degradation effects, related to their short channel length, thin gate oxide and their generally more complex structure and doping profiles. In strong inversion, with gate overdrive voltages ($V_{GS} - V_T$) of several hundred millivolts, the error in the transconductance predicted by square-law models with constant parameters is of the order of 20–60%.
- In moderate inversion, with gate overdrive voltages below 150 mV, the square-law model breaks down altogether and it may be in error by a factor of two or even more. This deficiency applies to all MOSFETs, regardless of channel length. However, the issue has become more pronounced with short channel devices, since moderate inversion represents a design "sweet spot" for a variety of circuits in these technologies [3]–[5].
- In weak inversion (subthreshold operation), the current flows by diffusion (like in a BJT) and the square-law model must be replaced with an exponential I–V relationship.

The above-stated issues are clearly visible in Figure 1.1, which shows the current density plot of a realistic 65-nm transistor, together with exponential and square-law approximations. The exponential provides a reasonably good fit for very low V_{GS} (weak inversion) and the quadratic approximation begins to make sense a few hundred millivolts above the device's threshold voltage (vertical dashed line). The transition from weak to strong inversion should ideally be smooth and continuous,

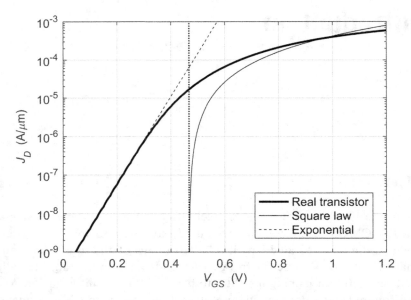

Figure 1.1 Current density of a minimum-length n-channel device in 65-nm CMOS technology versus V_{GS}. The dotted vertical line corresponds to the device's threshold voltage (see Chapter 2 for further details).

but finding a physical relationship that bridges the exponential and square-law approximations turns out to be non-trivial. In addition, at very large V_{GS}, the current density of the real device and the quadratic model diverge again due to the mentioned mobility degradation effects.

The above-stated modeling limitations are a great nuisance when it comes to design, since the square-law hand calculations described in textbooks typically won't match simulations for a classical flow (see Figure 1.2). Modern circuit simulation relies on complex device models such as BSIM6 [6] or PSP[1] [7], which are carefully crafted to reflect the "real" device characteristic in Figure 1.1. The result is a significant disconnect between hand analysis and simulation results, and consequently, designers tend to shy away from hand-calculations and resort to a design style built on iterative and time-consuming SPICE-based "tweaking."

There are several issues with the iterative simulation-based design of analog circuits. The problem is that the designer loses insight about the tradeoffs as well as the ability to sanity-check the results. While an equation-based design can reveal fundamental issues with a topology and help the designer advance his or her circuit architecture, it is difficult to gain knowledge about the fundamental limits of a circuit via repetitive sizing and simulation. What used to be design now resembles reverse engineering, which is highly undesirable for anyone who is interested in leading-edge innovation.

---

[1] The PSP compact MOSFET was developed by Philips Semiconductors and Penn State University.

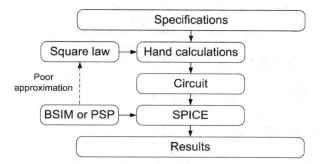

Figure 1.2 Typical analog circuit design flow based on square-law hand calculations SPICE simulation using advanced models.

The second issue is that highly iterative design based on SPICE "tweaking" is typically incompatible with the time-to-market pressure seen in today's IC develop-ments. As a response to this problem, universities and EDA vendors have created solutions that look to automate the iterative process, leveraging the vast amount of computing power available today. The work of [8] provides an extensive reference list of such programs and categorizes them as full design automation (FDA) tools. While an FDA approach can help overcome the design time issue, it comes with the same problem as manual tweaking: It is even more difficult for the designer to gain analytical insight and intuition, which is an important ingredient for topology selection and innovation.

Taking a step back, we note that the key problem is not the equation set that describes the circuit, which tends to be either amenable to manual derivation or available in standard textbooks and publications. The main issue lies in link-ing the device sizing parameters (geometries and bias currents) to the transis-tor's representation within the circuit, typically in form of a small-signal model. Therefore, while using FDA can be appropriate and justified in some cases, it goes one step further than required, providing full automation at the expense of analytical insight.

The design approach described in this book falls under the category of full design handcrafting (FDH) [8]. It builds on classical hand analysis methods and eliminates the gap between hand analysis and complex transistor behavior using SPICE-gen-erated lookup tables (see Figure 1.3). The tables contain the transistor's equiva-lent small-signal parameters (g_m, g_{ds}, etc.) across a multi-dimensional sweep of the MOSFET's terminal voltages. Since using the lookup table data closely captures the behavior of the SPICE model, the approximation issues of Figure 1.2 are elimi-nated and it is possible to achieve close agreement between the desired specs and the simulated performance without iterative tweaking. Though in some cases the cal-culations can literally be done by hand, it is usually more efficient to implement the design flow through a computer script. In this book, we chose the popular Matlab® environment for designing such scripts.

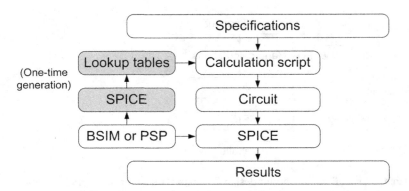

Figure 1.3 Analog circuit design flow used in this book.

It is worth noting that the outlined approach does not resemble the "SPICE in the loop" approach [9], [10] advocated in the 1980s. The main differences are: (1) The lookup tables are created once and stored permanently; they do not get updated with each circuit simulation run. (2) The design scripts tend to use abstract and simplified circuit models. This often means that the designer does not need to worry about auxiliary circuits that may be required to get a SPICE simulation to work. For example, it is possible to create a design script that evaluates the small-signal performance of an amplifier under the assumption that the bias point is perfectly set. How exactly that bias point is established can be determined later, after studying the first-order performance tradeoffs.

To implement the design flow of Figure 1.3, we need the following ingredients:

- A convenient way to generate and access the lookup table data. The generation of the proposed lookup table format is described in Appendix 2. Examples on how to access and use the stored data are given throughout this book (including an introductory example in Section 1.2.2).
- A suitable way to translate the design problem into a script that helps us study the key tradeoffs and ultimately computes the final device sizes. Most of this book is dedicated to this part of the flow. By means of examples, we study design problems of varying complexity and the derived scripts can form the basis for future design problems that the reader will encounter.

A key aspect of the proposed methodology is that we interpret and organize the lookup table data based on the transistor's inversion level, employing the transconductance efficiency g_m/I_D as a proxy, and key parameter for design. This metric captures a device's efficiency in translating bias current to transconductance and spans nearly the same range in all modern CMOS processes (~3...30 S/A). When combined with other figures of merit (g_m/C_{gg}, g_m/g_{ds}, etc.), thinking in terms of g_m/I_D allows us to study the tradeoffs between bandwidth, noise, distortion and power dissipation in a normalized space. The final bias currents and device sizes follow from a straightforward de-normalization step using the current density (I_D/W). We will take a first look at this normalized design approach in Section 1.2.2.

The idea of g_m/I_D-based design was first articulated by Silveira, Jespers et al. in 1996 [11]. Since then, the approach has been continuously refined through academic research (see e.g. [12]–[17]) and is being taught at various universities. Several companies known to the authors have integrated lookup table based design into their design environments. These efforts were driven by the first set of graduates being exposed to the methodology in school. Despite this growing popularity, much needs to be done to make the approach accessible to a broader community and specifically those engineers who have not acquired the material at the university. The goal of this book is to provide a comprehensive resource that will accomplish this.

It is important to note that several other authors have made contributions toward a design methodology that follows the spirit of full design handcrafting with bridges between hand analysis and simulation. Among them are the inversion coefficient (IC) based flows by Binkley [18], Enz [19], and Sansen [20] as well as the 2010 g_m/I_D-centric book by Jespers [21]. The main difference between these works and the present book is that they are still based on analytical device models. Instead of working with purely numerical lookup table data, these methodologies assume that the transistor characteristics can be fit to model equations (typically EKV [22]) that are more complex than the square-law model, but not too complex to be included in a design script environment. This approach is certainly workable for today's mainstream technologies, but we decided to go with a sizing approach that is agnostic to the increasingly complex physical behavior of nanoscale transistors. Despite this goal, we still make use of the EKV model to build intuition, but won't use it to compute the ultimate device sizes. This approach is made transparent in Chapters 2–4.

1.2 The Analog Circuit Sizing Problem and the Proposed Approach

Before outlining the remainder of this book, we feel that it is important to provide a short (and simplified) walk-through of the proposed design methodology. For this purpose, we assume that the reader is familiar with CMOS square-law design and we use the shortcomings of the square law to motivate our approach.

Generally, the types of analog circuits that we consider in this book fall into the class-A category, which means that they are operated with constant bias current. A basic example is the differential pair shown in Figure 1.4, which is usually part of a larger circuit (not shown for simplicity). Sizing the circuit in Figure 1.4 means that the designer must find suitable values for

- the bias current I_D;
- the device width W;
- the channel length L.

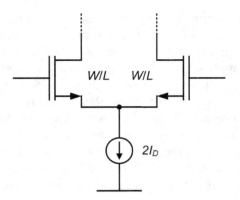

Figure 1.4 Differential pair.

For this introduction, we will assume that through some design process, we determined that the differential pair should implement a certain transconductance (g_m). How does this requirement translate into values for I_D, W and L? We will initially approach this question using simple square-law expressions and then refine our treatment to arrive at the proposed methodology.

1.2.1   Square-Law Perspective

Ideally, we would like to have an equation that relates all relevant parameters of the above example with one another. The square-law model used in standard textbooks provides such an expression [1]:

$$g_m = \sqrt{2\mu C_{ox} \frac{W}{L} I_D}. \tag{1.1}$$

Even though this formula is inaccurate for modern devices, it clarifies a basic, and generally important, point: there are an infinite number of choices for W, L and I_D that all lead to the design goal of realizing a certain value of g_m. To continue, we need a feel for the tradeoff that we are making by picking one of these many solutions.

To make progress, let us articulate what we would ideally like to achieve: We want to meet the design goal using the lowest possible current and the smallest possible transistor size. In absence of any other constraints (to be considered in later chapters), this immediately implies that we should use the smallest available channel length L (for example, $L_{min} = 60$ nm for the technology used in this book).

The remaining question is whether we should minimize the current or the device width. Note that achieving both simultaneously is not possible, since the product $W \times I_D$ is fixed. To think about this tradeoff systematically, we introduce two figures of merit that relate the design objective (g_m) to the variables that we want to minimize:

$$\frac{g_m}{I_D} \quad and \quad \frac{g_m}{W}. \tag{1.2}$$

Using the standard textbook square-law expressions [1], we can write these figures of merit as:

$$\frac{g_m}{I_D} = \frac{2}{V_{OV}}. \tag{1.3}$$

$$\frac{g_m}{W} = \frac{\mu C_{ox}}{L} V_{OV}. \tag{1.4}$$

Here, $V_{OV} = V_{GS} - V_T$ is the quiescent point gate overdrive voltage of the transistor. Physically, large V_{OV} means that the channel is more strongly inverted, i.e. more inversion charge is present underneath the gate.

The key observation from the above equations is that the gate overdrive controls how efficiently we employ current (I_D) or device width (W) to generate the desired transconductance. The designer can pick a large V_{OV} to arrive at a small device width or a small V_{OV} to minimize the current. Thus, the gate overdrive voltage can be viewed as a "knob" (see Figure 1.5) that fully defines the sizing tradeoff. Also, note that once V_{OV} has been chosen, the required device current (for a given g_m) follows directly from (1.3); no technology-specific parameters are needed (assuming the square law holds).

In addition, the choice of V_{OV} sets the minimum V_{DS} for which the transistor remains saturated ($V_{Dsat} = V_{OV}$ for a square-law device) and it also determines the

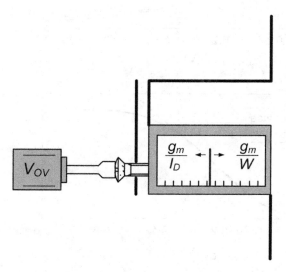

Figure 1.5 The gate overdrive voltage V_{OV} is a "knob" that lets us control the tradeoff between g_m/I_D and g_m/W.

circuit's linearity [23]. It is therefore not surprising that the gate overdrive is among the most important parameters used in square-law centric circuit optimization. For example, the seminal work by Shaeffer and Lee [24] studied the relationship between the gate overdrive voltage and the bandwidth, noise and linearity of a low-noise amplifier (LNA). It was found that the tradeoff between these performance metrics is fixed once a certain V_{OV} is chosen.

Unfortunately, and as already discussed in Section 1.1, the square-law model has become obsolete for design with modern MOS transistors. To see this, consider Figure 1.6, which plots the figures of merit of (1.2) for a minimum-length n-channel device in 65-nm CMOS. For reasons discussed in Chapter 2, the square-law expressions fit reasonably well only for a narrow range of gate overdrives in strong inversion (say $V_{OV} = 0.2...0.4$ V). Thus, (1.3) and (1.4) do not accurately link V_{OV} with g_m/I_D and g_m/W and the expressions are consequently unsuitable for design in the given 65-nm technology.

One way to solve this problem is to develop a more sophisticated equation set that can capture the physics of a modern device more accurately. However, as already explained, we want to eliminate the undesired tradeoff between algebraic model complexity and adequacy for design using a numerical approach.

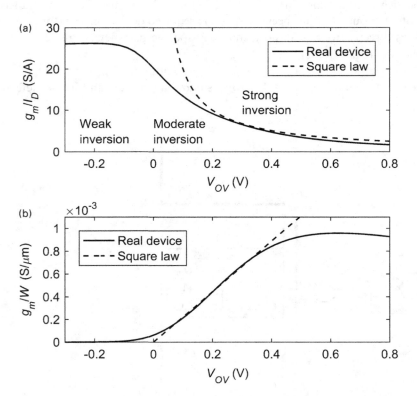

Figure 1.6 g_m/I_D and g_m/W for a minimum-length ($L = 60$ nm) n-channel in 65 nm CMOS. $V_{DS} = 1$ V and $V_{SB} = 0$ V.

1.2.2 Capturing the Tradeoffs Using Lookup Tables

This book advocates lookup tables to quantify the tradeoff between the relevant device figures of merit in each design (including, but not limited to g_m/I_D and g_m/W). The lookup tables can be generated (once) using a SPICE-like circuit simulator and the data can be stored in a file for future use (see Appendix 2). This is illustrated in Figure 1.7. Starting with a "well-calibrated" model file from the silicon foundry, we perform DC sweeps (and noise simulations) in four dimensions (L, V_{GS}, V_{DS} and V_{SB}) and tabulate all relevant device parameters along these sweeps for a fixed device width W. While one could in principle include the device width as a fifth sweep variable, this is not necessary since the parameters scale (approximately) linearly with W across the typical range encountered in analog design. We validate this important assumption of width independence for parameters like g_m/I_D, g_m/W, etc., in Appendix 3.

The sweeps can be repeated for all devices offered by a given foundry, leading to one lookup table per transistor type. With this flow, the quality of the lookup table data is of course directly linked to the quality of the foundry models. If the foundry models are poor, it will be challenging to produce a working circuit altogether, independent of which tool set and sizing methodology is used. Model quality assurance is therefore outside the scope of this book. However, proper inspection of the lookup table data (leveraging the physical intuition conveyed in Chapter 2) can sometimes expose modeling issues. For example, the shape of the g_m/I_D curve in Figure 1.6 may reveal discontinuities, improper location of the weak inversion plateau, etc.

To make the lookup table data easily accessible in Matlab, we have created a function (called "lookup") that allows us to read all transistor model parameters as a function of the applied bias voltages (with interpolation capabilities). Further details on this function are found in Appendix 2. To give the reader a basic feel for its usage, we provide two simple examples from the Matlab command line[2]:

```
>> ID = lookup(nch, 'ID', 'VGS', 0.7, 'VDS', 0.5, 'VSB', 0,
'L', 0.06)
ID =
9.3127e-04
>> Cgs = lookup(nch, 'CGS', 'VGS', 0.7, 'VDS', 0.5, 'VSB', 0,
'L', 0.06)
Cgs =
7.0461e-15
```

To see the device width (in microns) for which this data was tabulated, one can type:

```
>> nch.W
ans =
10
```

---

[2] In the given example, the gate length L is specified in microns. The argument "nch" specifies the device type and points to a Matlab structure containing the data. W is 10 μm, which is the width that was used to create the lookup table data; see Appendix 3 for more information on how this reference width was selected.

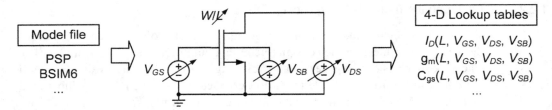

Figure 1.7 Lookup table generation using a four-dimensional SPICE sweep. The width W is set to 10 μm (5 fingers, 2 μm each) for the lookup tables used in this book.

Once these lookup tables have been generated, the remaining question is: how should the designer organize and use the data to gain insight into circuit sizing tradeoffs? To answer this question, we return to the differential pair example, which established a fundamental tradeoff between g_m/I_D and g_m/W.

One direction we could pursue with the lookup table data is to produce the "real device" curves of Figure 1.6, and pick a value for V_{OV} that dials in a suitable tradeoff between our figures of merit (g_m/I_D and g_m/W). However, since we are not dealing with a device that behaves per the square law, V_{OV} is just a remnant of an obsolete model. Hence, a key concept advocated in this book is to eliminate V_{OV} as a design variable altogether and instead link all design tradeoffs to the choice of g_m/I_D, which (like V_{OV}) can be viewed as a proxy for the device's inversion level.

As explained further in Chapter 2, g_m/I_D ranges from about 3...30 S/A in all modern CMOS technologies, where the lower end corresponds to strong inversion, the mid-range (around 12...18 S/A) amounts to moderate inversion, and the peak value is linked to weak inversion. In addition to indicating the inversion level, g_m/I_D is a useful figure of merit for another reason, as we have already discovered in our differential pair example. It directly quantifies the transconductance per unit of current invested in the device. Therefore, we advocate the unit of S/A (instead of 1/V).

With V_{OV} eliminated, the sizing tradeoff for our differential pair example is elegantly captured in a single plot, shown in Figure 1.8. Picking a small g_m/I_D means that we end up with a large g_m/W, implying a small device for a desired value of g_m. The opposite is true when we opt for a large value of g_m/I_D, where the device will be wider, at the benefit of reduced current. Note that the quantities plotted in Figure 1.8 can be extracted from the lookup table data introduced above. Each point of the tradeoff curve corresponds to a different V_{GS} value in the sweep. However, the exact value of V_{GS} for the desired tradeoff point (the chosen g_m/I_D) tends to be secondary from an optimization perspective. In our example, we only care about how much current and area we are investing to realize a certain value of g_m. More generally, we will in fact see throughout this book that width-independent parameters play a fundamental role in systematic tradeoff studies and circuit sizing.

To simplify the lookup of parameter ratios, the Matlab function introduced above has another usage mode that lets us directly fetch one ratio as a function of another. Below is a simple example that visualizes this functionality:

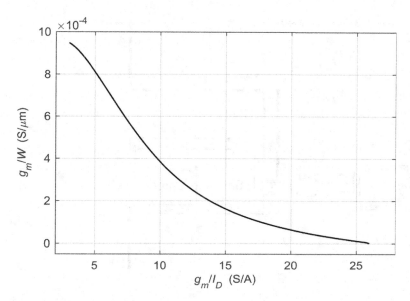

Figure 1.8 The tradeoff between g_m/I_D and g_m/W for an n-channel device with $L = 60$ nm, $V_{DS} = 1$ V and $V_{SB} = 0$ V.

```
>> gm_W = lookup(nch, 'GM_W', 'GM_ID', 10, 'VDS', 0.5, 'VSB', 0,
'L', 0.06)
gm_W =
  3.5354e-04
```

This example finds g_m/W for $g_m/I_D = 10$ S/A, $V_{DS} = 0.5$ V, $V_{SB} = 0$ V and $L = 0.06$ μm.

1.2.3   Generalization

In the above discussion, we have highlighted the explicit link and tradeoff between g_m/I_D and g_m/W. However, it turns out that g_m/I_D not only controls g_m/W, but also a variety of other width-independent quantities that analog circuit designers care about. We therefore promote g_m/I_D as "the" knob for analog design (see Figure 1.9), like the way V_{OV} has played a central role in square-law design.

Table 1.1 provides a (non-exhaustive) preview of the kinds of metrics that we will consider in this book. The first row in this table is g_m/W, which we have already discussed. Making g_m/I_D large means that g_m/W deteriorates. The second row is the current density of the transistor, I_D/W. The current density simply quantifies how many microamps per micron we must inject into the device to operate at the desired inversion level (represented by g_m/I_D). We will typically use this parameter in the other direction, i.e. given g_m/I_D, we can compute how wide the device must be to produce a certain g_m (see sizing example below). Obviously, g_m/W contains the same info, and so it is purely a matter of taste which parameter is used for sizing. Most

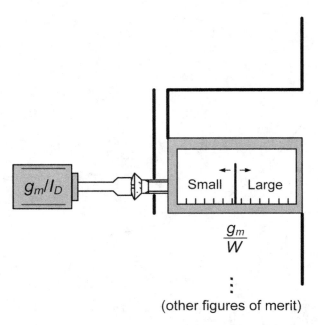

(other figures of merit)

Figure 1.9 The transconductance efficiency g_m/I_D replaces the gate overdrive voltage as the "knob" that lets us control g_m/W and several other figures of merit.

designers prefer to think in terms of current density ($J_D = I_D/W$), since this quantity also has important implications for long-term reliability (electromigration, etc.).

The last two rows offer performance parameters related to gain and bandwidth. The transistor's intrinsic voltage gain g_m/g_{ds} becomes larger as we push the device toward weak inversion. Unfortunately, at the same time, the device's transit frequency ($f_T = g_m/C_{gg}/2\pi$, defined more formally in Chapter 2) decreases significantly. In many circuits, operating the transistors at lower f_T translates into a bandwidth reduction. As a simple example, we will show in Chapter 3 that the gain-bandwidth product (GBW) of a gain stage scales directly with f_T, assuming a fixed "fan-out" (ratio of load to input capacitance). Similarly, we will see in the examples of Chapters 5 and 6 that the non-dominant poles of feedback amplifiers tend to occur at a fraction of f_T, thus also limiting the gain-bandwidth product via stability constraints.

From the above discussion, we can now identify a generic flow for transistor sizing:

1. Determine g_m (from design specifications).
2. Pick L:
   • short channel → high speed, small area;
   • long channel → high intrinsic gain, improved matching, ...
3. Pick g_m/I_D:
   • large g_m/I_D → low power, large signal swing (low V_{Dsat});
   • small g_m/I_D → high speed, small area.

Table 1.1 Figure-of-merit data for an n-channel device with $L = 60$ nm, $V_{DS} = 1$ V and $V_{SB} = 0$ V.

Figure of merit	g_m/I_D (S/A)		
	5	10	20
g_m/W (mS/µm)	0.811	0.384	0.064
J_D (µA/m)	162	38.4	3.20
g_m/g_{ds}	9.9	10.9	11.9
f_T (GHz)	122	64.2	12.7

4. Determine I_D (from g_m and g_m/I_D).
5. Determine W (from I_D/W).

This flow works for many basic circuit examples where the required transconductance is known a priori (see the examples in Chapter 3). In other cases, where g_m is not known explicitly and the specifications are more intertwined, one can devise a different sizing procedure for the specific scenario. The purpose of Chapters 4, 5 and 6 is to work through such scenarios.

For the time being, let us illustrate the generic sizing flow for our simple differential pair example, assuming that we wish to realize a transconductance of 10 mS with minimum-length transistors. Furthermore, let us assume that we consider two inversion levels: (a) $g_m/I_D = 5$ S/A (strong inversion), and (b) $g_m/I_D = 20$ S/A (moderate inversion). Using the data from Table 1.1, we can then complete the device sizing as follows.

Case (a):

$$I_D = \frac{g_m}{\left(\frac{g_m}{I_D}\right)} = \frac{10mS}{5S/A} = 2mA. \tag{1.5}$$

$$W = \frac{I_D}{\left(\frac{I_D}{W}\right)} = \frac{2mA}{149A/m} = 13.4\mu m. \tag{1.6}$$

Case (b):

$$I_D = \frac{g_m}{\left(\frac{g_m}{I_D}\right)} = \frac{10mS}{20S/A} = 0.5mA. \tag{1.7}$$

$$W = \frac{I_D}{\left(\frac{I_D}{W}\right)} = \frac{0.5mA}{2.8A/m} = 179\mu m. \tag{1.8}$$

The results are illustrated in Figure 1.10. The strong inversion design offers small area at the expense of larger current. The moderate inversion design draws less current, but employs a larger device (large area and large capacitance, lower f_T). In this context, it is interesting to note that the lowest possible current is achieved in weak inversion, and is given by

$$I_{Dmin} = \frac{g_m}{\left(\dfrac{g_m}{I_D}\right)_{max}} = nU_T g_m, \qquad (1.9)$$

where U_T is the thermal voltage (kT/q) and n is the device's subthreshold slope factor. The analysis leading to the maximum g_m/I_D value, seen as a "plateau" in Figure 1.6(a), is presented in Chapter 2. In the present context, we merely note that the peak value of g_m/I_D is mainly defined by physical quantities, once again underscoring the fundamental nature and significance of the transconductance efficiency.

Ultimately, the designer will need to pick an inversion level and design tradeoff that best fits his or her objectives. In some examples that we will consider in this book, the choice will be immediately obvious. In other examples, typically the ones involving multiple "critical" transistors, picking the proper tradeoff point is non-trivial. This is because the transistor parameters are often linked to the overall design objectives (e.g. bandwidth, power consumption, chip area, etc.) through a complex system of equations that does not have a closed-form solution. Thus, a large part of systematic design boils down to untangling the equation set using intuition about what matters, and where appropriate simplifications and immediate design choices can be made. Chapters 5 and 6 of this book provide many examples by which the reader can learn to approach and successfully tackle such problems.

The last point to emphasize in this section is that the device sizes shown in Figure 1.10 were obtained without any device modeling equations. Instead, we used pre-computed SPICE data in form of the lookup tables introduced in Section 1.2.2. Thus, the MOSFETS in Figure 1.10 are expected to achieve the targeted g_m value almost exactly when simulated in SPICE. However, small deviations from the ideal

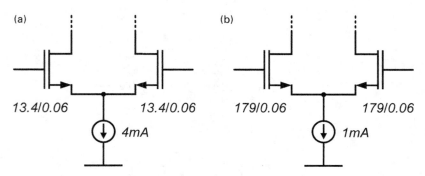

Figure 1.10 Differential pair sized for $g_m = 10$ mS using two options: (a) strong inversion $(g_m/I_D = 5$ S/A), (b) moderate inversion $(g_m/I_D = 5$ S/A). Device geometries are in microns.

value are still possible, and often stem from the fact that we do not know the final terminal voltages precisely. For example, we may not know the drain-source voltages in our differential pair during the initial sizing. We will discuss such second-order error sources throughout this book and will give the reader a feel for their magnitude. When the errors are significant, it is often possible to minimize them iteratively.

1.2.4 V_{GS}-agnostic design

In the previous sections, we have argued for a design flow that essentially ignores the gate-to source voltage (V_{GS}) and gate overdrive ($V_{OV} = V_{GS} - V_T$) for device sizing. In this section, we provide a few introductory remarks on why this is feasible (and indeed desired). The discussion will expand throughout the book, building on the basic points given here.

The first important point to note is that most MOS circuits are highly unsuitable for voltage biasing via some defined value of V_{GS} [25]. This is mostly because the threshold voltage of MOSFETS easily varies by $\pm100\ldots200$ mV over process and temperature. As a result, all voltages are either "computed" via replica circuits (example: high-swing current mirror biasing [1]) or set via a feedback mechanism. Consequently, the designer almost never picks some accurate value of V_{GS}, but instead works with a "reasonable feel" for V_{GS}, merely to ensure proper large-signal operation. We occasionally compute V_{GS} in this book mostly in the latter spirit, and not to drive the device sizing.

A second aspect concerns the drain-source saturation voltage, V_{Dsat}. In the square-law framework, $V_{Dsat} = V_{OV} = V_{GS} - V_T$. However, this expression does not hold in moderate and weak inversion and has therefore become relatively useless for design. We will show in Chapter 2 that V_{DSat} is well approximated by $2/(g_m/I_D)$, obviating once again the need to know V_{GS}.

Lastly, the gate overdrive V_{OV} has been traditionally used to predict circuit distortion [23]. However, as we will show in Chapter 4 (see also [17]), we can achieve a better and more universal modeling based on g_m/I_D.

1.2.5 Design in Weak Inversion

An argument that is sometimes made against using g_m/I_D as a design variable is that it saturates in weak inversion (see Figure 1.6(a)). Indeed, in this region, one can change the device current over a few orders of magnitude without seeing a significant change in g_m/I_D (see Chapter 2 for more detailed plots). Consequently, there is limited utility in designating g_m/I_D as the "knob" for design in weak inversion.

The design methodologies described in [19], [20] overcome this issue by designating the inversion coefficient (IC) as the design variable. The IC essentially corresponds to a normalized version of the current density [19]:

$$IC = \frac{I_D}{I_S} = \frac{I_D}{2nU_T^2\mu C_{ox}\frac{W}{L}} = \left(\frac{L}{2nU_T^2\mu C_{ox}}\right)J_D, \qquad (1.10)$$

where I_S is called the specific current. As seen from (1.10), IC varies continuously with current density (J_D) across all inversion levels. Despite this advantage, we note the following shortcomings:

- The IC-based methodology (as presented in [19], [20]) is not model agnostic, but assumes a simplified analytical function for the IV-characteristics. The characteristic depends on quantities like μC_{ox}, which must be extracted from a "real" reference transistor. However, since the value of μC_{ox} is bias dependent, this approach sacrifices numerical accuracy.
- Unlike g_m/I_D, the IC does not have a simple analytical link to V_{Dsat} and the transistor's distortion characteristics.

With these observations in mind, we opt for an approach that looks to combine the advantages of both approaches with g_m/I_D left as its core variable for analysis, intuition building and sizing in moderate and strong inversion. For design in weak inversion, we devise an alternative flow that enumerates the tradeoffs against the current density (see Chapter 3). For this purpose, we do not invoke an I-V curve template, but rely on lookup table data to maintain a model-agnostic approach.

Generally, it should be noted that most high-performance circuits operate either in moderate or strong inversion. Indeed, it is explained in [20] that moving deep into weak inversion (to the left of the g_m/I_D "knee" seen in Figure 1.6) makes little sense when both power and speed must be optimized jointly. To understand this, consider the simple example of Figure 1.11. Suppose you have 10 μA of drain current

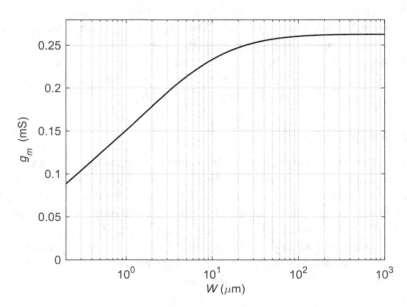

Figure 1.11 Transconductance versus device width for a n-channel device with $L = 60$nm and $I_D = 10$ μA.

available and you want to size a transistor to obtain large transconductance. As we increase the device width, g_m grows steadily and the required V_{GS} that sustains this current reduces along the sweep. As we enter the weak inversion plateau of Figure 1.6, g_m no longer increases and is given by (1.9). Thus, there is no incentive to increase W any further. Should we then always operate the device near the curve's saturation level? The answer is no, since it may be even more efficient to accept a smaller g_m, but in return enjoy the benefits of smaller device width and smaller capacitance, which may make it easier to achieve a given bandwidth. We will see throughout this book that this kind of tradeoff often points us toward design in moderate inversion.

Regardless of the above argument, operating deep in weak inversion can still be attractive for ultra-low power circuits that tend to have no significant speed constraints and, for example, operate in the nanoampere regime. The current density-driven sizing approach described in Chapter 3 covers this specific application scenario.

1.3 Content Overview

This book contains six chapters. The overview below outlines the content of the chapters that follow this introduction.

Chapter 2, Basic Transistor Modeling, familiarizes the reader with the physical mechanisms that control the drain current in a modern MOS transistor. The specific goal here is to illustrate why g_m/I_D is a meaningful proxy for the device's inversion level. To achieve this, we first invoke the charge sheet model (CSM), which paves the way toward the concept of inversion level. We then refine the discussion using a simplified version of the EKV model, which uses only three parameters to represent the drain current from weak to strong inversion. We make use of this model to illustrate how the inversion level controls MOS transistor characteristics and how second-order effects such as drain-induced barrier lowering (DIBL) connect to bias modifications affecting the parameters. To exemplify these effects, we compare drain currents, output conductances, transit frequencies, etc., predicted by the proxy to "real" transistor data. The latter consists of lookup tables (described above) generated from a well-calibrated 65 nm PSP model set.

Chapter 3, Basic Sizing Using the g_m/I_D Methodology, uses a common-source stage with ideal current source biasing (intrinsic gain stage, IGS) to introduce the reader to basic sizing problems. We determine the gate length, width and drain current that achieve a prescribed gain-bandwidth product (GBW), while optimizing for a predefined objective, as for example maximum gain, minimum power consumption, etc. The discussed examples are a natural extension of the simple sizing problem introduced in Section 1.2.3 and add other aspects step-by-step, including the addition of an active load device and the consideration of extrinsic capacitances. In addition, we consider sizing in weak inversion using a unified approach.

Each example concludes with SPICE simulations that validate the correctness of the sizing outcome.

Chapter 4, Noise, Distortion and Mismatch, reviews the fundamentals of noise, distortion and mismatch analysis using a g_m/I_D-centric framework. Specifically, we show that g_m/I_D plays a key role in quantifying and minimizing these non-idealities. For thermal noise, we highlight that the product of g_m/I_D and f_T is a useful figure of merit for circuits that demand low noise and wide bandwidth simultaneously. To quantify distortion, we leverage the simplified EKV model equations of Chapter 2 to derive nonlinear distortion models that hold for all inversion levels. We then compare the predicted distortion of a common-source stage and a differential pair to SPICE simulations and find good agreement. Finally, we review mismatch effects in current mirrors and differential pairs and formulate guidelines on how to pick the inversion level in these circuits given a mismatch constraint.

Chapter 5, Practical Circuit Examples I, takes the concepts established in Chapters 1–4 and applies them to more complex design problems. The main goal is to show that the script-based g_m/I_D methodology enables a systematic, hand analysis driven design flow for larger circuits of practical relevance. The specific examples include a constant-g_m bias circuit, a high-swing current mirror, a low-dropout voltage regulator (LDO), an RF low-noise amplifier (LNA) and a charge amplifier. For all examples, we establish a suitable design flow and validate the obtained results against SPICE simulations. At the end of the chapter, we address considerations for process corner aware design, using the charge amplifier as a specific example.

Chapter 6, Practical Circuit Examples II, provides additional design examples dealing with amplifiers for switched-capacitor (SC) circuits. We consider a generic SC gain stage and discuss the design of its constituent operational transconductance amplifier (OTA). We show how g_m/I_D-based design can be used to size the OTA circuit under noise and settling speed constraints. The investigation begins with the simplest possible OTA (a basic single-stage design) and subsequently considers the folded-cascode and two-stage topologies. Finally, we consider a strategy for using the lookup tables to size the switches that complete the SC circuit.

1.4   Prerequisites

The target audience for this book are MS and PhD students, as well as analog IC design practitioners at all levels of experience. Throughout the book, we assume familiarity with standard analog circuit design texts, such as Hurst, Lewis, Gray & Meyer [1] or Chan Carusone, Johns & Martin [2].

1.5   Notation

This book follows the notation for signal variables as standardized by the IEEE. Total signals are composed of the sum of DC quantities and small signals. For example, a total input voltage v_{IN} is the sum of a DC input voltage V_{IN} and an incremental voltage component v_{in}. The notation is summarized below.

* Total quantity has a lowercase variable name and uppercase subscript.
* DC quantity has an uppercase variable name and uppercase subscript.
* Incremental quantity has a lowercase variable name and lowercase subscript.

1.6   References

[1] P. R. Gray, P. Hurst, S. H. Lewis, and R. G. Meyer, *Analysis and Design of Analog Integrated Circuits*, 5th ed. Wiley, 2009.

[2] T. Chan Caruosone, D. A. Johns, and K. W. Martin, *Analog Integrated Circuit Design*, 2nd ed. Wiley, 2011.

[3] B. Toole, C. Plett, and M. Cloutier, "RF Circuit Implications of Moderate Inversion Enhanced Linear Region in MOSFETs," *IEEE Trans. Circuits Syst. I*, vol. 51, no. 2, pp. 319–328, Feb. 2004.

[4] A. Shameli and P. Heydari, "A Novel Power Optimization Technique for Ultra-Low Power RFICs," in *Proc. ISLPED*, 2006, pp. 274–279.

[5] T. Taris, J. Begueret, and Y. Deval, "A 60μW LNA for 2.4 GHz Wireless Sensors Network Applications," in *Proc. RF IC Symposium*, 2011, pp. 1–4.

[6] Y. S. Chauhan, S. Venugopalan, M.-A. Chalkiadaki, M. A. U. Karim, H. Agarwal, S. Khandelwal, N. Paydavosi, J. P. Duarte, C. C. Enz, A. M. Niknejad, and C. Hu, "BSIM6: Analog and RF Compact Model for Bulk MOSFET," *IEEE Trans. Electron Devices*, vol. 61, no. 2, pp. 234–244, Feb. 2014.

[7] G. Gildenblat, X. Li, W. Wu, H. Wang, A. Jha, R. Van Langevelde, G. D. J. Smit, A. J. Scholten, and D. B. M. Klaassen, "PSP: An Advanced Surface-Potential-Based MOSFET Model for Circuit Simulation," *IEEE Trans. Electron Devices*, vol. 53, no. 9, pp. 1979–1993, Sep. 2006.

[8] R. Iskander, M.-M. Louërat, and A. Kaiser, "Hierarchical Sizing and Biasing of Analog Firm Intellectual Properties," *Integr. VLSI J.*, vol. 46, no. 2, pp. 172–188, Mar. 2013.

[9] G. G. E. Gielen and R. A. Rutenbar, "Computer-Aided Design of Analog and Mixed-Signal Integrated Circuits," *Proc. IEEE*, vol. 88, no. 12, pp. 1825–1854, Dec. 2000.

[10] D. Han and A. Chatterjee, "Simulation-in-the-Loop Analog Circuit Sizing Method Using Adaptive Model-Based Simulated Annealing," in *4th IEEE International Workshop on System-on-Chip for Real-Time Applications*, 2004, pp. 127–130.

[11] F. Silveira, D. Flandre, and P. G. A. Jespers, "A gm/ID Based Methodology for the Design of CMOS Analog Circuits and Its Application to the Synthesis of a Silicon-on-Insulator Micropower OTA," *IEEE J. Solid-State Circuits*, vol. 31, no. 9, pp. 1314–1319, Sep. 1996.

[12] D. Flandre, A. Viviani, J.-P. Eggermont, B. Gentinne, and P. G. A. Jespers, "Improved Synthesis of Gain-Boosted Regulated-Cascode CMOS Stages Using Symbolic Analysis and gm/ID Methodology," *IEEE J. Solid-State Circuits*, vol. 32, no. 7, pp. 1006–1012, July 1997.

[13] F. Silveira and D. Flandre, "Operational Amplifier Power Optimization for a Given Total (Slewing Plus Linear) Settling Time," in *Proc. Integrated Circuits and Systems Design*, 2002, pp. 247–253.

[14] Y.-T. Shyu, C.-W. Lin, J.-F. Lin, and S.-J. Chang, "A gm/ID-Based Synthesis Tool for Pipelined Analog to Digital Converters," in *2009 International Symposium on VLSI Design, Automation and Test*, 2009, pp. 299–302.

[15] T. Konishi, K. Inazu, J. G. Lee, M. Natsui, S. Masui, and B. Murmann, "Design Optimization of High-Speed and Low-Power Operational Transconductance Amplifier Using gm/ID Lookup Table Methodology," *IEICE Trans. Electron.*, vol. E94-C, no. 3, pp. 334–345, Mar. 2011.

[16] H. A. Cubas and J. Navarro, "Design of an OTA-Miller for a 96dB SNR SC Multi-Bit Sigma-Delta Modulator Based on gm/ID Methodology," in *2013 IEEE 4th Latin American Symposium on Circuits and Systems (LASCAS)*, 2013, pp. 1–4.

[17] P. G. A. Jespers and B. Murmann, "Calculation of MOSFET Distortion Using the Transconductance-to-Current Ratio (gm/ID)," in *Proc. IEEE ISCAS*, 2015, pp. 529–532.

[18] D. M. Binkley, *Tradeoffs and Optimization in Analog CMOS Design*. John Wiley & Sons, 2008.

[19] C. Enz, M.-A. Chalkiadaki, and A. Mangla, "Low-Power Analog/RF Circuit Design Based on the Inversion Coefficient," in *Proc. ESSCIRC*, 2015, pp. 202–208.

[20] W. Sansen, "Minimum Power in Analog Amplifying Blocks: Presenting a Design Procedure," *IEEE Solid-State Circuits Mag.*, vol. 7, no. 4, pp. 83–89, 2015.

[21] P. Jespers, *The gm/ID Methodology, a Sizing Tool for Low-Voltage Analog CMOS Circuits*. Springer, 2010.

[22] C. C. Enz and E. A. Vittoz, *Charge-Based MOS Transistor Modeling: The EKV Model for Low-Power and RF IC Design*. John Wiley & Sons, 2006.

[23] W. Sansen, "Distortion in Elementary Transistor Circuits," *IEEE Trans. Circuits Syst. II*, vol. 46, no. 3, pp. 315–325, Mar. 1999.

[24] D. K. Shaeffer and T. H. Lee, "A 1.5-V, 1.5-GHz CMOS Low Noise Amplifier," *IEEE J. Solid-State Circuits*, vol. 32, no. 5, pp. 745–759, May 1997.

[25] B. Murmann, *Analysis and Design of Elementary MOS Amplifier Stages*. NTS Press, 2013.

2 Basic Transistor Modeling

The purpose of this chapter is to review physical aspects of MOS transistors and to consider a few models that describe their behavior. We begin with a physical model: the charge sheet model (CSM), which lays the foundation for understanding the concept of inversion level, as promoted in this book. Since the CSM is too complex for circuit design, we look for simplifications and therefore introduce the "basic" EKV model. We make use of the latter to construct characteristics that are compared with those of real transistors. Because the basic EKV model is a long-channel model (like the CSM), it does not fit the characteristics of modern transistors (short-channel devices) with high accuracy. However, the intuition that the model provides sets the stage for the g_m/I_D-based sizing methodology studied in this book.

2.1 The Charge Sheet Model (CSM)

The CSM was introduced by J.R. Brews [1] and F. Van de Wiele [2] in 1978 and 1979, respectively. Though it applies only to long-channel MOS transistors implemented in a uniformly doped substrate, the CSM is an invaluable tool for comprehending concepts like strong, weak and moderate inversion. Even though the model ignores many important physical aspects, it can predict drain currents that are very close to those of real transistors, even at very low levels of inversion. A short review of the physical background underlying the CSM is presented hereafter.

2.1.1 The CSM Drain Current Equation

Two distinct transport mechanisms define the drain current of MOS transistors: drift and diffusion. Their contributions are illustrated by the first and second term inside the brackets of the drain current expression below:

$$I_D = \mu W \left(-Q_i \frac{d\psi_S}{dx} + U_T \frac{dQ_i}{dx} \right). \tag{2.1}$$

The drift current is proportional to the electrical field represented by the derivative of the surface potential ψ_S with respect to the distance x along the channel. The diffusion current is proportional to the mobile charge density gradient represented by the derivative with respect to x of the mobile charge density Q_i. The factor U_T represents the thermal voltage kT/q.

The integration of the right part of (2.1) can be performed with respect to either the surface potential or the mobile charge density. In the CSM, the integration is carried out with respect to the surface potential. The mobile charge density Q_i is thus expressed as a function of ψ_S. This requires solving the Gauss and Poisson equations.[1] In the result given below, the three first terms between brackets represent the drift current, the remaining are for the diffusion current:[2]

$$I_D dx = \mu C_{ox} W \left[V_G - \gamma \sqrt{\psi_S} - \psi_S + U_T \left(1 + \frac{\gamma}{2\sqrt{\psi_S}} \right) \right] d\psi_S . \tag{2.2}$$

In the above result, γ is the same backgate parameter as defined in SPICE:

$$\gamma = \frac{1}{C_{ox}} \sqrt{2q\varepsilon_s N} . \tag{2.3}$$

After integration, the expression below is obtained where the indices D and S stand for the drain and the source terminals, respectively:

$$I_D = \beta \left[F(\psi_{SD}) - F(\psi_{SS}) \right] \tag{2.4}$$

and β is the current factor ($\mu C_{ox} W/L$). $F(\psi_{SX})$ is defined as:

$$F(\psi_{SX}) = -\frac{1}{2} \psi_{SX}^2 - \frac{2}{3} \gamma \psi_{SX}^{1.5} + (V_G + U_T) \psi_{SX} + \gamma U_T \psi_{SX}^{0.5} . \tag{2.5}$$

Equations (2.4) and (2.5) pave the way toward the drain current, but we still need an expression connecting ψ_S to the gate-to-bulk voltage, V_G. This link can be found by considering the Gauss equation, which bridges the gate-to-bulk voltage V_G to the surface potential ψ_S and the voltage drop across the gate oxide Q_i/C_{ox}:

$$V_G = -\frac{Q_i}{C_{ox}} + \psi_S . \tag{2.6}$$

After replacing Q_i by its Boltzmann approximation, the nonlinear implicit equation below is obtained:

$$V_G = \gamma \left(U_T \exp \left(\frac{\psi_S - 2\phi_B - V}{U_T} \right) + \psi_S \right)^{0.5} + \psi_S , \tag{2.7}$$

[1] All expressions reported in this section are documented in Chapter 2 of [12].
[2] Single indices are used when voltages are defined with respect to the bulk while double indices relate to well-defined references, e.g. V_G is the gate voltage with respect to the bulk whereas V_{GS} is the gate voltage with respect to the source.

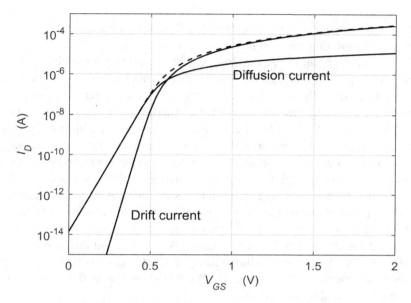

Figure 2.1 Contributions of drift and diffusion currents to the total drain current (dashed line). The transistor is in the common-source configuration and assumed to be saturated. The flat-band voltage is assumed to be equal to 0.6 V.

where V represents the non-equilibrium voltage along the channel, and ϕ_B is the bulk potential:

$$\phi_B = U_T \log(N / n_i). \tag{2.8}$$

Consider the common-source, saturated MOS transistor as an example. Assume that the doping concentration N of the substrate is equal to 10^{17} atoms/cm³, the gate oxide thickness is $t_{ox} = 5$ nm and $\mu C_{ox} = 3.45\ 10^{-4}$ A/V². Because the source is grounded, V_S is zero. We assume that the mobile charge density at the drain is also equal to zero since the transistor is saturated. Extracting ψ_{SS} and ψ_{SD} from (2.7), we find the drain current using (2.4) and (2.5).

Unfortunately, (2.7) can only be solved numerically. To overcome this problem, we make use of the surfpot(p, V, V_G) function,[3] which can be found in the Matlab toolbox accompanying this book. The obtained result is shown in Figure 2.1. The figure displays not only the current but also the contributions due to drift and diffusion. We see that in the right part of Figure 2.1, the drift current is larger than the diffusion current, meaning that strong inversion conditions prevail. Left, the diffusion current takes over, which corresponds to weak inversion. Of particular interest is the point where drift and diffusion currents are equal. We will show further that the gate voltage at the intersection can be viewed as the threshold voltage, though

---

[3] The p argument (vector or matrix) in the surfpot function coalesces the absolute temperature T, the substrate concentration N and the oxide thickness t_{ox}.

the CSM truly ignores the concept of a threshold for the simple reason that it is essentially not a meaningful physical parameter.

The changing nature of the drain current when the gate voltage is swept from small to large values brings important consequences. In weak inversion, the current varies exponentially with the gate voltage. In strong inversion, the drain current follows a quadratic law. The well-known quadratic and exponential drain current equations currently used by circuit designers represent acceptable approximations of these regions. It is important to be aware, however, that an increasing number of CMOS circuits do not operate in strong or weak inversion but in the so-called moderate inversion region, which lies in between. The trouble is that there aren't any simple models for this region. The CSM coalesces weak and strong inversion and describes the passage from one mode to another in a continuous manner. Unfortunately, because an analytical expression of the surface potential is lacking, the model is impractical for circuit design. A rigorous and simple continuous physical model that lends itself to hand-sizing of CMOS circuits throughout all possible modes of operation does not exist. We will show below that the so-called EKV model is a valuable alternative.

Example 2.1 Surface Potential Calculation

Plot the surface potential ψ_S versus V of the CSM model considered above. Assume $T = 300 \, °K$, $N = 10^{17}$ atoms/cm³, $t_{ox} = 5$ nm and $\mu C_{ox} = 3.45 \, 10^{-4}$ A/V².

SOLUTION

Two approaches are possible to obtain the plot shown in Figure 2.2:

1. The simplest approach is use the surfpot.m function to compute the surface potential ψ_S versus V, then plot V versus ψ_S.
2. Alternatively, extract V from (2.7). There is no need to use a solver; V can be expressed as an analytic function of ψ_S:

$$V = - U_T \log\left(\left(\left(\frac{V_G - \psi_S}{\gamma}\right)^2 - \psi_S\right)\frac{1}{U_T}\right) + \psi_S - 2\phi_B.$$

The values for ψ_S must be carefully chosen, owing to the trend of V to increase very rapidly right to the pinch-off voltage V_P, which lies at the middle breakpoints in Figure 2.2. Left of the breakpoint, the semiconductor surface is inverted so that ψ_S varies directly with V. Right, the surface is depleted and ψ_S is nearly constant.

What makes the calculation of V tricky is the need to take very small increments of the surface potential (on the hand side of the above equation) to control the fast increase of V. A simple way to circumvent the problem is to first

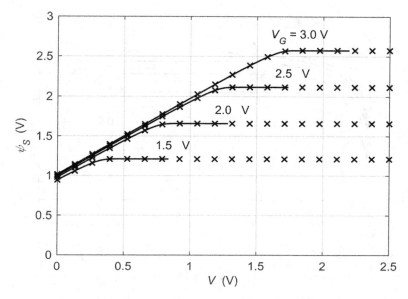

Figure 2.2 Evolution of the surface potential versus the non-equilibrium voltage V, considering four gate voltages. Below the pinch-off voltage (strong inversion), the surface potential increases with the non-equilibrium voltage V. Beyond pinch-off (weak inversion), the surface potential remains practically constant. Crosses illustrate the result from the surfpot function, while the solid lines are obtained by inverting the Gauss equation.

assess the upper limit of ψ_S. This is possible since (2.7) boils down to a simple quadratic expression once we null the mobile charge. The surface potential is given then by:

$$\psi_{Smax} = \left(-\frac{\gamma}{2} + \sqrt{\left(\frac{\gamma}{2}\right)^2 + V_G} \right)^2.$$

To prevent large steps of V to the right of V_P, we generated ψ_S by subtracting a decaying exponential from ψ_{Smax}.

Notice that changing the gate voltage V_G displaces the pinch-off voltage V_P by an almost equal amount. The fact that these two voltages are almost proportional to each other is leveraged in the EKV model that we will consider below. Also, notice that in strong inversion, the surface potential depends only weakly on the gate voltage, since all curves tend to coincide as V decreases.

2.1.2   The Dependence of the Drain Current on the Drain Voltage

The plot of Figure 2.3 shows drain currents versus the drain voltage as predicted by (2.4) and (2.5), considering a series of constant gate voltages encompassing

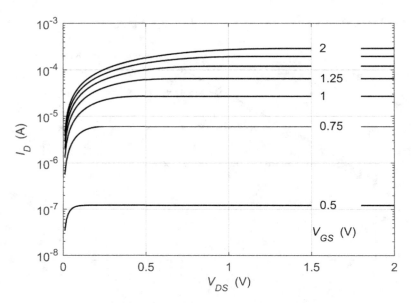

Figure 2.3 Drain currents versus the drain voltage considering the same transistor as in the previous figure. The gate-to-source voltages spans a wide range from weak to strong inversion.

all modes of operation (from weak to strong inversion). Every drain current characteristic starts from zero and grows until it approaches a constant value. The drain voltage that marks the onset of saturation is known as the drain saturation voltage, V_{Dsat}. Beyond this limit, the current remains constant since ψ_{SD} doesn't change any more. The transistor turns into a nearly ideal current source. Notice that V_{Dsat} decreases steadily as the gate voltage reduces until the transistor enters weak inversion. When this happens, the drain saturation voltage is stuck around 60 to 100 mV. This agrees with the approximate expression of the weak inversion drain current given by (2.9), which is obtained when the drift current is ignored, and only diffusion current is taken into consideration. As soon as V_D gets larger than approximately four times U_T, the drain voltage loses control over the drain current.

$$I_{Dwi} = I_o \exp\left(\frac{V_G}{n_{wi}U_T}\right) \cdot \left[\exp\left(-\frac{V_S}{U_T}\right) - \exp\left(-\frac{V_D}{U_T}\right)\right].$$

$$(2.9)$$

The prediction that the drain current remains constant when the drain voltage exceeds V_{Dsat} is a serious weakness of the CSM. Real (physical) transistors deliver currents that increase with the drain voltage in saturation. In other words, the output conductances are finite and not equal to zero. The reason for this non-physical behavior is that the CSM ignores many important second order effects, which are discussed in Section 2.3.

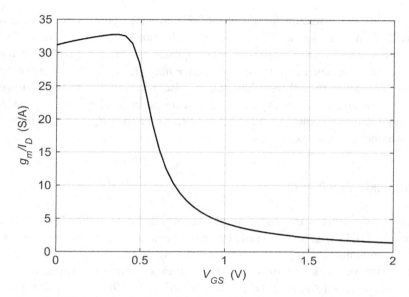

Figure 2.4 The g_m/I_D ratio obtained after taking the derivative of the semilog-scaled drain current shown in Figure 2.1.

2.1.3   The Transconductance Efficiency g_m/I_D

As discussed in Chapter 1 the transconductance efficiency g_m/I_D will be used as a key parameter throughout this book. We can find g_m/I_D for the drain current characteristic of Figure 2.1 using:

$$\frac{g_m}{I_D} = \frac{1}{I_D}\frac{\partial I_D}{\partial V_G} = \frac{\partial \log(I_D)}{\partial V_G}. \tag{2.10}$$

Thus, g_m/I_D is simply the slope of the semilog plot of $I_D(V_{GS})$. The result of this calculation is shown in Figure 2.4. Because the slope of $I_D(V_{GS})$ is largest and essentially constant in weak inversion, g_m/I_D plateaus in this region. We will show later that the maximum of g_m/I_D is equal to $1/(nU_T)$ where n is the subthreshold slope factor. Generally, n lies between 1.2 and 1.5 in a bulk technology, and is somewhat smaller for silicon-on-insulator (SOI) transistors.[4] The maximum possible g_m/I_D is obtained when n is equal to one, yielding $1/U_T = 38.46$ S/A at room temperature. This is what bipolar junction transistors (BJTs) achieve. Usually, the maximum transconductance efficiency of bulk MOS transistors lies between 20 and 30 S/A.

For larger V_{GS} beyond weak inversion, g_m/I_D decreases. Values of 20 to 10 S/A, correspond to moderate inversion, whereas smaller values, for example between 8 and 2 S/A, correspond to strong inversion. Since these ranges don't change significantly with the transistor technology, it is legitimate to use g_m/I_D as a proxy that

---

[4] SOI transistors are less sensitive to the body effect and exhibit steeper subthreshold characteristics. They can therefore attain g_m/I_D as high as 35 S/A in weak inversion, which is close to the BJT limit of $1/U_T$.

represents the inversion level of a MOS transistor. Thus, it is possible to target any desired inversion level by choosing g_m/I_D, and there is no need to know the gate-to-source voltage. Note that the opposite doesn't hold true: two transistors operated at the same gate-to-source voltage may have different inversion levels due to differences in their threshold voltages. Also, recall that threshold voltages necessarily bear the mark of a specific fabrication process. These are accounted for by distinct flat-band voltages. We don't have to care about V_{FB} when working with the transconductance efficiency.

2.2   The Basic EKV Model

The EKV model is the outcome of pioneering work at the Centre Electronique Horloger (CEH) in Switzerland. The name recognizes the contributions of C. Enz, F. Krummenacher and E. Vittoz. The starting point for this development was the derivation of weak and strong inversion drain currents, and the difficulty to combine these into a single representative equation. In 1982, Oguey and Cserveny [3] introduced a mathematical interpolation to mimic moderate inversion between the classical weak and strong inversion expressions. Though appealing, the model overestimated the transconductance in moderate inversion. In the eighties, several groups started investigating physical approaches to eliminate the mathematical interpolation. Cunha et al. [4] [5] from the University of Santa Catarina in Brazil, proposed an approximation that paved the way toward an analytical expression linking the mobile charge density to the surface potential in 1995. This led to what is called today the ACM (advanced compact model). The underlying theory of the ACM is similar to the model developed at the CEH (with regrettably distinct definitions to designate the same quantities).

   In 2003, Sallese et al. published a rigorous derivation of the EKV model [6]. In parallel, the group lead by C. Enz at EPFL began to extend the model to include second-order effects that are ignored by long-channel models. This turned the EKV model into a high-performance tool for circuit simulation [7]. The complexity it brings, however, moves us away from the possibility to employ it for sizing via hand calculations.

   In this book, we opt for a model that coalesces ideas developed at the CEH and the group of Galup-Montoro and M.C. Schneider at the University of Santa Catarina in Brazil. We refer to this model as the basic EKV model throughout the book. It should not be confused with more sophisticated versions of the EKV model, like those intended for circuit simulators.

2.2.1   The Basic EKV Equations

The basic EKV model is a one-dimensional long channel model (or "gradual channel" model) that closely approximates the CSM while covering all modes of operation. The model takes advantage of the observation that the factor associated with the diffusion current in (2.2) (the parentheses inside the square brackets) varies very little across strong, weak and moderate inversion. Thus, we approximate this term as:

$$d\left(-\frac{Q_i}{C_{ox}}\right) = -\left(1 + \frac{\gamma}{2\sqrt{\psi_S}}\right)d\psi_S = -n\,d\psi_S,\tag{2.11}$$

where n is the subthreshold slope factor, which was already introduced above. The merit of this simplification is that it enables us to integrate (2.1). Specifically, it allows us to substitute the mobile charge density Q_i to find the surface potential ψ_S. For this purpose, we introduce the normalized mobile charge density q, defined as:

$$q = -\frac{Q_i}{2nU_TC_{ox}}.\tag{2.12}$$

This lets us to rewrite (2.11) as:

$$d\psi_S = -2U_T dq\tag{2.13}$$

and we can now integrate (2.1) to find:

$$i = (q_S^2 + q_S) - (q_D^2 + q_D),\tag{2.14}$$

where we call q_S and q_D the normalized mobile charge densities at the source and drain,[5] respectively. The variable i is the normalized drain current, defined as:

$$i = I_D / I_S.\tag{2.15}$$

Here, I_S is the specific current, given by:

$$I_S = 2nU_T^2\mu C_{ox}\frac{W}{L} = I_{Ssq}\frac{W}{L}.\tag{2.16}$$

In this expression, the introduction of I_{Ssq}, which is the specific current of a square transistor ($W = L$), offers the possibility to separate what depends on the technology (n, μ and C_{ox}) and what is controlled by the circuit designer (W/L). Notice that I_S can also be written as the product of $2nU_T^2$ times β, the so-called current factor.

The basic EKV model makes use of a second equation that connects the non-equilibrium voltage V to q, the normalized mobile charge density along the channel. To find this equation, we make use of the Maxwell-Boltzmann approximation and leverage the fact that Q_i varies with the minority carrier concentration:

$$Q_i \propto \exp\left(\frac{\psi_S - \phi_F - V}{U_T}\right).\tag{2.17}$$

---

[5] Note that q^2 and q are the normalized counterparts of the drift and diffusion currents considered in the charge sheet model.

Taking the total differential of both sides establishes a link between the differentials of q, ψ_s and V:

$$\frac{dQ_i}{Q_i} = \frac{dq}{q} = \frac{d\psi_s - dV}{U_T}. \tag{2.18}$$

Using (2.13), we can now integrate (2.18) and we don't need the surface potential anymore. The result is shown below, where we call V_P the pinch-off voltage and the subscript x represents either the source S or the drain D, depending on the terminal that we consider:

$$V_p - V_x = U_T \left[2(q_x - 1) + \log(q_x) \right]. \tag{2.19}$$

The above expression relates the non-equilibrium voltage V_x along the channel to the local normalized mobile charge density q_x. At the source, V_x becomes V_S, and the normalized mobile charge density is q_S, while at the drain we have V_D and q_D. When $q_x = 1$, the non-equilibrium voltage V_x equals V_P. At this point, drift and diffusion currents are equal and we are in the middle of the moderate inversion region. When V_x gets smaller than V_P, we enter strong inversion. Conversely, when V_x is larger than V_P, we enter weak inversion.

Equations (2.14) and (2.19) let us link the normalized drain current i to the pinch-off voltage V_P as we turn the normalized mobile charge density into a parameter. When the source is grounded ($V_S = 0$) and the transistor is saturated ($q_D = 0$), the normalized drain current (q substituted by q_S) boils down to the semilog curve shown in Figure 2.5(a). Note that this curve does not use any other quantity than U_T.

Taking the derivative of the log-scaled normalized drain current i with respect to V_P yields the curve shown in Figure 2.5(b).

$$\frac{\partial \log(i)}{\partial V_P} = \frac{1}{U_T} \frac{1}{q+1}. \tag{2.20}$$

Note that when $q \ll 1$ (weak inversion), the expression boils down to the reciprocal of the thermal voltage U_T. When q is equal to one, (2.20) equals the reciprocal of $2U_T$, right in the middle of the moderate inversion region. On the other hand, making q much larger than one places the device in the strong inversion region. The fact that this curve is similar to the plot of Figure 2.4, and that it encompasses the whole range of inversion levels, supports the idea that the basic EKV model is a good approximation of the CSM.

What we need now is a connection linking V_P to the gate voltage V_G. We know already from the plot of Example 2.1 that the pinch-off voltage and the gate voltage vary together. Based on this observation, (2.21) establishes a linear link.

$$V_P = \frac{V_G - V_T}{n}. \tag{2.21}$$

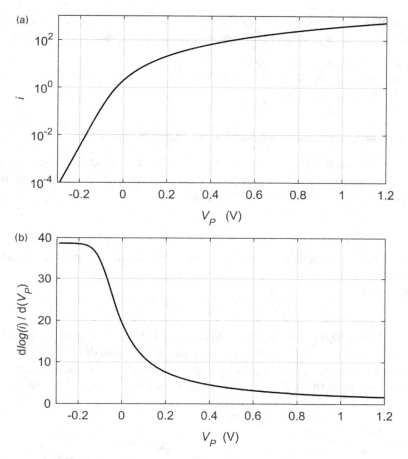

Figure 2.5 (a) The normalized drain current i of the grounded-source saturated transistor versus the pinch-off voltage V_P. (b) The derivative of the log-scaled normalized drain current with respect to V_P.

This expression incorporates two parameters, the subthreshold slope n (introduced earlier) and the threshold voltage V_T. A particularity of the latter is that it is defined with respect to the bulk, like V_P, V_S and V_D. The voltage difference $(V_T - V_S)$ is thus simply the commonly accepted concept of threshold voltage defined with respect to the source, while the difference $(V_G - V_T)$ represents the gate overdrive V_{OV}, also commonly used in the literature [8]. In this context, it should be noted once again that the threshold voltage is not a quantity of physical significance. It is merely a worthwhile notion that allows us to assess the behavior of MOS transistors. Ultimately, what matters is that the currents predicted by the model conform to the data that they were extracted from. This is possible using different definitions of V_T, and there is no unique standard defined in the literature.

Let us summarize. Two equations, (2.14) and (2.19), form the basis of basic EKV model. To reproduce the $I_D(V_{GS})$ characteristics, we use (2.15) and (2.21) to perform vertical and horizontal shifts of $i(V_P)$ (through I_S and V_T, respectively) as well as scaling (through n). This enables the modeling of I_D versus V_{GS} using only three parameters:

1. the subthreshold slope factor, n,
2. the specific current, I_S,
3. the threshold voltage, V_T.

2.2.2   The Basic EKV Model for a Grounded-Source MOS Transistor

For a saturated MOS transistor ($q_D = 0$) in a grounded-source configuration ($V_S = 0$), we can replace q_S with q, since the only terminal that matters is the source. Equations (2.14) and (2.15) then lead to:

$$I_D = I_S\left(q^2 + q\right).$$
(2.22)

Likewise, (2.19) modifies as shown below, since the non-equilibrium voltage at the source is zero:

$$V_P = U_T\left[2(q-1)+\log(q)\right].$$
(2.23)

All we need to do now to plot I_D versus V_{GS} is to eliminate q between (2.22) and (2.23). To find the actual gate-to-source voltage V_{GS} and drain current I_D, we make use of (2.21) and (2.15). Figure 2.6 shows the result for q ranging from 10^{-4} (deep weak inversion) to 10^1 (strong inversion), assuming n, I_S, and V_T equal to 1.3, 1 μA,

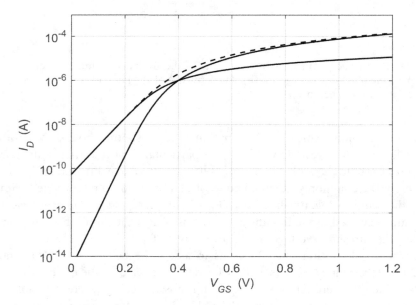

Figure 2.6 Drain current (dashed line) with the drift and diffusion contributions (solid lines) predicted by the basic EKV model. As in Figure 2.1, the drift current dominates in strong inversion, diffusion current in weak inversion.

and 0.4 V, respectively. The curves have the same general shape as in Figure 2.1. The figure also shows the contributions of the drift and diffusion currents, proportional to q^2 and q, respectively, while the sum is the drain current I_D. Note that at the point where drift and diffusion currents are equal, q is equal to one, thus V_p is equal to zero. Hence, V_G equals V_T at this point.

2.2.3   Strong and Weak Inversion Approximations of the EKV Model

In strong and weak inversion, the EKV model of the grounded-source saturated transistor boils down to well-known expressions. For example, in strong inversion, since $q \gg 1$, the drift current, which is proportional to q^2, dominates over the diffusion current. Hence:

$$V_P \approx 2U_T q \approx 2U_T \sqrt{i} = 2U_T \sqrt{\frac{I_D}{I_S}}.$$
(2.24)

This approximation leads to the square-law model after replacing V_P using (2.21):

$$I_D = \mu C_{ox} \frac{W}{L} \frac{(V_G - V_T)^2}{2n} = \beta \frac{(V_G - V_T)^2}{2n}.$$
(2.25)

The somewhat unusual subthreshold slope factor in the denominator of (2.25) reflects the fact that the constant n affects the threshold voltage in the basic EKV model (back-gate effect).

In weak inversion, where $q \ll 1$, we find:

$$V_P \approx U_T \left(-2 + \log(q)\right) \approx U_T \left(-2 + \log(i)\right) = U_T \left(-2 + \log\left(\frac{I_D}{I_S}\right)\right).$$
(2.26)

This leads to the well-known weak inversion exponential approximation:

$$I_D = I_S \exp\left(2 - \frac{V_T}{nU_T}\right)\exp\left(\frac{V_G}{nU_T}\right) = I_0 \exp\left(\frac{V_G}{nU_T}\right).$$
(2.27)

This expression corroborates the fact that the subthreshold factor n sets the slope of the $I_D(V_{GS})$ characteristic on a semilog plot.

Figure 2.7 illustrates the strong and weak inversion approximations using the same EKV parameters as in Figure 2.6, indicating satisfactory asymptotic trends. To define the boundaries between weak, moderate and strong inversion, the plot contains markers for the normalized mobile charge density q. We see that the basic EKV model approaches the exponential (weak inversion) and quadratic (strong inversion) approximations around 0.2 and 5, respectively. Between these landmarks, the device operates in moderate inversion. The center of moderate inversion is signified by $q = 1$ and $V_G = V_T$. We will utilize these boundaries throughout this book to distinguish between the three possible inversion levels.

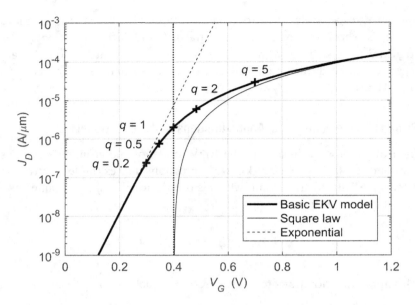

Figure 2.7 Basic EKV model with square-law and exponential approximations. The vertical line marks the threshold voltage (V_T).

2.2.4   Basic EKV Model Expressions for g_m and g_m/I_D

The transconductance of the saturated grounded-source MOS transistor can be derived from (2.22) and (2.23). We see that it is proportional to the normalized mobile charge density q and I_S, which contains W/L.

$$g_m = \frac{\partial I_D}{\partial V_G} = \frac{\partial I_D}{\partial q}\frac{\partial q}{\partial V_G} = I_S\left(2q+1\right)\frac{1}{nU_T}\frac{q}{2q+1} = \frac{I_S}{nU_T}q. \tag{2.28}$$

As stated in Section 2.2.1, the transconductance efficiency g_m/I_D is a useful proxy for the inversion level. It boils down to a simple function of the normalized mobile charge density q, obtained by dividing (2.28) by (2.22):

$$\frac{g_m}{I_D} = \frac{1}{nU_T}\frac{1}{q+1}. \tag{2.29}$$

Deep in weak inversion, where $q \ll 1$, the transconductance efficiency reaches a maximum equal to:

$$M = \max\left(\frac{g_m}{I_D}\right) = \frac{1}{nU_T}. \tag{2.30}$$

Dividing (2.29) by (2.30) leads to the concept of normalized transconductance efficiency ρ:

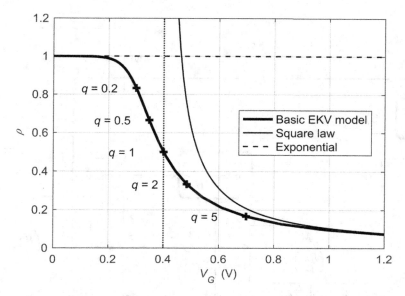

Figure 2.8 Plot of normalized transconductance efficiency ρ for the basic EKV model and its square-law and exponential approximations. The vertical line marks the threshold voltage (V_T).

$$\rho = \frac{g_m / I_D}{\max(g_m / I_D)} = \frac{1}{q+1}. \tag{2.31}$$

ρ varies between 0 (deep in strong inversion) and 1 (deep in weak inversion). In the middle of moderate inversion, ρ is equal to 0.5, since q is equal to one.

An alternative expression for g_m/I_D, which can be helpful when the current is given, is obtained when the normalized mobile charge density in (2.29) is replaced by the normalized drain current i, after inverting (2.22):

$$\frac{g_m}{I_D} = \frac{1}{nU_T} \frac{2}{\sqrt{1 + 4\dfrac{I_D}{I_{Ssq}}\dfrac{L}{W}} + 1}. \tag{2.32}$$

Finally, the counterpart of (2.30) for the square-law approximation in strong inversion (SI) is given by the well-known expression below [8]:

$$\left(\frac{g_m}{I_D}\right)_{SI} = \frac{2}{V_G - V_T}. \tag{2.33}$$

Figure 2.8 plots the normalized transconductance efficiency for the basic EKV model (2.31), along with its square-law (2.33) and exponential (2.30) approximations. The (+) marks denote the same range of normalized mobile charge densities q as in Figure 2.7. We see that moderate inversion ($q = 0.2...5$) corresponds to the steepest section of the curve.

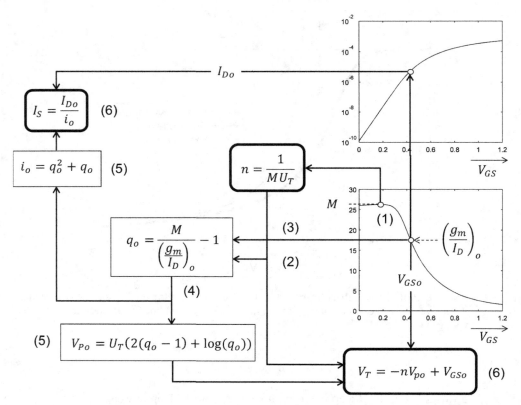

Figure 2.9 Extraction procedure for the parameters n, V_T and I_S from the saturated I_D and g_m/I_D characteristics.[6]

2.2.5   EKV Parameter Extraction

In this section, we ask the question: How can we extract the basic EKV model parameters from a given MOS transistor's characteristics? The suggested procedure is as follows:

1. We compute (numerically) the derivative of $\log(I_D)$ with respect to V_{GS}, i.e. we plot the $g_m/I_D(V_{GS})$ characteristic shown in the lower right corner of Figure 2.9. We extract the subthreshold slope n from the maximum M of the transconductance efficiency (see (2.30)).

2. To find V_T and I_S, we select a second point on the g_m/I_D curve (as $V_{GS} = V_{GSo}$) and assess the corresponding drain current and transconductance efficiency, respectively (I_{Do} and $(g_m/I_D)_o$).

3. Since we already know the subthreshold slope, we find the normalized mobile charge density q_o by inverting (2.29).

---

[6] The extraction method described in this figure is discussed in more detail in Appendix 1 and implemented in the XTRACT.m Matlab function.

4. Having q_o, we compute the normalized drain current i_o and the pinch-off voltage V_{Po} using (2.22) and (2.23).
5. Finally, making use of (2.15) and (2.21), we find the specific current I_S and the threshold voltage V_T.

   How effective is this extraction method? Can we reconstruct drain currents and transconductance efficiencies matching the characteristics where from the parameters were derived? Let us consider an example to investigate.

Example 2.2 EKV Parameter Extraction for the CSM Device

Reconstruct the CSM drain current and transconductance efficiency curves shown in Figure 2.1 and Figure 2.4, making use of the EKV parameters extracted from the original CSM characteristics via the procedure of Figure 2.9.

SOLUTION

1. This step is done, since the g_m/I_D curve is already available.
2. Using the maximum of g_m/I_D in Figure 2.4 we find $n = 1.18$.
3. Select the second reference point on the transconductance efficiency characteristic. Any point is in principle satisfactory. Being too close to the first point, however, could jeopardize accuracy. We fix the point in strong inversion,[7] e.g. $(g_m/I_D)_o = 3$ S/A, making $V_{GS_o} = 1.212$ V and $I_{D_o} = 57.55$ μA.
4. We evaluate the normalized mobile charge density q_o and find that it is equal to 9.90.
5. Having q_o, we compute the pinch-off voltage V_P and the normalized drain current i using (2.22) and (2.23). These are equal to 0.520 V and 107.8, respectively.
6. We evaluate the threshold voltage V_T and the specific current I_S using (2.15) and (2.21) and find $V_T = 0.5988$ V and $I_S = 0.5336$ μA.

With these values, we can now reconstruct the EKV drain current repeating the steps described in Section 2.2.2 and compare the result to the original CSM current. The outcome is depicted in Figure 2.10, and shows a nearly perfect agreement.

   In Figure 2.11, we compare the EKV and CSM transconductance efficiencies using (2.29). The matching is quite good in moderate and strong inversion, but less so in weak inversion. The difference is due to the assumption made earlier, when we approximated the subthreshold slope factor n as a constant (see (2.11)). In reality, n varies deep in weak inversion, causing the discrepancy

---

[7] We will see later that strong inversion is not the best choice when identifying EKV parameters of real transistors. Mobility degradation affects the current in this region. Considerations for selecting the second reference point are discussed in Appendix 1.

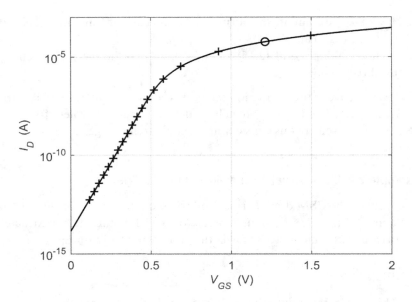

Figure 2.10 The reconstructed drain current predicted by the EKV model (crosses) is compared to the CSM characteristic of Figure 2.1 (solid line). The circle marks the second reference point for the parameter extraction.

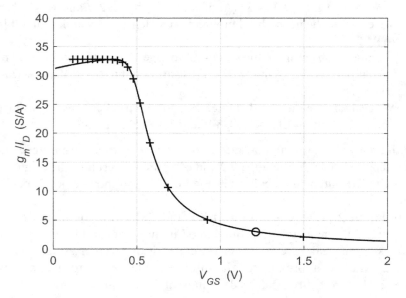

Figure 2.11 The reconstructed transconductance efficiency predicted by the basic EKV model (crosses) is compared to the CSM g_m/I_D of Figure 2.4. The circle marks the second reference point used for the parameter extraction.

we see on the left side of the graph. This is not a significant drawback, since the data to the left of the maximum relate to inversion levels that are encountered only in very low-power/low-speed applications (see Chapter 3 for some examples).

2.3 Real Transistors

Can we use the basic EKV model to describe the behavior of "real" modern transistors? The direct answer is no, since the simple EKV model that we considered does not include mobility degradation or important second-order effects that play a significant role in short-channel devices. Capturing these effects would require a substantial increase in model complexity, and consequently contradict our objective of using only simple expressions when sizing analog CMOS circuits.

In the context of this book, we solve this problem by employing lookup tables containing drain currents, small-signal parameters, capacitances, etc., all extracted either from physical devices or from advanced models like those used in circuit simulators like SPICE (BSIM6 or PSP). In the first case, the data are the result of measurements carried out on large numbers of physical devices considering some range of gate lengths and widths as well as bias conditions. In the second case, the same data are derived from "virtual" devices, represented by high-level models that are believed or known to be accurate. The data considered throughout the remainder of this book belong to the second category. They are the result of 65-nm PSP-modeled transistors simulated using the Cadence Spectre simulator. PSP models are based on surface potential representations like the CSM and are known to be very accurate when properly extracted.

The simulation data used throughout this book consists of a multi-dimensional DC sweep and steps V_{GS} and V_{DS} from 0 to the nominal process supply voltage (1.2 V). The data (including many small-signal parameters across the sweep) were exported from Cadence Spectre to MATLAB using the Spectre Matlab Toolbox. The created data files (65nch.mat and 65pch.mat) can be read into Matlab to look up various device parameters against one another. To simplify this task, we created the Matlab function lookup.m, which provides most of the functionality needed for design. If the function resides in the current Matlab path, a short description of its usage can be echoed by typing "help lookup". Further details on the inner workings of this function and how the lookup data can be created are provided in Appendix 2.

2.3.1 Real Drain Current Characteristics $I_D(V_{GS})$ and g_m/I_D

Figure 2.12 shows several drain current characteristics versus V_{GS} extracted from the lookup tables for a 10 μm wide n-channel transistor. We consider three gate

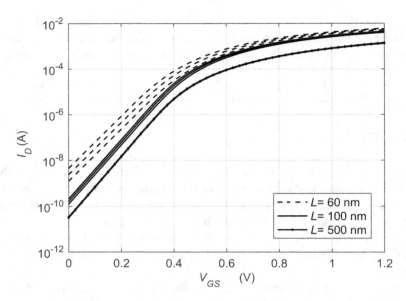

Figure 2.12 Drain currents of a 10 μm wide n-channel transistors considering three gate lengths: 60, 100 and 500 nm, and three drain-to-source voltages, 0.6, 0.9 and 1.2 V, respectively (from bottom to top).

lengths: 60, 100 and 500 nm, and three drain-to-source voltages: 0.6, 0.9 and 1.2 V. The Matlab code used to read in the currents is shown below:

```
L   = [.06 .10 .50];                          % μm
VDS = .6: .3: 1.2;                            % V
ID  = lookup(nch, 'ID', 'VDS', VDS, 'L', L);  % A
```

The arguments passed to the lookup function are:

```
nch    % structure that contains the data
ID     % the desired output variable
VDS    % the range of drain voltages
L      % the range of gate lengths
```

If V_{SB} is not passed to the function, it assumes that this voltage is zero. Similarly, we don't specify V_{GS}, and the function then assumes that we are interested in all values along the sweep vector nch.VGS of the lookup table (equal to 0:0.025:1.2 in this book). The drain current output is an array whose size is defined by the lengths of the gate-to-source, drain-to-source and gate length vectors.

Looking at the representation of Figure 2.12, one of the differences with respect to earlier plots is that the current of real transistors depends on the drain voltage. Two mechanisms explain this. channel length modulation (CLM) and drain-induced barrier lowering (DIBL). To explain these effects, consider the region near the drain junction. As V_{DS} increases, the width of the depleted region surrounding the drain junction tends to increase. The result is a slight decrease of the effective length of the inversion layer. Therefore, because L is getting smaller, W/L increases, which in turn increases the drain current. This is what is meant by channel length modulation.

The second effect, DIBL, becomes relevant once the channel length falls below approximately 0.5 μm. The depletion region surrounding the drain then takes over a significant share of the fixed space charge region under the channel. Thus, the amount of fixed charge controlled by the gate reduces. Smaller gate voltages are therefore needed to induce the same amount of drain current. This is equivalent to a slight decrease of the threshold voltage, which enhances the gate overdrive $(V_{GS} - V_T)$, and thus also increases the drain current. This phenomenon is visible in the 100-nm weak inversion characteristic of Figure 2.12 and more pronounced in the 60-nm characteristic. We see that DIBL shifts the weak inversion drain current characteristic to the left as the drain voltage grows. The effect becomes very significant when L is equal to 60 nm but on the other hand almost vanishes for $L = 500$ nm, where CLM causes only minor changes, making the three curves controlled by the drain voltage almost coincide. Notice also the significant reduction of the drain current slope in weak inversion, when L is equal to 60 nm.

Notice that every single characteristic of Figure 2.12 resembles the CSM current represented in Figure 2.1. The current grows exponentially in weak inversion until we enter moderate inversion. Here, the slope (or g_m/I_D) starts to decrease, reducing further in strong inversion.

An interesting question to pose at this point is whether we can extract basic EKV parameters for the real drain current characteristics in Figure 2.12 and achieve a good match. Clearly, this won't be possible unless we make the parameters a function of V_{DS}, L and V_{SB}. In other words, each curve in Figure 2.12 will require a different set of parameters. In the following example, we will first determine how good of a match we can achieve with this approach. Later in this chapter, we will show that despite this apparent complication, the basic EKV modeling approach is in fact useful for understanding and quantifying gate length and drain voltage dependencies in modern transistors.

Example 2.3 EKV Parameter Extraction for Real Transistors

Extract the EKV parameters for the characteristics shown in Figure 2.12 and compare the reconstructed curves to original drain current and transconductance efficiency.

SOLUTION

Since the task is the same as in Example 2.2, the procedure does not deserve further discussion. The parameters are extracted from lookup functions with appropriate gate lengths and drain voltages using the XTRACT.m function. The reconstructed characteristics are compared to the original data in Figure 2.13. The agreement is generally satisfactory, but not in strong inversion, since the model does not take into account the impact of increasing electrical fields on mobility [8]. Recall that the basic model ignores mobility degradation and therefore overestimates the drain currents in strong inversion. The point is clearly visible in Figure 2.13(b) where the original and reconstructed characteristics for L equal to 60 nm diverge increasingly when V_{GS} gets large. If I_S, which contains μ, were expressed as a function of the electrical field, mobility degradation could in principle be considered. But, the complexity that this brings makes the model too complex and does not justify the loss of flexibility.

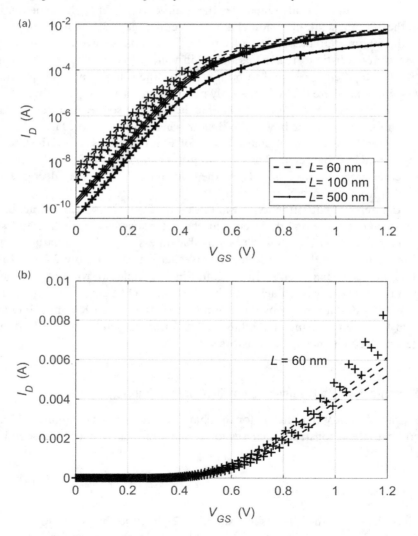

Figure 2.13 The EKV-reconstructed drain currents (+ markers) are compared to the currents of Figure 2.12 (solid and dashed lines). (a) Logarithmic scale, and (b) linear scale for L = 60 nm, W = 1 μm and V_{DS} = 0.6, 0.9 and 1.2 V.

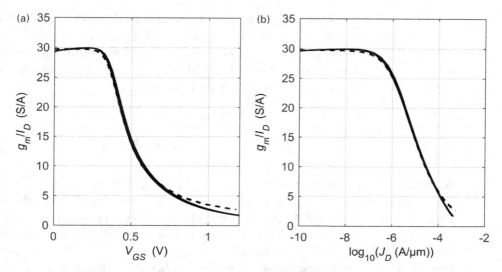

Figure 2.14 Comparison of g_m/I_D considering the real transistor curves of Figure 2.12 (solid lines) and the basic EKV counterpart (dashed lines) derived from (2.29) for $L = 100$ nm. Each plot has overlaid curves for $V_{DS} = 0.6$, 0.9 and 1.2 V.

The observation on mobility degradation brings us back to the parameter extraction algorithm, and in particular the location of the second reference point. We must avoid the strong inversion data because it is affected by mobility degradation. On the other hand, we prefer to be as far away from weak inversion as possible, to avoid a loss of accuracy. Therefore, an optimum exists in moderate inversion and this is discussed further in Section A.1.4 (Appendix 1). We recommend choosing the point such that the normalized transconductance efficiency ρ is in the range between 0.5 and 0.8. The XTRACT function has one optional variable ("rho") that specifies this, the default value being 0.6.

Figure 2.14 compares the transconductance efficiency curves. We use (2.29) to evaluate the transconductance efficiency predicted by the basic EKV model and compare the result to the lookup table curves, considering the real transistors characteristics above with $L = 100$ nm ($V_{DS} = 0.6$, 0.9 and 1.2 V). As we see in from the plots, the agreement is very good from strong to weak inversion, regardless of whether we plot g_m/I_D versus V_{GS} (Figure 2.14(a)) or versus J_D (Figure 2.14(b)). Also, note that the influence of V_{DS} is very small; the three curves for different V_{DS} are very close to one another and are almost indistinguishable. Finally, we note again that the reconstructed characteristics depart significantly from the real g_m/I_D only in strong inversion and deep in weak inversion (for the reasons explained earlier).

In the above example, we saw that the g_m/I_D characteristics versus V_{GS} and J_D are relatively independent of V_{DS} (see Figure 2.14). This is a very welcome feature, since V_{DS} may not always be known for initial circuit sizing. We will now take a closer look at this aspect, to get a numerical feel for the deviations.

First consider a scenario where we would like to operate the device at $g_m/I_D = 15$ S/A and want to assess the required current density. Graphically, we would find J_D by drawing a horizontal line at 15 S/A in Figure 2.14(b) and looking up the abscissa values for the each of the three curves ($V_{DS} = 0.6$, 0.9 and 1.2 V). Using Matlab, we can find the numerical values using the following code:

```
VDS = [0.6  0.9  1.2];
for k = 1:length(VDS),
   JD1(k,1) = interp1(gm_ID(:,k), JD(:,k), 15);
end
```

We find that J_{DI} is equal to 9.09, 9.86 and 10.32 μA/μm. The corresponding variation from the mid-point is −7.8% and +4.7%, which is rather small, given the significant change in V_{DS} and the relatively short channel length (100 nm).

This insensitivity can be explained by inspecting the basic EKV model equations. We see from (2.29) that g_m/I_D is directly linked to the normalized charge density q and the subthreshold slope factor n. Constant g_m/I_D implies that q doesn't change significantly with V_{DS}, since n is almost constant. Consequently, the normalized drain current i doesn't change either, so that the net impact of V_{DS} on J_D is only due to changes in the specific current I_S. We will consider the dependence of I_S on V_{DS} in Section 2.3.3 (Figure 2.19), and will see that it is relatively weak.

Figure 2.15 focuses on another aspect of the transconductance efficiency. It concerns the decaying behavior of the transconductance efficiency left to the circle that marks its maximum. We encountered a declining trend already in the CSM plot of Figure 2.4, which is due to the term between brackets in (2.11). Since the decline is usually very modest, it is typically ignored, like in Figure 2.8, where the subthreshold factor n is approximated as a constant for the basic EKV model.

What we see in Figure 2.15 when $|V_{GS}|$ approaches 0.1 V is due to other effects that lead to a substantial departure from the weak inversion exponential model. One possible mechanism that could lead to the drop of g_m/I_D is junction leakage. However, we have removed the junction leakage component from this plot. What we see is in fact due to band-to-band tunneling (BT-BT) and gate-induced drain leakage (GIDL); see [9] for a detailed discussion. These effects become more visible with larger threshold voltage (hence smaller diffusion current in the channel), and we therefore apply a backgate bias in Figure 2.15 to highlight their impact more clearly.

We will never bias a transistor in the region where BT-BT and GIDL play a role, but we still need to pay attention to these effects for practical reasons related to table-lookup. Suppose that we want to find V_{GS} when $g_m/I_D = 25$ S/A in Figure 2.15. What we are looking for in this case is the intersect with the right side of the curve, and not the GIDL/BT-BT induced drop. The most convenient measure to prevent issues is to systematically disregard data to the left of the maximum every time we use the lookup function to find some quantity for a given value of g_m/I_D. Section A.2.4 provides more information on how we address this problem, which also applies to other quantities, like the device's transit frequency.

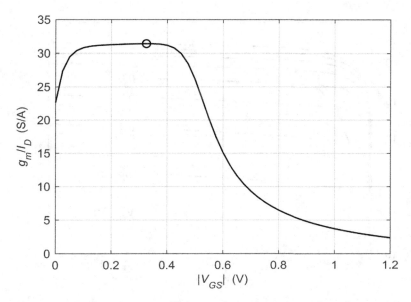

Figure 2.15 Transconductance efficiency of a p-channel transistor for $L = 0.2$ μm, $V_{SB} = -0.6$ V, $V_{DS} = -0.6$V.

2.3.2 The Drain Saturation Voltage V_{Dsat} of Real Transistors

So far, we have compared reconstructed to original drain current characteristics. Does the basic EKV model allow us to model other important parameters? Consider the drain saturation voltage V_{Dsat} that was briefly introduced in Section 2.1.2. Defining the drain saturation voltage of real transistors is less obvious than with the CSM model, since the drain current continues to increase with V_{DS} in saturation. The expression of the drain saturation voltage below, which we advocate in this book, links V_{Dsat} to the reciprocal of the transconductance efficiency, and thus the inversion level of the transistor.

$$V_{Dsat} = \frac{2}{\dfrac{g_m}{I_D}}. \tag{2.34}$$

In weak inversion, since g_m/I_D is equal to $1/(nU_T)$, (2.34) boils down to $2nU_T$, typically near 50 mV at room temperature. In strong inversion, V_{Dsat} is equal to the gate overdrive[8] since according to (2.33), g_m/I_D is equal to $2/(V_G - V_T)$, While constant in weak inversion, the drain saturation voltage increases with the gate-to-source voltage in strong inversion. This is illustrated in Figure 2.16 by the asterisks, which

---

[8] An alternative approximation of V_{Dsat} is obtained by dividing (2.34) with the subthreshold factor n. The weak and strong inversion approximations are then equal to $2U_T$ and $(V_G - V_{To})/n$, respectively.

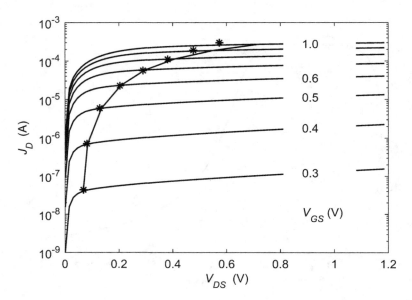

Figure 2.16 Evolution of V_{Dsat} from weak to strong inversion ($L = 100$ nm). The asterisks relate to (2.34), whereas the solid lines come from (2.35) and the basic EKV model.

mark the drain saturation voltage on every $I_D(V_{DS})$ characteristic. The data were obtained by running the Matlab code below:

```
L    = .1;
VGS  = (.3: .1: 1)';
gm_ID  = lookup(nch,'GM_ID','VGS',VGS,'L',L);
VDsat = 2./gm_ID; 9
JDsat = diag(lookup(nch,'ID_W','VDS',VDsat,'VGS',VGS,'L',L));
```

Notice that the V_{Dsat} locus conforms nicely with the heuristic image of the saturation drain voltage. We see that in weak inversion, the predicted V_{Dsat} is almost constant between 50 and 100 mV, while in strong inversion, V_{Dsat} increases rapidly according to the approximate square-law expression above.

In terms of the EKV model, (2.34) boils down to a simple analytical expression after applying (2.29):

$$V_{Dsat} = 2nU_T\left(1+q_S\right). \tag{2.35}$$

In weak inversion, since $q_S \ll 1$, we see that V_{Dsat} approaches $2nU_T$. For strong inversion, we find:

$$V_{Dsat} \approx 2nU_Tq_S \approx 2nU_T\sqrt{i} = 2nU_T\sqrt{\frac{I_D}{I_S}} = 2nU_T\sqrt{\frac{J_D}{J_S}} = \left(V_G - V_T\right). \tag{2.36}$$

---

[9] Note that there is no need to re-evaluate g_m/I_D for $V_{DS} = V_{Dsat}$; the transconductance efficiency varies very little with the drain voltage.

The solid curve of Figure 2.16, illustrates the basic EKV V_{Dsat} loci that were computed using the following code:

```
y   = XTRACT(nch,L,VDsat,0);
n = y(:,2);
VT = y(:,3);
JS = y(:,4);
VP = (VGS - VT)./n;
UT = .026;
qs = invq(VP/UT);10
VDsat_EKV = 2*n*UT.*(1 + qs);
JDsat_EKV = JS.*(qs.^2 + qs);
```

The result conforms rather well with the apparent V_{Dsat} of the real transistor, except for deep in strong inversion, because the model ignores mobility degradation.

2.3.3   Impact of Bias Conditions on EKV Parameters

In this section, we examine the impact that the gate length and drain voltage have on the basic EKV parameters. We take advantage of the XTRACT function to assess n, V_T and J_S, considering a grounded-source n-channel transistor. The same procedure can be used to identify EKV parameters across process corners (Slow/Nominal/Fast) as illustrated in Section A.1.5 (see Appendix 1).

Figure 2.17 shows the threshold voltage versus the gate length of the n-channel transistor when the drain voltage is stepped from 0.5 to 1 V. If the gate length is large (typically 1 μm), the threshold voltage doesn't vary significantly with the drain voltage. As we move toward submicron transistors, this isn't true any more. We see the threshold voltage increasing gradually at first. The dependence on V_{DS} grows progressively until a rapid collapse of V_T appears once the gate length gets smaller than 100 nm. This marks the so-called threshold voltage roll-off, which sets the limit of the technology (60 nm for the technology used in this book). Manufacturers make use of local ion implantations (called pocket implants) to combat the threshold voltage roll-off. This is done generally at the price of a slight increase of the threshold voltage that is visible in Figure 2.17 before the roll-off takes place. This is called reverse short channel effect.

While pocket implants enable us to increase the achievable operating frequencies by making shorter gate lengths feasible, the anisotropy they create unfortunately impacts the sensitivity of V_T regarding the drain voltage. This is clearly visible in Figure 2.17 and emphasized further in Figure 2.18, which displays similar data, exchanging L and V_{DS}. Notice the approximately linear drop of the threshold voltage with increasing drain voltage. It appears that the slope dV_T/dV_D is nearly constant. For the 60-nm gate length, we see that the threshold voltage drops by almost 80 mV per Volt increase in V_{DS}. Going from 60 to 100 nm suffices to reduce this rate by a factor larger than two. This behavior is a consequence of DIBL (drain-induced

---

[10] The "invq" function extracts q_S from V_P/U_T by inverting (2.23).

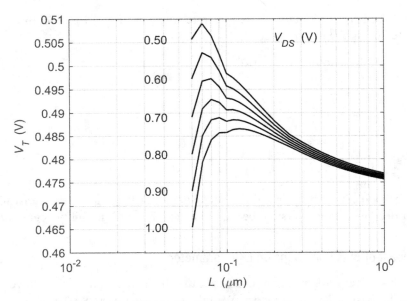

Figure 2.17 The extracted EKV threshold voltage versus the gate length, considering equally spaced drain voltages from 0.5 V to 1 V.

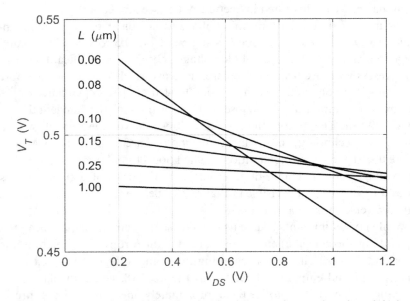

Figure 2.18. Dependence of the threshold voltage on the gate length and drain voltage.

barrier lowering). As mentioned earlier, DIBL concerns the amount of fixed charge below the channel that is controlled by the drain instead of the gate. The depletion layer surrounding the drain junction is gradually absorbing the depleted space charge region controlled by the gate. Thus, a lower gate voltage suffices to yield the

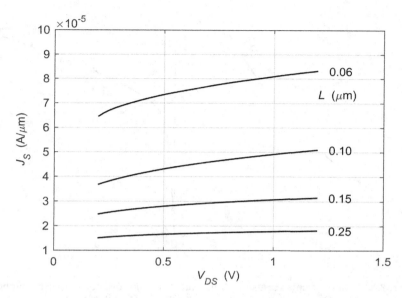

Figure 2.19 Dependence of the specific current density on the drain voltage.

same drain current. We will see later that this has a strong impact on the transistor's intrinsic voltage gain. Adding only 20 to 40 nm to the minimum gate length reduces DIBL significantly.

The extracted specific current is shown in Figure 2.19. Generally, I_S increases with V_{DS} due to CLM, since the specific current is proportional to W/L. Note that the impact of CLM grows as we move toward shorter gate lengths.

2.3.4   The Drain Current Characteristic $I_D(V_{DS})$

As already mentioned, V_{DS}-induced changes of the basic EKV parameters enable us to predict the evolution of the drain current. Figure 2.20 shows the saturated drain current of the CS (common-source) n-channel MOS transistor considering two gate lengths, 60 nm in (a) and 200 nm in (b). The gate-to-source voltage V_{GS} is kept constant and equal to 0.4 V, which corresponds to g_m/I_D values of 19.8 and 24.3 S/A, respectively (moderate and weak inversion). The Matlab code below shows how to compute the parameters as a function of the drain voltage to reconstruct the EKV drain current characteristic of Figure 2.20(a).

```
L   = .06;
VDS = 0.2: 0.1: 1.2;
y   = XTRACT(nch,L,VDS,0);[11]
n   = y(:,2);
VT  = y(:,3);
```

---

[11] The EKV parameters are automatically updated for every V_{DS}, since the drain-to-source voltages used in the XTRACT function form a column vector.

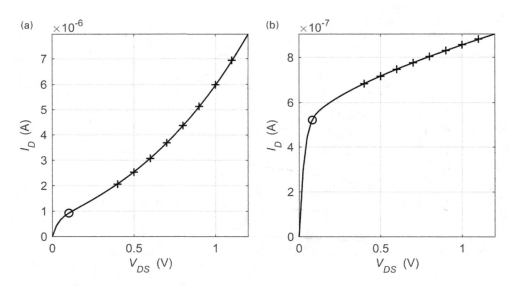

Figure 2.20 Drain currents versus the drain voltage of the common source n-channel transistor ($W = 10$ μm and $V_{GS} = 0.4$ V). In (a) the gate length is 60 nm, in (b) it is 200 nm. Solid lines = lookup table, crosses = basic EKV model. Circles mark the drain saturation voltage V_{Dsat} introduced in the previous section.

```
JS  = y(:,4);
VP  = (VGS - VT)./n;
UT  = .026;
qS  = invq(VP/UT);
IDEKV = W*JS.*(qS.^2 + qS);
```

The crosses in Figure 2.20 illustrate the drain currents of the n-channel transistor predicted by the EKV model, while the solid lines show the lookup table data. For the 60-nm transistor, the drain current shows a rapid increase with the drain voltage, caused by the threshold voltage reduction that is due to DIBL. The almost linear decrease of the threshold voltage caused by the increasing drain voltage produces a nearly linear increase of V_P, which, in turn, causes a nearly exponential increase of q (and thus also current). For the 200-nm transistor, we don't see this effect since DIBL is insignificant, as illustrated in Figure 2.18. The soft hyperbolic increase of the drain current seen instead reflects mainly the impact of CLM.

In Figure 2.21, we compare the drain currents of the 60-nm gate length transistor when the inversion level of the transistor changes from weak to strong inversion. V_{GS} is equal to 0.3 V in (a) ($g_m/I_D \sim 25$ S/A, weak inversion), and equal to 0.8 V in (b) ($g_m/I_D \sim 5$ S/A, strong inversion). Even though the threshold voltage decreases with V_{DS} in both figures, the exacerbated impact of DIBL on the drain current is visible only in (a) but not in (b). The reason is that in case (b), the impact on V_P on the nonlinear $\log(q)$ term of (2.23) is overwhelmed by the linear term in the same expression.

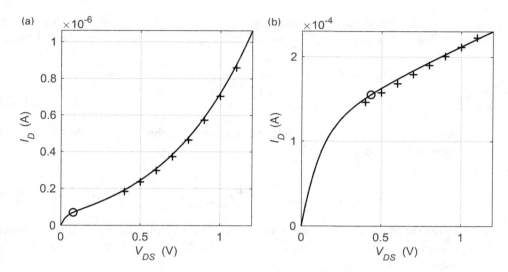

Figure 2.21 Drain currents versus the drain voltage of the common source n-channel. In (a), V_{GS} is equal to 0.3 V ($g_m/I_D \sim 25$ S/A), in (b) it is 0.8 V ($g_m/I_D \sim 5$ S/A). Solid lines = lookup table, crosses = EKV model. Circles mark the drain saturation voltage V_{Dsat}. $W = 10$ μm, $L = 60$ nm.

2.3.5 The Output Conductance g_{ds}

The fact that the CSM ignores second-order effects such as CLM and DIBL explains the absence of output conductance in the saturation region. Saturated real transistors exhibit non-zero output conductances, which vary significantly from weak to strong inversion as evident from the drain current plots of the previous section. To reproduce drain currents in saturation using the basic EKV model, we must add parameters that reflect the influence of V_{DS}. We therefore introduce the derivatives of n, V_T and $\log(I_S)$ with respect to V_{DS}, respectively. We call these the sensitivity parameters S_n, S_{VT} and S_{IS}:

$$S_n = \frac{dn}{dV_D}; \quad S_{VT} = \frac{dV_T}{dV_D}; \quad S_{IS} = \frac{d \log(I_S)}{dV_D}. \tag{2.37}$$

The XTRACT function evaluates these derivatives and outputs them either as a row-vector or a matrix versus the input drain voltage (scalar or column vector). The output parameter order of XTRACT is: V_{DS}, n, V_T, I_S, S_n, S_{VT} and S_{IS}.

Since the output conductance g_{ds} is the derivative of the drain current with respect to the drain-to-source voltage, we start from the basic EKV expression of I_D given in (2.15):

$$g_{ds} = \frac{dI_D}{dV_D} = i \frac{dI_S}{dV_D} + I_S \frac{di}{dV_D}. \tag{2.38}$$

Using the specific current sensitivity factor S_{IS}, which represents the slope of the semilog curves of Figure 2.19, we can now express the first term on the right-hand side of (2.38) as:

$$i \frac{dI_S}{dV_D} = I_D \frac{d \log(I_S)}{dV_D} = I_D S_{IS}. \tag{2.39}$$

Per (2.22) and (2.23), and approximating n as constant ($S_n \approx 0$), the second contributor to g_{ds} becomes:

$$I_S \frac{di}{dV_D} = I_S \frac{di}{dq} \frac{dq}{dV_D} = I_S \left(\frac{q}{U_T} \frac{dV_P}{dV_D} \right) = I_S \frac{q}{nU_T} \left(-\frac{dV_T}{dV_D} \right). \tag{2.40}$$

Using the transconductance expression of (2.28) and introducing the threshold voltage sensitivity factor S_{VT} (the slope of the threshold voltage curves of Figure 2.18) we turn (2.40) into:

$$I_S \frac{q}{nU_T} \left(-\frac{dV_T}{dV_D} \right) = -g_m S_{VT}. \tag{2.41}$$

Finally, combining (2.39) and (2.41) we get the simple analytical expression below:

$$g_{ds} = -g_m \cdot S_{VT} + I_D \cdot S_{IS}. \tag{2.42}$$

In this expression, S_{VT} primarily captures DIBL effects and is usually a negative number (V_T decreases approximately linearly with drain voltage). S_{IS} mainly accounts for CLM.

The two plots shown in Figure 2.22 compare predicted and lookup output conductances versus the drain voltage when the gate length takes two distinct values: 60 nm (a) and 200 nm (b). We consider the same n-channel transistor as in Figure 2.20, which has $V_{GS} = 0.40$ V ((a) $g_m/I_D = 19.8$ and (b) 24.3 S/A). The solid lines represent the lookup table results, crosses are for the basic EKV model and circles mark V_{Dsat}. The nearly exponential increase of the drain current due to DIBL that is visible in Figure 2.20(a) also explains the steady increase of g_{ds} that is visible in Figure 2.22(a). In fact, the term $g_m S_{VT}$ overwhelms $I_D S_{IS}$ so that the output conductance varies almost directly with g_m. On the other hand, for the 200-nm curve in Figure 2.22(b), the $I_D S_{IS}$ term dominates because the DIBL term is small.

The Matlab code yielding the g_{ds} curve of Figure 2.22(a) is listed below for reference.

```
VDS = 0.1*(3:12)';              % V
VGS = .4;                       % V
L   = .06;                      % micron
% gds from look-up tables =========
gds = lookup(nch,'GDS','VDS',nch.VDS,'VGS',VGS,'L',L);
% extract EKV param =================
y   = XTRACT(nch,L,VDS,0);
```

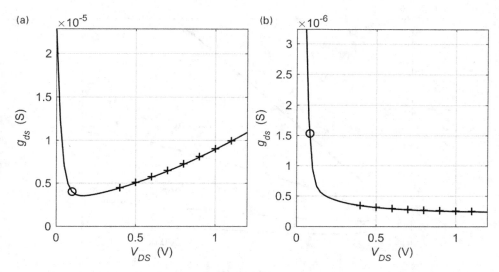

Figure 2.22 Output conductances versus the drain voltage of the common source n-channel transistor with $W = 10$ μm and $V_{GS} = 0.4$ V. (a) $L = 60$ nm, (b) $L = 200$ nm. Solid lines = lookup table, crosses = EKV model. Circles mark the drain saturation voltage V_{Dsat}.

```
n   = y(:,2);
VTo = y(:,3);
JS = y(:,4);
SVT = y(:,6); SIS = y(:,7);
% EKV drain current
VP    = (VGS - VTo)./n;
qS    = invq(VP/UT);
IDEKV = W*JS.*(qS.^2 + qS);
UT    = .026;
gm = IS./(n*UT).*qS;
gdsEKV = - gm.*SVT + IDEKV.*SIS;
```

2.3.6 The g_{ds}/I_D Ratio

Dividing the two sides of (2.41) by I_D, we get an expression linking g_m/I_D to g_{ds}/I_D:

$$\frac{g_{ds}}{I_D} = -S_{VT}\frac{g_m}{I_D} + S_{IS}. \tag{2.43}$$

Per the above expression and if the gate length and the drain voltage remain constant, g_{ds}/I_D and g_m/I_D are linearly related. Figure 2.23 shows that this indeed holds true for real transistors over a wide range of transconductance efficiencies, regardless of the gate length.

Recall that the reciprocal of g_{ds}/I_D is called also the Early voltage, V_{Early}. To visualize the concept, consider a set of $I_D(V_{DS})$ characteristics. Suppose that we select a quiescent point drain-to-source voltage V_{DS} and a drain current I_D. The slope of the $I_D(V_{DS})$ characteristic at the selected point signifies the output conductance g_{ds}. To

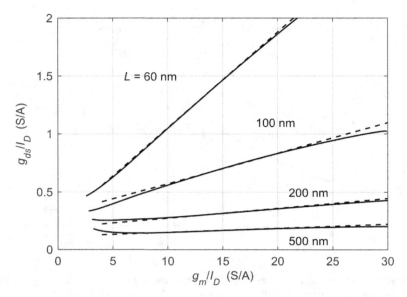

Figure 2.23 The ratio g_{ds}/I_D versus g_m/I_D. Solid lines = real transistor, dashed lines = basic EKV model. $V_{DS} = 0.6$ V and $V_{SB} = 0$ V.

illustrate V_{Early} graphically, we prolong the tangent until it intersects the horizontal axis. The Early voltage is the voltage difference between the intersection point and V_{DS}.

2.3.7   The Intrinsic Gain

We now consider the so-called intrinsic gain of the transistor, which is the ratio of g_m and g_{ds}:

$$A_{intr} = \frac{g_m}{g_{ds}} = \frac{g_m / I_D}{g_{ds} / I_D} = \frac{1}{S_{IS} \left(\dfrac{g_m}{I_D} \right)^{-1} - S_{VT}}. \tag{2.44}$$

The result on the right-hand side of this expression is obtained by substituting (2.43). In the circuit domain, the intrinsic gain represents the magnitude of the low-frequency voltage gain (A_{v0}) in a common-source, open-drain[12] transistor configuration (see small-signal model in Figure 2.24):

$$A_{v0} = \frac{v_{out}}{v_{in}} = -\frac{g_m}{g_{ds}} = -A_{intr}. \tag{2.45}$$

In addition, the intrinsic gain is also relevant in more complex circuits, such as the cascode stage and operational amplifiers, which typically have a voltage gain that is proportional to $A_{intr}{}^n$, where n is an integer.

---

[12] Open drain means that the drain terminal of the transistor is connected to an ideal current source. In other words, the only impedance seen by the drain is its own internal resistance $1/g_{ds}$.

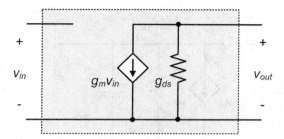

Figure 2.24 Small-signal transistor model with output conductance g_{ds}.

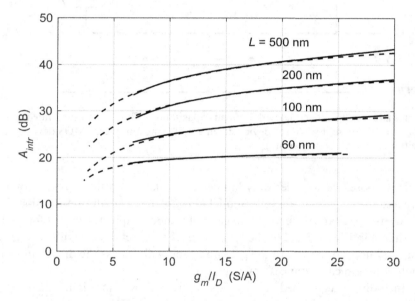

Figure 2.25 Intrinsic gain of an n-channel transistor. Solid lines = real transistor, dashed lines = basic EKV model. $V_{DS} = 0.6$ V and $V_{SB} = 0$ V.

When the transistor operates in weak inversion (g_m/I_D is large) and the channel is very short (S_{VT} is large), A_{intr} becomes essentially constant and is approximately equal to the reciprocal of $-S_{VT}$. Thus, DIBL determines the voltage gain. For example, in the 60-nm transistor, where S_{VT} equals –0.080, the intrinsic gain is approximately 22 dB. It does not vary substantially with g_m/I_D as illustrated by the lower curve of Figure 2.25, which plots A_{intr} versus g_m/I_D. When the gate length increases, S_{VT} decreases in magnitude, and consequently A_{intr} increases, while the S_{IS} term enhances the sensitivity to g_m/I_D.

2.3.8 MOSFET Capacitances and the Transit Frequency f_T

So far, we have taken only low-frequency parameters into consideration. However, device capacitances must be introduced to model a circuit's frequency response. For

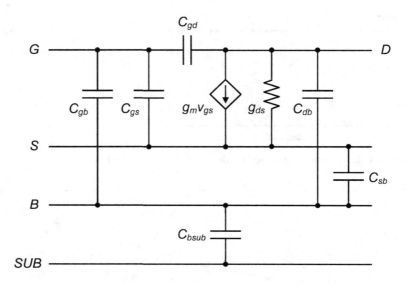

Figure 2.26 MOSFET model with lumped capacitances. The bulk-to substrate capacitance C_{bsub} is only present when the device sits in a well (for example, a PMOS device in an n-well within a p-type chip substrate).

this purpose, we consider only lumped capacitances in this book, in other words, we assume a quasi-static (QS) model and don't deal with non-quasi-static (NQS) or transcapacitance effects.[13] The general lumped-capacitor model for a MOSFET is shown in Figure 2.26. In the remainder of this section, we will consider only the capacitances connected to the gate node and will discuss the drain- and source-to-bulk junction capacitances in Chapter 3.

Including the lumped device capacitances shown in Figure 2.26 extends the applicability of our model to include QS behavior. However, the required capacitance parameters do not follow simple analytic expressions like those used so far to model the drain current. For example, the gate-to-drain capacitance (C_{gd}) follows from the device's three-dimensional structure, and depends on complex fringe fields at the drain contact. Hence, we will not attempt to construct an analytical model and instead rely on lookup table data only.

In total, there are three capacitances connected to the gate: from gate to source (C_{gs}), from gate to bulk (C_{gb}), and gate to drain (C_{gd}). The sum of these capacitances is commonly called the total gate capacitance, C_{gg}:

$$C_{gg} = C_{gs} + C_{gb} + C_{gd}. \tag{2.46}$$

Figure 2.27 shows lookup table results for the dependence on the inversion level of the three components making up C_{gg} for two gate lengths: 0.1 μm (a) and 1 μm (b).

---

[13] Transcapacitance effects are signified in modern devices by the fact that C_{gd} is not exactly equal to C_{dg}. Such effects are covered in the advanced modeling literature.

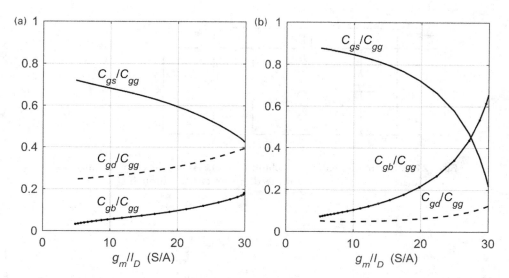

Figure 2.27 The relative parasitic capacitances C_{gx}/C_{gg} versus g_m/I_D. In (a), the gate length L is 0.1 μm, and in (b) it is 1 μm ($V_{SB} = 0$ V, $V_{DS} = 0.6$ V).

To clarify commonalities and differences, it helps to understand the role of the inversion layer. When the transistor is saturated, a large share of the capacitive gate current is diverted through the inversion layer to the source. The gate-to-source capacitance is therefore larger than the gate-drain capacitance. While true in strong inversion, this gets less so in moderate inversion and does not hold deep in weak inversion. Here, the source and drain gate capacitances are mostly gate overlaps and fringe fields, owing to the disappearance of the channel. Since nothing differentiates the source from the drain anymore, C_{gs} and C_{gd} tend to equalize. The gate moreover "sees" the substrate directly without the screen formed by the channel in strong inversion. Hence, the gate-to-bulk capacitance (C_{gb}) is now maximized.

The striking difference between the gate-to-drain and gate-to-bulk capacitances when we go from long to the short channels deserves clarification. In long channel devices, the larger the gate length, the larger the associated gate-to-bulk capacitance while the gate-to-drain capacitance remains unchanged. This is clearly visible in Figure 2.27(b), where the relative gate-to-bulk capacitance in weak inversion dominates over the gate-to-source and gate-to-drain capacitances. For the short channel transistor in Figure 2.27(a) the situation is different. The proximity of the source and drain junctions helps hide the bulk from the gate to some extent. The gate-to-bulk capacitance consequently remains relatively small for all inversion levels. While the sum of the relative gate-to-source and gate-to-drain capacitances represents 80% of the total gate capacitance in the short channel transistor, it is only 30–40% in the long-channel device.

The total gate capacitance is typically used to define an important figure of merit: the angular transit frequency.

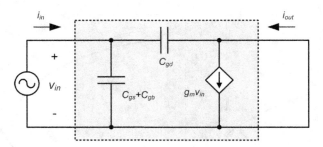

Figure 2.28 Test circuit used in the definition of ω_T. At ω_T, the current gain $|i_{out}/i_{in}|$ is approximately unity (neglecting the feedforward current through C_{gd}).

$$\omega_T = 2\pi f_T = \frac{g_m}{C_{gg}}$$

(2.47)

In the literature, the physical meaning of the transit frequency is often interpreted by showing that the current gain of a common-source stage (ratio of i_{out} and i_{in} in Figure 2.28) with ac-shorted drain becomes unity at ω_T. Therefore, the transit frequency can be viewed as a proxy for the maximum useful operating frequency of a MOSFET. Alternatively, one can consider ω_{max}, the maximum frequency of oscillation, defined at unity power gain [10].

In practice, MOSFETs are usually not operated at or near ω_T. This is mostly because the actual device behavior becomes difficult to model and is strongly affected by NQS effects and other parasitic elements such as gate resistance. The circuit simulations and parameters assumed in this book are based on the PSP model, which can incorporate such effects in principle [11]. However, unless great care is taken in populating the respective model parameters based on actual measurements, there is always uncertainty about the physical behavior near the transit frequency.

Figure 2.29 shows a SPICE simulation of the current gain for a 60-nm n-channel transistor. The setup is the same as in Figure 2.28; the gate is driven by a voltage source and we measure the magnitude of i_{out}/i_{in}. The solid line shows the response of the plain transistor as modeled by the PSP model. We see that the impact of C_{gd} is significant, and the feedforward zero that it introduces is clearly visible. To reveal the response with C_{gd} taken out, we connect a capacitance of $-C_{gd}$ between gate and drain and obtain the dashed line shown in the plot. As we can see, the intrinsic device without C_{gd} shows a much more complex response around 2–3 times ω_T, which may or may not be well calibrated relative to the actual physical behavior. Part of this behavior is due to gate resistance and other second-order effects (such as transcapacitance).

In light of the modeling uncertainties around ω_T, a conservative rule of thumb is to assume that most circuits operate predictably up to $\omega_T/10$ [10]. It therefore makes sense to view ω_T as an extrapolated quantity, merely defining the ratio g_m/C_{gg}, i.e. the

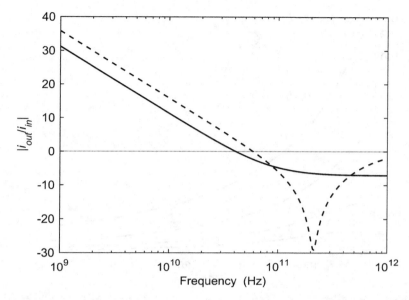

Figure 2.29 Simulated current gain for a 65-nm n-channel device (g_m/I_D = 9.6 S/A). Solid line: Transistor without any modifications. Dashed line: A neutralization capacitor of size $-C_{gd}$ is connected between gate and drain to reveal the response of the intrinsic transistor.

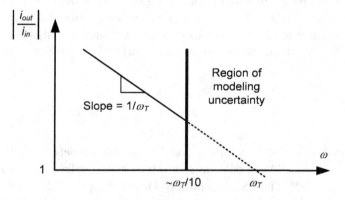

Figure 2.30 Interpretation of ω_T as an extrapolated quantity.

amount of transconductance per total gate capacitance presented by the transistor. This is illustrated in Figure 2.30.

For most examples in this book, we will assume that ω_T is ten times larger than the maximum angular frequency of interest to constrain the choice of channel length. Later, in Chapters 5 and 6, we will encounter cases where we may occasionally model the circuit behavior up to $\sim\omega_T/3$. This is common practice and reasonable in the modeling of non-dominant poles that define the phase margin of feedback amplifiers.

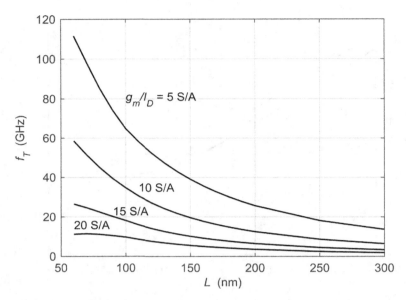

Figure 2.31 Transit frequency versus channel length for an n-channel device. Using short channel lengths and operating in strong inversion substantially improves the transit frequency.

In any case, the fact that f_T can be as high as 100 GHz in modern technology (see Figure 2.31) leaves plenty of room to operate transistors at high frequencies without having to rely on NQS models and complex passive model extensions. As the gate lengths continue to shrink, f_T tends to improve proportionally to $1/L^\alpha$, where α is close to one for velocity saturated devices [8] (as opposed to $\alpha = 2$ for the ideal square law).

2.4   Summary

After the introduction of two long-channel transistor models, the charge sheet model and the basic EKV model, the connection with low-frequency parameters like g_m, g_{ds}, etc. is made. The CSM is a powerful tool for apprehending the operation modes of MOS transistors: weak, moderate and strong inversion. It is too complex, however, to be used for circuit design and sizing. The basic EKV model is a simpler alternative that approximates the CSM. Both models are continuous, since they encompass weak to strong inversion with the same equations. Continuity is important, because many modern CMOS circuits operate nowadays in the moderate inversion region. We will see in later chapters that this region offers a good tradeoff between high performance and low power consumption.

The impact of second-order effects like DIBL and CLM is ignored, however, by the two models. Making the parameters dependent on the source and drain voltages as well as the gate length, extends the validity of the basic EKV model to short-channel devices. This paves the way toward an intuitive interpretation of second-order effects and the possibility to assess their relative importance using a simple model extension.

Most importantly, this chapter refined our understanding and physical under-pinnings of the transconductance efficiency g_m/I_D, which will be at the core of the sizing methodology developed in the remaining chapters.

2.5   References

[1] J. R. Brews, "A Charge-Sheet Model of the MOSFET," *Solid. State. Electron.*, vol. 21, no. 2, pp. 345–355, 1978.

[2] F. Van de Wiele, "A Long Channel MOSFET Model," *Solid. State. Electron.*, vol. 22, no. 12, pp. 991–997, 1979.

[3] H. Oguey and S. Cserveny, "Modèle du transistor MOS valable dans un grand domaine de courant," *Bull. SEV/VSE*, vol. 73, no. 3, pp. 113–116, 1982.

[4] A. I. A. Cunha, M. C. Schneider, and C. Galup-Montoro, "An Explicit Physical Model for Long Channel MOS Transistors including Small-Signal Parameters," *Solid. State. Electron.*, vol. 38, no. 11, pp. 1945–1952, 1995.

[5] "An MOS Transistor Model for Analog Circuit Design," *IEEE J. Solid-State Circuits*, vol. 33, no. 10, pp. 1510–1519, Oct. 1998.

[6] J.-M. Sallese, M. Bucher, F. Krummenacher, and P. Fazan, "Inversion Charge Linearization in MOSFET Modeling and Rigorous Derivation of the EKV Compact Model," *Solid. State. Electron.*, vol. 47, no. 4, pp. 677–683, 2003.

[7] C. C. Enz and E. A. Vittoz, *Charge-Based MOS Transistor Modeling: The EKV Model for Low-Power and RF IC Design*. John Wiley & Sons, 2006.

[8] P. R. Gray, P. Hurst, S. H. Lewis, and R. G. Meyer, *Analysis and Design of Analog Integrated Circuits*, 5th ed. John Wiley & Sons, 2009.

[9] Y. Taur and T. H. Ning, *Fundamentals of Modern VLSI Devices*, 2nd ed. Cambridge University Press, 2013.

[10] T. H. Lee, *The Design of CMOS Radio-Frequency Integrated Circuits*, 2nd ed. Cambridge University Press, 2004.

[11] G. Gildenblat *et al.*, "PSP: An Advanced Surface-Potential-Based MOSFET Model for Circuit Simulation," *IEEE Trans. Electron Devices*, vol. 53, no. 9, pp. 1979–1993, Sep. 2006.

[12] P. Jespers, *The gm/ID Methodology, a Sizing Tool for Low-Voltage Analog CMOS Circuits*. Springer, 2010.

3   Basic Sizing Using the g_m/I_D Methodology

This chapter introduces the reader to the concepts underpinning the g_m/I_D sizing methodology. To simplify this initial treatment, we focus on elementary circuits with only a few transistors and leave the sizing of more complex circuits for later chapters.

3.1   Sizing an Intrinsic Gain Stage (IGS)

The starting point of our discussion is the intrinsic gain stage (IGS), shown in Figure 3.1. The IGS can be viewed as an idealized version of a common-source stage with active load, which is frequently used in linear amplifiers. It also represents the small-signal half-circuit model of an actively-loaded differential pair, which is typically used as the input stage of a differential amplifier. The term "intrinsic" reflects the fact that no external components (other than the load capacitance C_L) are considered. For simplicity, we also assume that the stage is driven by an ideal voltage source and defer more realistic scenarios to later chapters (and Example 3.13).

In the circuit of Figure 3.1, the drain current is set up using an ideal current source (I_D) and the input bias voltage (V_{BIAS}) is assumed to be adjusted such that the transistor operates in saturation at some desired output quiescent point (e.g. $V_{OUT} = V_{DD}/2 = 0.6$ V). We will leave out details on how the bias voltage is generated, since such discussions are much more meaningful in the context of larger circuits. For basic biasing considerations, the reader may refer to introductory textbooks on CMOS transistor stages [1].

3.1.1   Circuit Analysis

The first step in any systematic design flow is to perform a suitable circuit analysis. For this purpose, we briefly review the frequency response of the IGS, considering its small-signal model shown in Figure 3.2. This model contains the gate capacitances C_{gs}, C_{gb} and C_{gd}, whichwere already introduced in Section 2.3.8, as well as the drain-to-bulk junction capacitance C_{db}.

To simplify further, we will initially neglect the junction capacitance and assume $C_{db} \ll C_L$. Additionally, it is usually true that C_{gd} is much smaller than C_L. Under

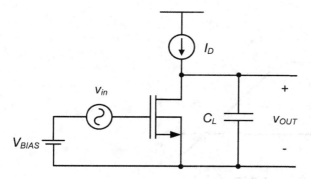

Figure 3.1 Circuit schematic of an intrinsic gain stage.

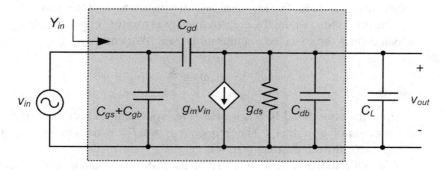

Figure 3.2 Small-signal model of the IGS (the transistor's bulk is assumed to be connected to the source).

these conditions, the frequency response is well-approximated by the following expression:

$$A_v(j\omega) = \frac{v_{out}}{v_{in}} \cong \frac{A_{v0}}{1 + j\dfrac{\omega}{\omega_c}}, \tag{3.1}$$

where A_{v0} is the low-frequency (LF) voltage gain (equal in magnitude to the intrinsic gain of the transistor):

$$A_{v0} = -\frac{g_m}{g_{ds}} = -A_{intr} \tag{3.2}$$

and ω_c is the circuit's angular corner frequency, given by

$$\omega_c = \frac{g_{ds}}{C_L}. \tag{3.3}$$

Figure 3.3 shows a straight-line approximation of the frequency response magnitude. The gain is constant at low frequencies and falls with −20 dB/decade beyond

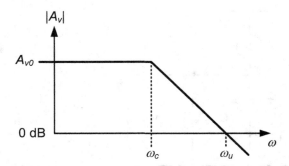

Figure 3.3 Magnitude response of the IGS.

the corner frequency. The low- and high-frequency asymptotes meet at ω_c. Another significant point is where the high-frequency asymptote crosses the horizontal axis, which defines the so-called angular unity gain frequency ω_u:

$$\omega_u \cong \frac{g_m}{C_L}. \tag{3.4}$$

Since the considered IGS is a first-order system, ω_u is (approximately) equal to the product of ω_c and A_{v0}. The frequency $f_u = \omega_u/2\pi$ is therefore commonly called the gain-bandwidth product (GBW).

Another important figure of merit of the IGS is how much output capacitance it drives relative to the capacitance it presents at its input. To quantify this ratio, we define the stage's fan-out (*FO*) as:

$$FO = \frac{C_L}{C_{gs} + C_{gb} + C_{gd}} = \frac{C_L}{C_{gg}}. \tag{3.5}$$

While the fan-out is known as a significant metric in the sizing of digital logic gates, it tends to be an underappreciated concept in analog design. However, as we shall see throughout this book, some analog design problems can be elegantly framed using *FO*.

Using *FO* to express the stage loading in a normalized manner is motivated by the fact that the load is often just another transistor stage (or general circuit) whose input capacitance is linked to the global specifications in a similar way as the stage under consideration. In many circuits, the capacitances of all components scale up and down together when different combinations of gain, bandwidth and noise specifications are being explored in the overall design space. For example, we will see in Chapter 4 that *FO* plays a significant role in the tradeoff between GBW, supply current and noise performance of the IGS.

The fan-out metric defined in (3.5) also coincides with a frequency ratio that is important from a modeling perspective:

$$\frac{\omega_T}{\omega_u} = \frac{f_T}{f_u} = \frac{g_m/C_{gg}}{g_m/C_L} = FO. \tag{3.6}$$

As we have argued in Chapter 2, the quasi-static transistor model used in Figure 3.2 becomes inaccurate as the frequencies of interest approach about 1/10th of f_T in moderate or strong inversion. Hence, for the IGS model to hold near f_u, the fan-out should ideally be larger than 10.

As a final note, it should be mentioned that the exact input admittance of the IGS is neither purely capacitive, nor exactly equal to $j\omega C_{gg}$. Using the circuit model in Figure 3.2, and applying the Miller theorem [1], it follows that:

$$Y_{in}(j\omega) = j\omega(C_{gs} + C_{gb}) + j\omega C_{gd}(1 - A_v(j\omega)). \tag{3.7}$$

This expression reduces asymptotically to $j\omega C_{gg}$ as the circuit loses its voltage gain at high frequencies (or if the drain is ac-grounded). The fan-out metric that we defined in (3.5) should therefore be viewed as an asymptotic metric that was intentionally simplified to be useful without getting into the complexities of accurate admittance modeling.

3.1.2 Sizing Considerations

With the expressions derived in the previous section, we are now ready to size the IGS per a given set of specifications. Specifically, for a given load capacitance C_L, we typically want to establish a methodology that enables us to determine the drain current, the device width and gate length for a given unity-gain frequency (f_u) target. For this purpose, we can consider the generic sizing flow that was already introduced in Chapter 1 (repeated here for convenience):

1. Determine g_m (from design specifications).
2. Pick L:
   - short channel → high speed, small area;
   - long channel → high intrinsic gain, improved matching, …
3. Pick g_m/I_D:
   - large g_m/I_D → low power, large signal swing (low V_{Dsat});
   - small g_m/I_D → high speed, small area.
4. Determine I_D (from g_m and g_m/I_D).
5. Determine W (from I_D/W).

Step 1 of this flow is straightforward for the given IGS circuit, since g_m is fixed per (3.4) and must be equal to $\omega_u C_L$. However, additional constraints are needed to decide on the best choice for the channel length L and the transconductance efficiency g_m/I_D. For example, suppose that we want to minimize the drain current, which is given by:

$$I_D = \frac{g_m}{g_m/I_D}. \tag{3.8}$$

Based on this expression alone, it would make sense to operate the transistor at the maximum possible g_m/I_D, which is obtained in weak inversion. Unfortunately, the problem with operating the device at high g_m/I_D is that the transistor's f_T is small, leading to large gate capacitance and a fan-out that may be smaller than the recommended bound of 10.

The take-home from this initial discussion is that finding the optimal inversion level is not straightforward, and that all applicable design constraints must be considered jointly. In the remainder of this chapter and the next, we will therefore gradually introduce a variety of design constraints, and subsequently show how they impact the choice of L and g_m/I_D. However, an important prerequisite for this exploration is the ability to size a device once g_m/I_D and L have been chosen (by some design/optimization criterion). This procedure is thus the topic of the next section.

3.1.3 Sizing for Given L and g_m/I_D

If L and g_m/I_D are known, the sizing procedure simplifies to the flow illustrated in Figure 3.4. As discussed earlier, g_m follows from the design specifications, while I_D can be directly computed using g_m and g_m/I_D via (3.8). The last unknown parameter is the device width, which follows from the ratio of drain current and drain current density $J_D = I_D/W$ (expressed in A/μm):

$$W = \frac{I_D}{J_D} = \frac{I_D}{I_D/W}.$$

(3.9)

We refer to the step of computing the device width as "de-normalization," as it marks the transition from normalized quantities like g_m/I_D and I_D/W to absolute geometries.

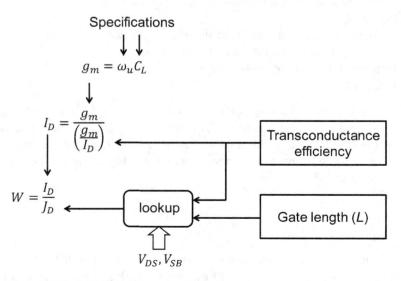

Figure 3.4 Sizing for given L and g_m/I_D.

Equation (3.9) assumes that the drain current scales strictly proportional to the transistor width, which requires that the device is large enough to make narrow-width effects negligible. Fortunately, this is a condition that is met in most analog circuits. Appendix 3 takes a closer look at this assumption.

Since there is a one-to-one mapping between transconductance efficiency and current density for given L, V_{DS} and V_{SB} (see Section 2.3.1), we can find J_D using the lookup table data for our technology and thereby complete the sizing. The mechanics of this approach are illustrated through the following example.

Example 3.1 A Basic Sizing Example

Size the circuit of Figure 3.1 so that that $f_u = 1$ GHz when $C_L = 1$ pF. Assume $L = 60$ nm, $g_m/I_D = 15$ S/A (moderate inversion), $V_{DS} = 0.6$ V and $V_{SB} = 0$ V (default values). Find the low-frequency voltage gain and the Early voltage of the transistor. Validate the results through SPICE simulations.

SOLUTION

Following the above-discussed procedure, we start by computing the transconductance using (3.4):

```
fu = 1e9;
CL = 1e-12;
gm = 2*pi*fu*CL;
```

Since g_m/I_D is given and equal to 15 S/A, we can find I_D via (3.8):

```
ID = gm/gm_ID
```

This yields $I_D = 419$ μA. To find the width W, we divide I_D by the drain current density J_D per (3.9). Conceptually, to find J_D, we can consider a plot like that of Figure 2.14(b) (for $L = 60$ nm), and look for the drain current density at $g_m/I_D = 15$ S/A. Equivalently, this is accomplished using the following lookup command:

```
JD = lookup(nch,'ID_W','GM_ID',gm_ID,'VDS',VDS,'VSB',VSB,'L',L);
```

where V_{DS}, V_{SB} and L correspond to the drain-to-source, source-to-bulk voltage and gate length of the transistor. We find that J_D is equal to 10.05 μA/μm. Hence, dividing I_D/J_D yields $W = 41.72$ μm and we have thus completed the sizing procedure.

We see from this flow that the drain current I_D is known as soon as we fix the inversion level (via g_m/I_D). Fixing the gate length then also determines W. These steps not only fix I_D and W, but also V_{GS}, A_{v0} and f_T, since these are the outcome of similar lookup operations involving the same variables. For example, we can find V_{GS} using the lookupVGS companion function (see Appendix 2 for a description):

```
VGS = lookupVGS(nch,'GM_ID',gm_ID,'VDS',VDS,'VSB',VSB,'L',L);
```

This yields $V_{GS} = 0.4683$ V. Using similar commands, we can also find the low-frequency voltage gain and the device's transit frequency:

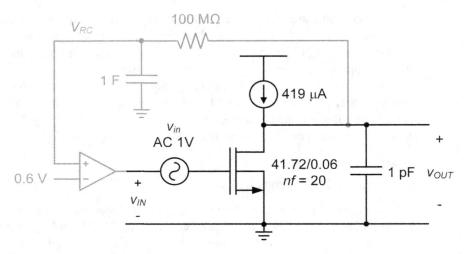

Figure 3.5 Schematic for SPICE simulation.[1]

```
Av0 = -lookup(nch,'GM_GDS','GM_ID',gm_ID,'VDS',VDS,'VSB',VSB,'L',L)
fT  = lookup(nch,'GM_CGG','GM_ID',gm_ID,'VDS',VDS,'VSB',VSB, ...
'L',L)/2/pi
```

This yields A_{vo} = −10.25 and f_T = 26.46 GHz. Note that f_T is much larger than the desired unity gain frequency. This implies that the *FO* is sufficiently large (*FO* = 26.46 > 10), and that the equivalent circuit of Figure 3.2 is valid. Finally, we compute the Early voltage V_A using:

$$V_A = \frac{I_D}{g_{ds}} = \frac{\left(\dfrac{g_m}{g_{ds}}\right)}{\left(\dfrac{g_m}{I_D}\right)}$$

This yields the rather low value of V_A = 0.683 V, due to the strong DIBL effect for L = 60 nm (see Sections 2.3.5 and 2.3.6).

We turn now to verification, and run SPICE simulations using the setup shown in Figure 3.5. The transistor is partitioned into 20 fingers (*nf* = 20), each 2.086 µm wide (see Appendix 3 for a discussion on finger partitioning). The quiescent point gate voltage is set using an auxiliary feedback circuit (drawn in gray) that computes V_{GS} such that V_{DS} = 0.6 V. For any meaningful frequency above DC, the feedback loop is open, and the circuit is evaluated across frequency as intended. In a practical implementation of this circuit, the gate voltage is sometimes set via an actual feedback circuit (similar to what is shown), or computed using a replica circuit [1]. More commonly, the transistor

---

[1] The AC amplitude of 1 V is conveniently chosen so that the transfer function follows directly from the output voltage. Note that one is free to choose the test amplitude, since the circuit is perfectly linear in the performed small-signal AC analysis.

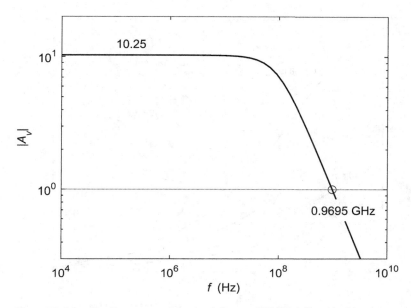

Figure 3.6 Magnitude response obtained from a SPICE AC analysis.

is operated within a differential pair, where the bias current is drawn from the source terminal, obviating the need for an explicit computation of V_{GS} (see Section 3.3).

We simulate this circuit and first inspect the DC operating point output:

V_{GS} = 468.119 mV,
V_{DS} = 600.468 mV,
I_D = 419.004 uA,
g_m = 6.28284 mS,
g_{ds} = 612.939 μS.

We thus have g_m/I_D = 6.28 mS/419 μA = 14.99 S/A, which is very close to the desired value. Also, the quiescent point gate voltage and the simulated g_m/g_{ds} ratio agree with the predicted values.

Next, we run a small-signal AC analysis and obtain the plot shown in Figure 3.6. Again, the result is very close to our expectation. This is not surprising, because we already saw from the operating point data that g_m is set almost exactly as desired. The gain-bandwidth error of approximately 3% can be explained by the fact that we have neglected the extrinsic capacitances (C_{gd} and C_{db}) when using (3.4); we will address this issue in Section 3.1.7.

The above example gave us a first feel for g_m/I_D-based design, taking advantage of pre-computed lookup tables. We arrived at the desired result without any iterations and "tweaking" in SPICE.

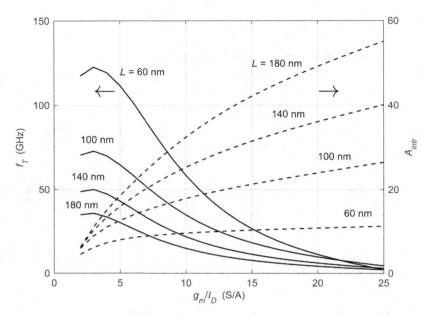

Figure 3.7 Transit frequency f_T and intrinsic gain A_{intr} versus g_m/I_D, considering four equally spaced gate lengths from 60 to 180 nm ($V_{DS} = 0.6$ V, $V_{SB} = 0$ V).[2]

3.1.4   Basic Tradeoff Exploration

In the previous section, we assumed that L and g_m/I_D were known, and this led to a straightforward sizing procedure. We will now begin to explore tradeoffs that will constrain and ultimately define the choice of these parameters in a practical design. We will focus in this section on first-order metrics (gain and bandwidth), and leave the inclusion of more advanced specifications (noise, linearity and mismatch) to Chapter 4.

We already know from Chapters 1 and 2 that g_m/I_D and L affect parameters that tradeoff with one another: the transit frequency and the intrinsic gain. This point is illustrated in Figure 3.7, which shows f_T and $A_{intr} = g_m/g_{ds}$ plotted against g_m/I_D. The data for this plot was obtained using:

```
Avo = lookup(nch,'GM_GDS','GM_ID',gmID,'L',L);
fT  = lookup(nch,'GM_CGG','GM_ID',gmID,'L',L)/(2*pi);
```

where gmID and L are vectors defining the shown sweep range.

The key observations from this plot are summarized as follows:

• The transit frequency is largest in strong inversion (small g_m/I_D), and gradually decays as we approach weak inversion (large g_m/I_D). Unfortunately, this means that we can either make the transistor fast or efficient, but not both.

---

[2] Note that the values of V_{DS} and V_{SB} won't affect the general tradeoffs shown in this figure (as long as the device remains saturated). See also Section 2.3.1.

- The intrinsic gain is large for long channels, but long channels have an adverse effect on the transit frequency. We can either achieve large gain (using large L) or high transit frequency (using short L), but not both.

Our job as circuit designers is to manage these tradeoffs in accordance with the overall design goal, which may vary widely (see the examples in Chapters 5 and 6). To untangle this problem further without losing generality, we will now consider three examples that each include one constraint.

The first example assumes that g_m/I_D is fixed and that we are free to choose L. This allows us to see the connection between channel length, voltage gain and other design parameters more clearly. In practice, the case of constant g_m/I_D may reflect a scenario where the circuit is limited by distortion. We will see in Chapter 4 that linearity requirements place upper bounds on g_m/I_D. Another scenario is a low-voltage, high-dynamic range circuit where stringent bounds on signal swing and V_{Dsat} may exist. For example, requiring $V_{Dsat} = 150$ mV, means $g_m/I_D = 2/V_{Dsat} = 13.33$ S/A. We will see an example of this in Chapter 6.

Example 3.2 Sizing at Constant g_m/I_D

Consider an IGS with $C_L = 1$ pF and g_m/I_D of 15 S/A. Find combinations of L, W and I_D, that achieve $f_u = 100$ MHz and compute the corresponding low-frequency gain and fan-out. Assume $V_{DS} = 0.6$ V and $V_{SB} = 0$ V.

SOLUTION

Graphically, the problem boils down to tracing the thick vertical line in the plot of Figure 3.8(a) and to collect the intersecting transit frequencies and intrinsic gains for every gate length. The result is plotted in Figure 3.8(b), showing opposing trends in $|A_{v0}|$ and f_T as L is increased.

To compute the drain current that meets the desired f_u, we follow the same approach used in Section 3.1.3:

```
gm = 2*pi*fu*CL;
ID = gm/15;
```

This yields $I_D = 41.89$ μA. Note that since g_m/I_D is constant and g_m is fixed by the specifications, the drain current is also constant, regardless of the chosen L.

To find the device widths, we need the drain current densities, which depend on L. In the calculation below, we consider the entire L vector stored in the lookup tables (nch.L = [(0.06:0.01:0.2) (0.25:0.05:1)]):

```
JD = lookup(nch, 'ID_W', 'GM_ID', 15, 'L', nch.L);
W  = ID./JD;
```

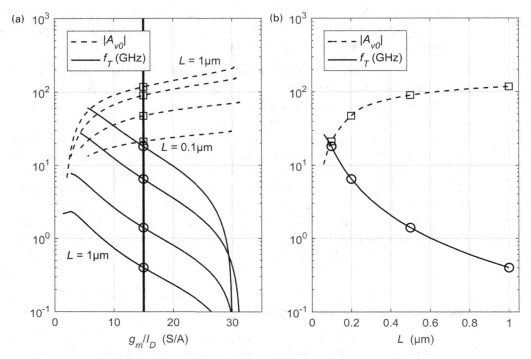

Figure 3.8 (a) Plot of LF gain and transit frequency versus g_m/I_D. The bold line marks the given value of $g_m/I_D = 15$ S/A. (b) Plot of $|A_{vo}|$ and f_T that intersect with the bold line in (a).

This completes the sizing and we can now compute the fan-out for each choice of L. We can do this using:

```
fT = lookup(nch, 'GM_CGG', 'GM_ID', 15,'L', nch.L)/(2*pi);
FO = fT/fu;
```

and find that f_T ranges from 0.4 to 26.5 GHz. Since we do not consider transit frequencies less than 10 times f_u (100 MHz), we find an index vector M for the usable gate length range and the corresponding maximum value:

```
M    = fT >= 10*fu;
Lmax = max(nch.L(M));
```

This yields $L_{max} = 0.60$ μm. Note that this value can also be read from Figure 3.8(b). The f_T curve crosses 1 GHz near $L = 0.6$ μm.

   Figure 3.9 plots the above-computed data for the voltage gain, device width and fan-out for $L \leq L_{max}$. We see that the largest achievable voltage gain is about 100 for $L = L_{max}$. Note that this number would reduce if we increased the f_u requirement, since it would lead to a higher f_T requirement and a correspondingly shorter channel. Conversely, larger L and hence larger voltage gains would be possible if we relaxed the f_u requirement. However, we can see

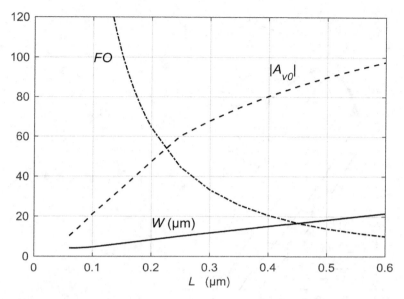

Figure 3.9 LF voltage gain, device width and fan-out versus gate length. Parameters: $f_u = 100$ MHz, $C_L = 1$ pF, $g_m/I_D = 15$ S/A, $V_{DS} = 0.6$ V, $V_{SB} = 0$ V.

from Figure 3.8(b) that the gain curve saturates for long channels, and not much improvement is possible.

The next scenario that we will consider is sizing at constant transit frequency. This case is of practical relevance, for instance, in amplifiers that target a fixed gain-bandwidth product. Another example is a cascode configuration, where we often want to size the common-gate device such that the non-dominant pole lies at or above some given frequency. Such cases will be studied in more detail in Chapter 6.

Example 3.3 Sizing at Constant f_T

Consider an IGS with $C_L = 1$ pF and f_u of 1 GHz. Find combinations of L and g_m/I_D, that achieve (i) maximum low-frequency gain and (ii) minimum current consumption. Assume $FO = 10$, $V_{DS} = 0.6$ V and $V_{SB} = 0$ V. Validate the results using SPICE simulations.

SOLUTION

As before, we start with a plot of $|A_{v0}|$ and f_T versus g_m/I_D, shown in Figure 3.10(a). This time, however, we look for the intersects of f_T and the bold gray line, marking the target of 10 GHz. This yields corresponding g_m/I_D values that we can use to find A_{v0} across a range of gate lengths. The result of this collection is shown in Figure 3.10(b). We see that there is a value pair of L and g_m/I_D that maximizes the gain.

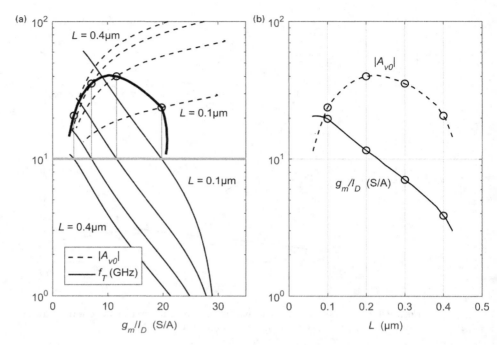

Figure 3.10 (a) Plot of transit frequency and intrinsic gain versus g_m/I_D. The bold black line marks the $|A_{vo}|$ values for which $f_T = 10$ GHz (bold gray line). (b) Plot of the corresponding $|A_{vo}|$ and g_m/I_D versus gate length L.

To investigate further, Figure 3.11 shows an expanded view of (a) the LF gain and (b) the gate length versus g_m/I_D. Plot (b) shows that L must decrease to keep f_T unchanged as we increase g_m/I_D. This explains the decline of $|A_{vo}|$ for large g_m/I_D beyond the maximum (marked by a circle). Before the maximum is reached, the dependence of A_{v0} on g_m/I_D dominates and leads to a positive slope. This is explained by the strong positive slope in $|A_{vo}|$ seen in Figure 3.10(a) for $g_m/I_D < 10$ S/A.

Interestingly, the plots of Figure 3.11 also show a trend reversal near their tail end and as we reach the minimum L offered by the technology. This is due to the reduced slope in the I-V characteristic for large g_m/I_D and near-minimum L (see Section 2.3.1). Physically, this effect is related to reduced gate control, or equivalently, increased drain-induced barrier lowering (DIBL) for short channels. The largest g_m/I_D value on these curves is marked with an asterisk and represents the minimum current design.

The design data for maximum gain (option (i)) and minimum current (option (ii)) are summarized in Table 3.1. It is important to note that that these design points were obtained only through the manipulation of parameter ratios in a normalized space. Both f_u and C_L have not yet entered the design process and become only important once we want to compute the device widths, which is the next step in this problem.

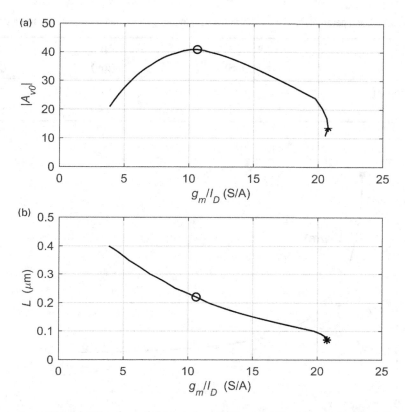

Figure 3.11 (a) LF gain and (b) gate length versus g_m/I_D. The circles mark the maximum gain points (option (i)) and the asterisks mark the design with minimum current (option (ii)). Parameters: $f_T = 10$ GHz, $V_{DS} = 0.6$ V, $V_{SB} = 0$ V.

Table 3.1 Design parameters that maximize the gain (option (i)) or minimize the drain current (option (ii)).

	$|A_{v0}|$	g_m/I_D (S/A)	L (nm)	V_{GS} (V)
Option (i)	40.88	10.62	220	0.5786
Option (ii)	13.75	20.76	70	0.4103

Since we know both L and g_m/I_D at this stage, all further calculations are carried out as in Example 3.1. Given $FO = 10$, we know that $f_u = f_T/FO = 1$ GHz and arrive at the sizing parameters listed in Table 3.2.

To compare the two sizing options, Figure 3.12 plots W and $|A_{v0}|$ versus the drain current I_D (the maximum gain and minimum current designs are marked with a circle and asterisk, respectively). Note that for the minimum current design, increasing the drain current only slightly would lead to a significant increase in voltage gain. Operating the design at the absolute minimum current (and correspondingly short L) may therefore not be desirable.

Table 3.2 Sizing parameters that maximize the gain (option (i)) or minimize the drain current (option (ii)).

	g_m (mS)	W (μm)	I_D (μA)
Option (i)	6.283	45.92	591.6
Option (ii)	6.283	114.2	302.7

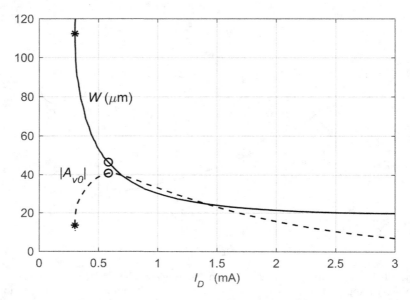

Figure 3.12 LF gain and device width versus drain current. The circles mark the maximum gain design option (i) and the asterisks mark the design with minimum current (ii).

Table 3.3 Simulation result summary.

	$|A_{v0}|$ SPICE	Error (%)	f_u (GHz) SPICE	Error (%)
Option (i)	41.0	+2.9	0.967	−3.3
Option (ii)	13.75	0	0.933	−6.7

To conclude, we validate the maximum gain and minimum current design options in SPICE and obtain the results listed in Table 3.3. We observe that the low-frequency voltage gains are almost exactly as predicted, whereas the unity gain frequencies are somewhat smaller than the design target. As already mentioned in Example 3.1, the small discrepancy is due to the parasitic drain capacitance that we have not yet considered in our design flow.

As a final example, we consider a scenario where we fix the circuit's low-frequency voltage gain. Such a situation may arise in the design of operational amplifiers, where we typically want to achieve a certain target for the feedback circuit's loop gain. We will encounter this situation in some of the examples in Chapter 6.

Example 3.4 Sizing at Constant $|A_{vo}|$

Consider an IGS with $C_L = 1$ pF and constant $|A_{vo}| = 50$. Find combinations of L and g_m/I_D, that achieve (i) maximum unity gain frequency, and (ii) minimum current consumption for a design that achieves 80% of the maximum unity gain frequency. Assume $FO = 10$, $V_{DS} = 0.6$ V and $V_{SB} = 0$ V. Validate the results using SPICE simulations.

SOLUTION

The flow of our solution is very close to Example 3.3. We start with a plot of $|A_{vo}|$ and f_T versus g_m/I_D, shown in Figure 3.13(a), but this time we look for the

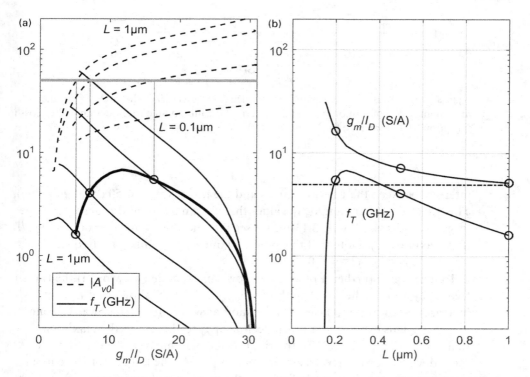

Figure 3.13 (a) Plot of transit frequency and LF gain versus g_m/I_D. The bold black line marks the f_T values for which $|A_{vo}| = 50$ (bold gray line). (b) Plot of the corresponding f_T and g_m/I_D versus gate length L.

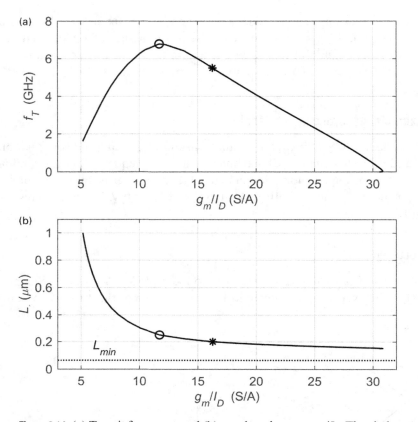

Figure 3.14 (a) Transit frequency and (b) gate length versus g_m/I_D. The circles mark the maximum f_T design (option (i)) and the asterisks the minimum current design at 20% loss in f_u (option ii). Parameters: $|A_{vo}| = 50$, $V_{DS} = 0.6$ V, $V_{SB} = 0$ V.

intersect between the LF gain curves and the target value of 50 (bold gray line). This yields corresponding g_m/I_D values that we can use to find f_T across a range of gate lengths (see Figure 3.13(b)). We see that there is a value pair of L and g_m/I_D that maximizes f_T. Note that this point maximizes the unity gain frequency as well, since $f_u = f_T/FO = f_T/10$.

To investigate further, Figure 3.14 shows an expanded view of (a) the transit frequency and (b) the gate length versus g_m/I_D. Plot (b) shows that L must increase significantly to keep $|A_{vo}|$ constant as we reduce g_m/I_D toward strong inversion. This explains the sharp decline of f_T to the left of the maximum that we see in Figure 3.13. To the right of the maximum, f_T decreases due to the reduced inversion level as we increase g_m/I_D. At the tail end of the curve (onset of weak inversion), we find that the minimum channel length for which an LF gain of 50 can be achieved is about 150 nm. This lower bound explains the sharp drop of the f_T curve in Figure 3.13(b) as the designs to the left of the curve are unfeasible.

Table 3.4 Design and sizing parameters that maximize the unity gain frequency (option (i)) or minimize the drain current for a 20% f_u reduction (option (ii)).

	g_m/I_D (S/A)	L (nm)	f_u (MHz)	I_D (μA)	W (μm)	V_{GS} (V)
Option (i)	11.7	250	679	363.8	41.95	0.5533
Option (ii)	16.3	200	543	209.5	54.24	0.4818

Table 3.5 Simulation result summary.

	$\|A_{v0}\|$ SPICE	Error (%)	f_u (MHz) SPICE	Error (%)
Option (i)	49.9	−0.2	661	−2.7
Option (ii)	50.0	0	525	−3.3

At the maximum transit frequency point, we find $f_u = f_T/10 = 679$ MHz and $L = 0.25$ μm; this is the solution for option (i). Although there are two options for designs with 20% reduction in f_u, the point to the right of the maximum (marked with an asterisk) achieves minimum current. Both options require the same g_m, but the point with larger g_m/I_D will require a smaller drain current. The channel length for option (ii) is thus 0.2 μm. Since we now know L, g_m/I_D and f_T for both designs, the sizing parameters are readily computed (see Table 3.4).

Finally, we validate the maximum f_u and minimum I_D designs in SPICE and obtain the results listed in Table 3.5. Once again, we observe good agreement with the Matlab prediction.

In all the prior examples, we have assumed a unity gain frequency constraint in the range of 100 MHz to 1 GHz. However, there exist applications that operate at much lower frequencies. For example, biomedical and sensor interface circuits often operate in the kilohertz range. In such circuits, the transit frequency typically won't appear as a significant constraint that will affect the sizing.

Consider for example the tradeoff between f_T and g_m/I_D for a fixed value of LF gain in Example 3.4. From Figure 3.14(a), we know that we can move toward larger values of g_m/I_D when the f_T requirements are reduced, and this can help us save current. If the f_T constraint is removed entirely, the preferred design point would be at the tail end of the curve, namely at the maximum possible g_m/I_D. In this case, the transistor will operate in weak inversion and g_m/I_D is essentially constant. Finding and optimizing the device width in this specific scenario requires a different approach, which we will investigate in the following section.

3.1.5   Sizing in Weak Inversion

As an introduction to sizing in weak inversion, consider the amplifier circuit used in the ultra-low power sensor of [2] as an example. The node is designed to dissipate

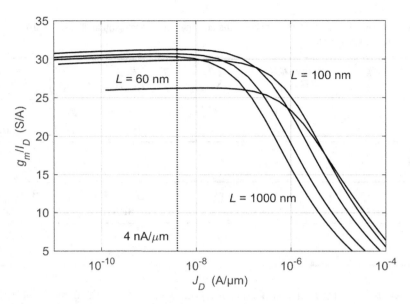

Figure 3.15 Transconductance efficiency (g_m/I_D) versus drain current density (J_D) of an n-channel device with gate lengths $L = 60$, 100, 200, 500 and 1000 nm. ($V_{DS} = 0.6$ V, $V_{SB} = 0$V).

only 3 nW, so that it can be powered from a thin-film battery for about 10 years. The differential front-end amplifier of the sensor node consumes only 1 nW from a 0.6 V supply and thus the differential pair transistors have a bias current of only about 0.8 nA each. With such low currents, the current densities will be correspondingly low. Even if we assume that the MOSFET has a width of 200 nm, which is the minimum value allowed in a typical 65-nm process, the current density will not exceed 0.8 nA/0.20 μm, which is 4.0 nA/μm. Considering the plot of g_m/I_D versus current density in Figure 3.15, we can therefore conclude that the transistor must operate in weak inversion.

In weak inversion, g_m is equal to $I_D/(nU_T)$. In this expression, the subthreshold slope parameter n is only a weak function of the channel length, provided that L is somewhat larger than minimum length (avoiding significant DIBL, as explained in Section 2.3.1). This is evident from Figure 3.15, which shows that $(g_m/I_D)_{max}$ is nearly constant for $L = 100...1000$ nm. Therefore, the required drain current in weak inversion is to first order gate-length independent.

If no other constraint applies, it would seem natural to design for minimum area, i.e. minimum device width and length. However, this is usually not a good choice in analog circuits. Minimum length implies low intrinsic gain, and small gate area ($W \cdot L$) leads to poor device matching and flicker noise (see Chapter 4). This means that for a real-world problem, we typically need to invoke some of these additional constraints to complete the device sizing.

To investigate further, let us evaluate the intrinsic gain as a function of channel length, shown in Figure 3.16(a). We use the current density instead of g_m/I_D on the

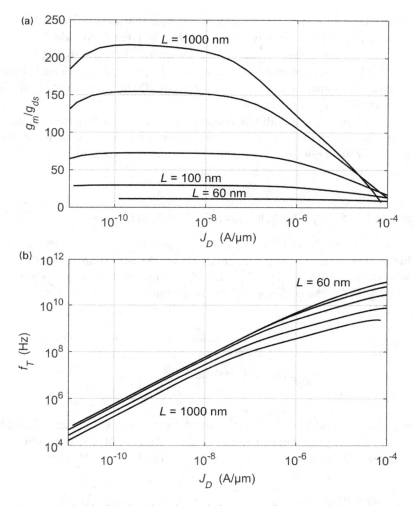

Figure 3.16 Intrinsic gain (a) and transit frequency (b) versus drain current density (J_D) of an n-channel device with gate lengths $L = 60$, 100, 200, 500 and 1000 nm. ($V_{DS} = 0.6$ V, $V_{SB} = 0$ V).

x-axis since the latter is essentially constant in the design region of interest. Also, we show the device f_T in Figure 3.16(b), to get a feel for the resulting numbers. As expected, the intrinsic gain increases significantly with longer channel lengths. The device f_T drops as we increase the channel length, but even at a current density of 4.0 nA/μm it is still above 1 MHz, and thus significantly larger than f_u in the kilohertz range. With this specific example at hand, we could make the channel length even larger than 1000 nm and continue to benefit from increasing intrinsic gain. At which point should we stop?

From a theoretical standpoint, it would make sense to continue to increase L until the given current density (4.0 nA/μm in our example) corresponds to the weak inversion "knee" (see trend in Figure 3.15). Moving any further would mean that

g_m/I_D and thus g_m drops. For the extreme example considered here (0.8 nA of drain current), this design point is reached at extremely large L, for which the transistor model may no longer be accurate. The designer will therefore stop at a point where a reasonable amount of gain is achieved and the f_T is still significantly higher than the operating frequencies. With a gate length of 500 nm, we already have an intrinsic gain of almost 150 per Figure 3.16(a).

As far as the device width is concerned, there is no incentive to increase W beyond the minimum value of 200 nm, unless we include further constraints on matching or flicker noise (see Chapter 4). We consider an example below that takes a minimum-width constraint into account.

Example 3.5 Sizing in Weak Inversion Given a Width Constraint

Size an ultra-low power IGS with $C_L = 1$ pF and $I_D = 0.8$ nA. Determine L to achieve a low-frequency voltage gain of about 150, and assume that the minimum desired device width is 5 μm. Compute V_{GS}, f_T and the circuit's unity gain frequency. Verify the design using a SPICE simulation. Assume $V_{DS} = 0.6$ V and $V_{SB} = 0$ V.

SOLUTION

We begin by computing the current density:

```
JD = ID/W
```

This yields 0.16 nA/μm. To find the gate length, we look at Figure 3.16(a), which indicates that $L = 500$ nm will suffice to achieve a gain of 150 at the computed current density. We can now use the lookup function to find g_m/I_D and f_T:

```
gm_ID = lookup(nch, 'GM_ID', 'ID_W', JD,'L', 0.5)
fT  = lookup(nch, 'GM_CGG', 'ID_W', JD, 'L',0.5)/2/pi
```

This leads to $g_m/I_D = 30.5$ S/A and $f_T = 445$ kHz. The transconductance and f_u now follow mechanically:

```
gm = gm_ID*ID;
fu = gm/(2*pi*CL)
```

Finally, to compute the gate-to-source voltage we use:[3]

```
VGS = lookupVGS(nch, 'ID_W', JD, 'L', 0.5, 'METHOD', 'linear')
```

Table 3.6 summarizes the results. The above calculations and SPICE simulation data are in good agreement. Note that the circuit's fan-out is very large ($FO = 445$ kHz/3.82 kHz = 116).

---

[3] The default interpolation method of lookupVGS is "pchip." In the shown computation, the interpolation method is changed to "linear" to avoid numerical issues that can arise near the endpoints of the stored lookup data.

Table 3.6 Result summary.

	$|A_{v0}|$	g_m/I_D (S/A)	f_T (kHz)	g_m (nS)	f_u (kHz)	V_{GS} (mV)
Calculation	~150	30.5	445	24.4	3.82	128
SPICE	145	30.5	442	24.4	3.87	128

From the above example, we see that sizing in weak inversion is relatively straight-forward, since g_m/I_D is essentially constant and does not play a role in the optimization. The gate length follows directly from the gain requirement. Furthermore, we do not need to worry about fan-out from a modeling perspective, since the ratios of f_T and f_u will almost always be larger than the desired bound of 10.

3.1.6   Sizing Using the Drain Current Density

Throughout Section 3.1.4 we used the sizing flow of Figure 3.4, which takes g_m/I_D as the main "knob" to define the transistor's inversion level. The advantage of this approach is that g_m/I_D spans a well-defined range that is nearly independent of technology, thus serving as a very intuitive numerical proxy for the inversion level and the associated tradeoff between gain, speed and efficiency. In addition, once g_m/I_D is known, one can immediately compute the current that is required for a certain g_m.

However, as we saw in Section 3.1.5, a problem with the sizing flow of Figure 3.4 is that g_m/I_D no longer uniquely defines the drain current density once the transistor enters weak inversion. In other words, a wide range of current densities (and thus designs) map to nearly the same value of g_m/I_D. In Section 3.1.5, our solution to this problem was to first choose the current density, and then work with the resulting g_m/I_D to complete the design (see Example 3.5). In this section, we explore a generalization of this approach as shown in Figure 3.17. Since there is a one-to-one mapping between J_D and g_m/I_D, we can always begin by selecting J_D, then look up g_m/I_D and complete the design as usual. As already mentioned in Section 1.2.5, this approach is similar to the EKV-based sizing method of [2] which works with a normalized representation of the current density.

The flow of Figure 3.17 is applicable or advantageous in the following scenarios:

- We know a priori that the circuit will operate in weak inversion. In this case, we have no choice but to design based on current density (see Example 3.5).
- We have no a-priori information or intuition about the device's inversion level and hence want to search across all possibilities, from weak inversion to strong inversion.

Since the latter scenario occurs infrequently in real-world design, we advocate the g_m/I_D-driven flow of Figure 3.4 for the practitioner working on moderate to high-speed circuits, and only resort to current density-based design when the

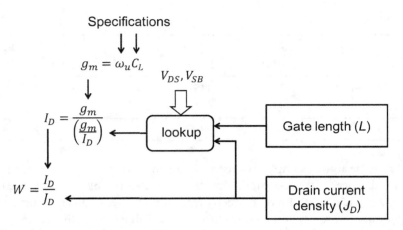

Figure 3.17 Current density-based sizing, appropriate for all inversion levels.

circuit is best operated in weak inversion. It is usually straightforward to make this decision.

Despite this stated preference, we dedicate the remainder of this section to a tradeoff exploration using the current density as the design variable. This lets us appreciate the sizing tradeoffs seen in the previous sections from another angle, and allows the reader to see the overall design space in a unified representation.

We begin by examining Figure 3.18. Plots (a), (b) and (c) show contours of constant intrinsic gain (A_{intr}), constant transconductance efficiency (g_m/I_D), and constant transit frequency (f_T), respectively. All quantities are plotted against L and J_D, which are the "knobs" in the design flow of Figure 3.17. The contours in (a), which are based on the same data as in Figure 3.15, exhibit a large "empty" area delineated by the 30 S/A locus (weak inversion). In this region, the choice of J_D and L does not impact g_m/I_D significantly. To enter moderate inversion, J_D must be made larger than 10 to 100 nA/μm. Figure 3.18(b) shows the corresponding evolution of the intrinsic gain. In weak inversion, L sets A_{intr}, corroborating the data of Figure 3.16(a), while in moderate and especially in strong inversion, the drain current density becomes the defining variable. Lastly, (c) conveys the same trend as seen in Figure 3.16(b). The constant transit frequency loci are nearly parallel and almost vertical (independent of J_D) in weak inversion, but more dependent on the current density in strong inversion.

It is now interesting to locate and contrast our two previous design examples in the plots of Figure 3.18. The strong-inversion design of Example 3.3 is marked by a circle in each subplot, while the weak-inversion design of Example 3.5 is marked using a plus sign.

We start with the weak inversion design. Inside the region delineated by the 30 S/A contour in Figure 3.18(a), g_m/I_D does not change significantly. Since g_m is fixed by the specifications, the drain current remains almost constant, too. However, we

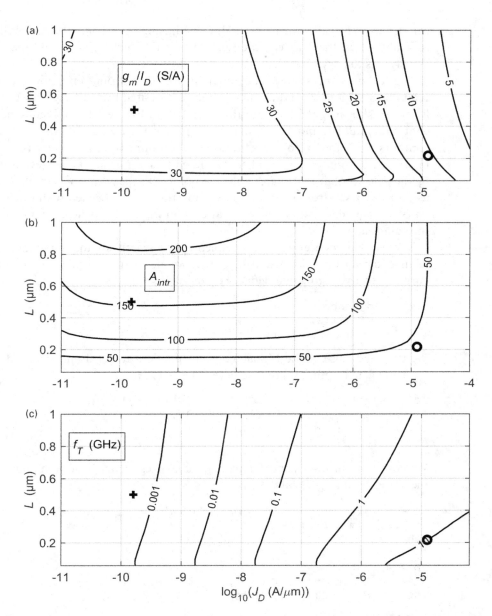

Figure 3.18 Constant contour plots representing (a) the transconductance efficiency, (b) intrinsic gain and (c) transit frequency versus the drain current density J_D and the gate length L ($V_{DS} = 0.6$ V, $V_{SB} = 0$ V). The data from Example 3.3 and Example 3.5 are marked by "o" and "+", respectively.

can still vary the gain moving up or down (adjusting L) keeping the current density constant. Thus, the gate length essentially sets the intrinsic gain, and the current density sets the device width.

The circumstances are very different for the strong inversion design of Example 3.5. Here, J_D and L pairs meeting the desired transit frequency (10 GHz) lie on the shown f_T contour with the label "10" in Figure 3.18(c). Picture now this contour of valid J_D and L pairs in the gain plot of Figure 3.18(b). Some of the gain contours in Figure 3.18(b) cross the target f_T contour twice. As we increase gain, the intersections come closer together. The largest gain that we can achieve is defined by the gain contour that shares only one point with the f_T contour. In other words, as we move away from the tangential point, the gain associated with the intersect pairs can only decrease. We saw this already in Figure 3.11 where for a given g_m/I_D the gain is largest.

The Matlab code given below implements a search for the maximum voltage gain, assuming $V_{DS} = 0.6$ V and $V_{SB} = 0$ V. We first set up the gate length and drain current density vectors that define the axes of Figure 3.18:

```
JDx  = logspace(-10,-4,100);
Ly   = .06:.01:1;
[X Y] = meshgrid(JDx,Ly);
```

We then make use of the Matlab contour function[4] to get drain current densities (J_{D1}) and gate lengths (L_1) that achieve a desired f_T value:

```
fTx   = lookup(nch,'GM_CGG','ID_W',JDx,'L',Ly)/(2*pi);
[a1 b1] = contour(X,Y,fTx,fT*[1 1]);
JD1 = a1(1,2:end)';
L1  = a1(2,2:end)';
```

We now look for the J_D and L pair that maximizes the intrinsic gain and evaluate the corresponding g_m/I_D and V_{GS}.

```
Av = diag(lookup(nch,'GM_GDS', 'ID_W', JD1, 'L', L1));
[a2 b2] = max(Av);
Avo = a2;
L  = L1(b2);
JD = JD1(b2);
gm_ID = lookup(nch, 'GM_ID', 'ID_W', JD, 'L', L);
VGS = lookupVGS(nch, 'GM_ID', gm_ID, 'L', L);
```

Table 3.7 summarizes the results obtained across a wide range of f_T. Note that for the lowest f_T values, we converge to the longest channel length available in our lookup tables (1 μm).

In the following example, we consider another illustration of the current density-based sizing approach of Figure 3.17, and reinforce some of the above-discussed observations.

---

[4] The Matlab function contour(x, y, F(x, y), C*[1 1]) finds x1 and y1 vectors that make F(x1, y1) equal to C.

Table 3.7 Summary of maximum gain parameters.

f_T (GHz)	g_m/I_D (S/A)	J_D (μA/μm)	max($|A_{vo}|$)	L (μm)	V_{GS} (V)
0.02	29.93	0.0125	206.4	1.00	0.2861
0.05	28.78	0.0369	196.0	1.00	0.3228
0.10	26.54	0.0986	179.0	0.99	0.3572
0.20	24.85	0.1874	155.1	0.79	0.3773
0.50	22.21	0.4746	124.0	0.57	0.4058
1.00	**20.00**	**0.9371**	**102.0**	**0.44**	**0.4300**
2.00	16.42	2.105	81.6	0.37	0.4718
5.00	13.19	5.397	57.6	0.27	0.5242
10.00	10.62	12.90	40.9	0.22	0.5787
20.00	8.75	28.48	27.4	0.17	0.6310
50.00	6.66	75.64	15.5	0.11	0.7050

Example 3.6 Sizing Using Contours in the J_D and L Plane

Consider an IGS with $C_L = 1$ pF and $f_u = 100$ MHz, which requires $f_T \geq 1$ GHz. Per Table 3.7, the largest voltage gain magnitude that we can achieve is 102. In this example, consider reduced gain values of $|A_{vo}| = 50$ and 80 and size the circuit for maximum fan-out and minimum current for each case (four design options total). Use a current density sweep to find the required g_m/I_D and L values. Assume $V_{DS} = 0.6$ V and $V_{SB} = 0$ V.

SOLUTION

For illustrative purposes, Figure 3.19 combines the contours of Figure 3.18 into one plot, focusing on the case of $f_T = 1$ GHz (bold line). The solid thin black line marks the contour with the maximum $|A_{vo}| = 102$. Note that these two curves have only one intersection point, as discussed previously. Also shown are gray contours for constant g_m/I_D. The contour for 20 S/A passes through the point with maximum gain, as expected from the result in Table 3.7.

Assume now a reduced value of $|A_{vo}| = 80$. The J_D and L pairs tracing this gain locus are illustrated by the dashed contour and are obtained using:

```
Av = lookup(nch,'GM_GDS','ID_W',JD,'L',L);
[a3 b3] = contour(X,Y,Av,80*[1 1]);
JD3 = a3(1,2:end)';
L3  = a3(2,2:end)';
```

The gain locus has two intersections with the bold $f_T = 1$ GHz locus, and the pertaining design values are given in Table 3.8. Note that the upper intersect (A) lies in strong inversion, while the lower (B) is in weak inversion (within ~20% of maximum g_m/I_D).

To the right to the bold line in Figure 3.19, the transit frequency increases, leading to $FO > 10$. We select these points by running:

```
M   = FO >= FOmin;
FO  = diag(lookup(nch,'GM_CGG', 'ID_W', JD3, 'L', L3))/(2*pi*fu);
FO4 = FO(M);
JD4 = JD3(M);
L4  = L3(M);
```

As a last step, we find the corresponding transconductance efficiencies, drain currents and device widths:

```
gm_ID4 = diag(lookup(nch,'GM_ID','ID_W',JD4,'L',L4));
gm = 2*pi*fu*CL;
ID = gm./gm_ID4;
W  = ID./JD4;
```

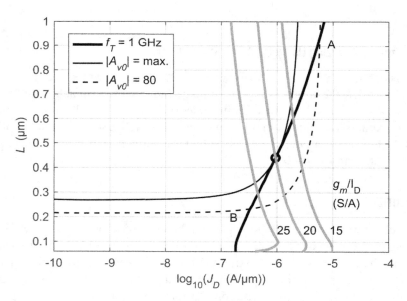

Figure 3.19 Contours for constant transit frequency (1 GHz), constant gain magnitude (102 and 80) and constant g_m/I_D (15, 20 and 25 S/A).

Table 3.8 Data for intersects A and B in Figure 3.19.

	g_m/I_D (S/A)	L (µm)	V_{GS} (V)
Point A	8.78	0.930	0.6120
Point B	27.02	0.233	0.3720

The results are plotted in Figure 3.20 considering not only $|A_{vo}| = 80$ but also 50. Notice that we meet the design requirements for a wide range of transconductance efficiencies, across which the currents change by about 3–5x. This stands in contrast with a circuit operating at maximum gain, which narrows the space to one single point. Note that the range of feasible g_m/I_D values widens for the design with lower gain.

In Table 3.9, we summarize the sizing data considering designs aiming at maximum fan-out and compare these to SPICE simulations. We see from the table

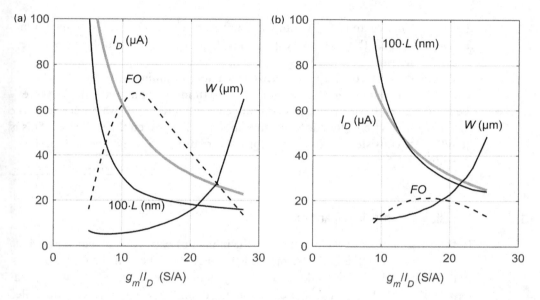

Figure 3.20 Plot of drain currents, device widths, gate lengths and fan-outs (a) $|A_{vo}| = 50$, and (b) $|A_{vo}| = 80$. ($V_{DS} = 0.6$ V and $V_{SB} = 0$ V).

Table 3.9 Result summary.

	(a)	SPICE verification	(b)	SPICE verification
$|A_{vo}|$	50	50.5	80	80.45
L (µm)	0.240	–	0.340	–
W (µm)	6.479	–	17.41	–
I_D (µA)	50.74	–	36.59	–
FO	68.0	–	21.2	–
g_m/I_D (S/A)	12.38	12.31	17.17	17.03
f_u (MHz)	100	99.0	100	98.0
V_{GS} (V)	0.541	0.544	0.465	0.465
C_{gg} (fF)	14.7	14.6	47.0	46.2

Table 3.10 Sizing data for minimum drain current.

	50	80
$|A_{vo}|$	50	80
g_m/I_D (S/A)	28.53	27.02
L (µm)	0.158	0.233
I_D (µA)	22.02	23.26
W (µm)	88.90	71.02

that the calculated and SPICE-simulated numbers are very close. Since maximum fan-out minimizes the total gate capacitance $C_{gg} = C_L/FO$, this can be a desirable design choice.

As a final step, Table 3.10 summarizes the data for minimum I_D, which is achieved for maximum g_m/I_D and thus $FO = 10$ (minimum allowed value). As we see from the result, this yields design points in weak inversion. We observe that the currents and device widths don't change appreciably between the two cases. This is qualitatively consistent with our previous observations on weak inversion design in Section 3.1.5.

3.1.7 Inclusion of Extrinsic Capacitances

In this section, we take a closer look at the impact of extrinsic capacitances (which were ignored so far), and devise methods for including them in the sizing process. Extrinsic capacitances include the device's junction capacitances (C_{sb} and C_{db}), as well as the gate-to-drain fringe capacitance (C_{gd}). These capacitances are called extrinsic, since unlike the intrinsic gate capacitance (C_{gs}), they are not essential to the operation of the MOSFET. In other words, the device would still function if these where eliminated.

Figure 3.21(a) shows the small-signal model of the IGS with the relevant extrinsic capacitances C_{db} and C_{gd} included. Relative to Figure 3.2, we omitted C_{gs} and C_{gb} since these capacitances are shorted by the input voltage source, which is still assumed to be ideal for the sake of simplicity. With this assumption, the circuit is well approximated by the model in Figure 3.21(b), where the total drain capacitance $C_{dd} = C_{db} + C_{gd}$ appears in parallel to the load. This is an approximation since C_{gd} contributes a feedforward current that is being neglected. However, it can be easily shown that this current is relevant only beyond the transit frequency of the transistor, which is outside the range of our analysis.

Especially for high-speed designs with large g_m and correspondingly large device width, the added capacitance C_{dd} can result in a noticeable error in the gain-bandwidth product. For instance, in the high-speed, low-power design of Example 3.3, option (ii), we saw an error of 6.7% between the Matlab calculation and the SPICE simulation. The situation worsens once additional devices (such as active loads in Section 3.2.1) are connected to the output. We call this self-loading and will refer to the sum of these unwanted capacitances C_{self}.

We would like to establish a sizing process in which C_{dd} (and other self-loading parasitics) are accounted for. A fundamental issue with factoring C_{dd} into the design is that we don't know its value before the sizing is completed (i.e. the device width W is known). One way to deal with this problem is to use a three-step design procedure:

1. Design the IGS while ignoring the total drain capacitance C_{dd} (as done in all prior examples).
2. Find C_{dd} of the design obtained in step 1, call this value C_{dd1}.
3. Now scale the device width and current by the following factor to arrive at the final design:

$$S = \frac{1}{1 - \dfrac{C_{dd1}}{C_L}}.$$

(3.10)

To see why this will work (perfectly), consider the scaling behavior of the device and how it affects the unity gain frequency. When we scale the device current and

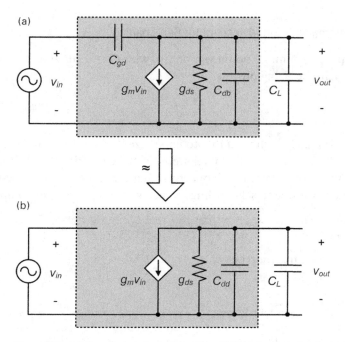

(a)

(b)

Figure 3.21 (a) Small-signal model of the IGS including the drain junction capacitance C_{db} and the gate-drain fringe capacitance C_{gd}. (b) Approximate model that adds $C_{dd} = C_{db} + C_{gd}$ in parallel to C_L.

width together, the current density and g_m/I_D stay constant, and both C_{dd} and g_m scale linearly with S. Therefore:

$$\omega_u = \frac{Sg_m}{C_L + SC_{dd1}} = \frac{g_m}{C_L}. \tag{3.11}$$

Solving (3.11) for the scaling factor S yields (3.10).

While this method works very well for the simple examples considered so far, it is restricted to cases where the parasitic capacitance scales linearly with the device transconductance that sets the unity gain frequency. As we shall see in Chapter 6, this is not the case for all circuits encountered in practice. We therefore devise a second solution that finds the correct sizing iteratively, and without any analytical assumptions (such as (3.11)). This approach is outlined as follows:

1. Start by assuming $C_{dd} = 0$.
2. Size the circuit to meet the GBW spec for $C + C_{dd}$. (For the first iteration, this means that we are ignoring C_{dd}.)
3. Estimate C_{dd} for the obtained design (using the device width from step 2).
4. Go to step 2 using the new C_{dd} estimate.
5. Repeat until convergence.

We will now illustrate this approach using an example.

Example 3.7 Iterative Sizing to Account for Self-Loading

Repeat Example 3.3, minimal power option (ii), with $C_L = 1$ pF and $f_u = 1$ GHz. Account for C_{dd} in the sizing process.

SOLUTION

In the quoted example, we showed that to maximize the intrinsic gain, g_m/I_D and L should equal 20.76 S/A and 70 nm, respectively. SPICE verifications showed good agreement as far as the gain is concerned, but the GBW was 6.7% less than expected. The error is due to self-loading. To account for C_{dd}, we pre-compute:

```
JD    = lookup(nch,'ID_W','GM_ID',gm_ID,'L',L);
Cdd_W = lookup(nch,'CDD_W','GM_ID',gm_ID,'L',L);
```

and then run the iterative loop below:

```
Cdd = 0;
for m = 1:5,
    gm = 2*pi*GBW*(CL + Cdd);
    ID(m,1) = gm./gm_ID;
    W(m,1)  = ID(m,1)./JD;
    Cdd = W*CDD_W;
end
```

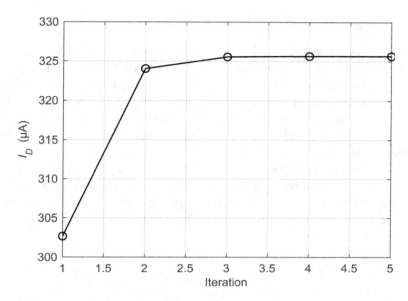

Figure 3.22 Progression of I_D as the extrinsic capacitance is factored in.

Figure 3.22 shows the evolution of the drain current as the number of iterations increases. After step 1, we get a first estimate of C_{dd} that we add to C_L in step 2. The current increases from 302.7 μA to 324.0 μA. We reach 325.6 μA after step 3 and observe that additional iterations lead to negligible increments. The final cumulative increase of the drain current is 7.60%. The width increases by the same percentage, from 114.2 to 122.8 μm.

The reader can verify that the scale factor S in (3.10) is equal to 1.076, predicting the same amount of current and width increase analytically.

Having laid the grounds of the g_m/I_D sizing methodology, we now turn the IGS progressively into a more realistic amplifier stage. In the following sections, we introduce active loading (Section 3.2.1) and resistive loading (Section 3.2.2). Finally, we consider a differential pair arrangement (Section 3.3).

3.2 Practical Common-Source Stages

To turn the IGS into a more practical common-source (CS) stage, we replace the ideal bias current source by a saturated p-channel transistor (Figure 3.23(a)) or by a resistor (Figure 3.23(b)). The following subsections discuss the sizing of these circuits.

3.2.1   Active Load

The presence of M_2 in Figure 3.23(a) introduces an extra conductance g_{ds2} in the circuit's small-signal model. Since the conductance lies in parallel with g_{ds1} (see Figure 3.2), it decreases the low-frequency voltage gain but does not impact the gain-bandwidth product. However, the additional extrinsic capacitance introduced by the active load can have a noticeable impact on the GBW if it is a significant fraction of C_L. We will consider this capacitance in some of the examples that follow below.

The drain current of M_2 is fixed per the unity gain frequency that the signal path device M_1 must realize (approximately g_{m1}/C_L). The only degrees of freedom for M_2 are its inversion level $(g_m/I_D)_2$ and gate length L_2. To develop guidelines on selecting these parameters, we begin by inspecting the circuit's low-frequency gain expression:

$$A_{v0} = -\frac{g_{m1}}{g_{ds1} + g_{ds2}} = -\frac{\left(\dfrac{g_m}{I_D}\right)_1}{\left(\dfrac{g_{ds}}{I_D}\right)_1 + \left(\dfrac{g_{ds}}{I_D}\right)_2} = -\frac{\left(\dfrac{g_m}{I_D}\right)_1}{\dfrac{1}{V_{EA1}} + \dfrac{1}{V_{EA2}}}. \qquad (3.12)$$

When the Early voltages of the n- and p-channel transistors are equal, the CS stage exhibits a gain loss of 50% relative to the IGS. We can reduce this loss by increasing V_{EA2}, or equivalently, by reducing $(g_{ds}/I_D)_2$. Thus, to investigate further, Figure 3.24 illustrates the evolution of $(g_{ds}/I_D)_2$ versus gate lengths, with $(g_m/I_D)_2$ swept from strong inversion toward weak inversion.

We see that minimizing $(g_{ds}/I_D)_2$ requires the transconductance efficiency of the load transistor to be small, nearly 5 S/A, regardless of the gate length. This corresponds to strong inversion and per (2.34) yields a large saturation voltage V_{Dsat2} of nearly 0.4 V. With only 1.2V supply voltage, this loss in headroom is typically not acceptable, and one may instead pick a compromise with $(g_m/I_D)_2 = 10\ldots12$ S/A,

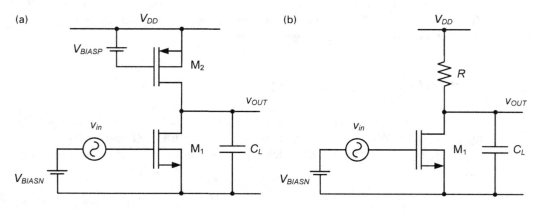

Figure 3.23 (a) Common source stage with (a) active p-channel load and (b) resistive load.

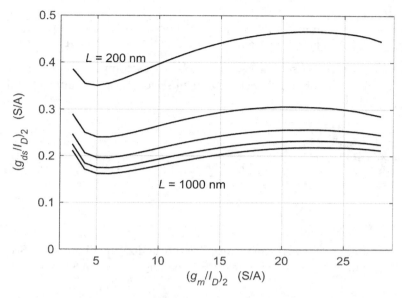

Figure 3.24 Plot of $(g_{ds}/I_D)_2$ versus g_m/I_D of a p-channel load device for various channel lengths ($V_{DS} = 0.6$ V, $V_{SB} = 0$ V).

leading to $V_{Dsat2} = 0.20...0.17$ V. Picking the "optimum" $(g_m/I_D)_2$ requires further knowledge about the specific design tradeoffs, and the extent to which gain is preferred over signal swing. In Section 4.1.1, we re-visit this question from a perspective of output dynamic range (ratio of maximum signal power and noise power). There, we find that g_m/I_D values near 10 S/A are indeed a good choice, and we therefore pick this value for the remainder of the discussion.

The final decision to be made concerns the gate length L_2. The choice is not obvious, since the benefits of large L_2 (for high gain) are offset by larger widths needed to sustain the same current. We will therefore study the tradeoffs in the following example.

Example 3.8 Sizing a CS Stage with Active Load

Consider the IGS of Example 3.3 with $f_u = 1$ GHz, $C_L = 1$ pF and $FO = 10$, but now add a p-channel load as shown in Figure 3.23(a). The supply voltage V_{DD} is equal to 1.2 V and the quiescent output voltage is $V_{OUT} = V_{DD}/2$. Assuming $(g_m/I_D)_2 = 10$ S/A, evaluate the impact of L_2 on all other device geometries and the required drain current. Compare the results for $L_2 = 0.3$, 0.5, and 1 μm with the data from Example 3.3 and perform a SPICE validation for $L_2 = 0.5$ μm.

SOLUTION

We begin by defining a suitable sweep range for L_1 and then compute the corresponding $(g_m/I_D)_1$ vector per the required transit frequency ($f_T = f_u \cdot FO$). This now also lets us compute $(g_{ds}/I_D)_1$.

```
L1 = .06: .001: .4;
gm_ID1  = lookup(nch,'GM_ID','GM_CGG',2*pi*fu*FO,'L',L1);
gds_ID1 = diag(lookup(nch,'GDS_ID','GM_ID',gm_ID1,'L',L1));
```

Next, we perform a similar sweep, which lets us compute $(g_{ds}/I_D)_l$ and A_{v0} using (3.12). From the obtained vector, we then select the maximum gain value, called $|A_{v0max}|$ below.

```
gm_ID2 = 10;
L2 = [.06 .1*(1:10)];
for k = 1:length(L2)
   gds_ID2 = lookup(pch,'GDS_ID','GM_ID',gm_ID2,'L',L2(k))
   Av0(:,k) = gm_ID1./(gds_ID1 + gds_ID2);
end
[a b] = max(Av0);
gain  = a';
```

The remaining steps perform the usual de-normalization discussed in Section 3.1.3, while simultaneously accounting for self-loading as done in Section 3.1.7.

```
Cself = 0;
for k = 1:10,
   gm = 2*pi*fu*(CL + Cself);
   ID = gm./gm_ID1(b);
   W1 = ID./diag(lookup(nch,'ID_W','GM_ID',gm_ID1(b),...
   'L',L1(b)));
   Cdd1 = W1.*diag(lookup(nch,'CDD_W','GM_ID',gm_ID1(b),,...
   'L',L1(b)));
   W2 = ID./lookup(pch,'ID_W','GM_ID',gm_ID2,'L',L2);
   Cdd2 = W2.*lookup(pch,'CDD_W','GM_ID',gm_ID2,'L',L2);
   Cself = Cdd1 + Cdd2;
end
```

Figure 3.25 shows the transistor geometries, maximum gain values and the drain current when L_2 is swept from 0.1 to 1 μm. We see that there is no reason to push L_2 beyond 0.5 μm, since the gain doesn't increase appreciably and the width of the load transistor grows rapidly. Also, notice the significant increase of the drain current for large L_2. To first order, I_D is fixed by M_1. However, as the self-loading increases with larger L_2, a larger g_{ml} is needed to maintain the desired unity gain frequency.

Table 3.11 compares sizes and performance data of the IGS from Example 3.3 and the two-transistor stage. We see that the gain loss caused by the active load is beyond 40%, even when L_2 is largest.

Table 3.12 investigates the impact of small changes in $(g_m/I_D)_2$ while keeping L_2 constant at 0.5 μm. We observe that the geometries of M_1 are nearly unaffected, while the width of the active load (W_2) varies strongly. We see that I_D increases for large W_2, due to the increased self-loading.

To conclude, we compare the data for $L_2 = 0.5$ μm to a SPICE simulation in Table 3.13. The agreement is better than in the earlier examples, since we included self-loading effects in the sizing process.

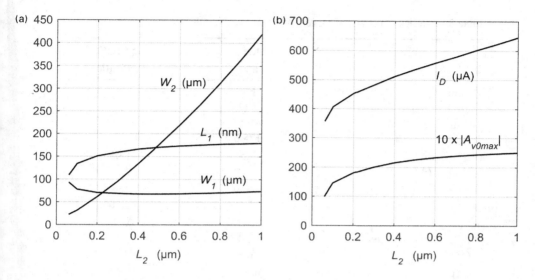

Figure 3.25 (a) Device geometries and (b) drain current and maximum gain of the actively loaded CS stage versus the gate length of the active p-channel load.

Table 3.11 Result comparison. IGS and p-channel load CS stages designed for maximum LF gain and $f_u = 1$ GHz, $C_L = 1$ pF and $FO = 10$.

	IGS	CS stage with p-channel load		
		$L_2 = 1$ μm	$L_2 = 0.5$ μm	$L_2 = 0.3$ μm
$|A_{v0max}|$	40.82	25.0	22.6	20.1
$(g_m/I_D)_1$	10.62	12.92	13.55	14.38
I_D (μA)	592	645	537	483
L_1 (nm)	220	179	170	159
W_1 (μm)	45.9	73.4	67.2	68.5
W_2 (μm)	–	419.3	176	95.6
C_{self} (pF)	–	0.327	0.157	0.104

Table 3.12 Impact of small changes in $(g_m/I_D)_2$ for fixed $L_2 = 0.5$ μm.

$(g_m/I_D)_2$ (S/A)	$|A_{v0max}|$	$(g_m/I_D)_1$ (S/A)	L_1 (nm)	W_1 (μm)	W_2 (μm)	I_D (μA)
9	22.85	13.48	171	65.32	135.1	526.8
10	22.61	13.55	170	67.21	175.6	536.6
11	22.37	13.63	169	69.52	225.6	548.8

Table 3.13 Design values and SPICE verification data.

	Design values	SPICE verification
L_1 (nm)	170	–
L_2 (nm)	500	–
W_1 (μm)	67.21	–
W_2 (μm)	175.6	–
$(g_m/I_D)_1$ (S/A)	13.55	13.55
$(g_m/I_D)_2$ (S/A)	10.00	9.99
I_D (μA)	536.6	536.7
$|A_{vo}|$	22.61	22.60
V_{GS1} (V)	0.5213	0.5212
V_{GS2} (V)	0.5857	–
f_u (MHz)	1000	1013
C_{self} (pF)	0.157	0.151

As a final example, we take now a closer look at the large-signal characteristic of an actively loaded CS stage. The large-signal characteristic is important as it defines the output signal swing that the stage can accommodate. The example will show that we can obtain an accurate characteristic using our Matlab lookup tables.

Example 3.9 Large-Signal Characteristic of a CS Stage with Active Load

Construct the transfer characteristic (v_{OUT} versus v_{IN}) of the CS stage from Table 3.13. Compare the small-signal voltage gain to the slope of the transfer characteristic at the quiescent point ($V_{OUT} = 0.6$ V) and assess the available output voltage swing. Compare the results to a SPICE simulation.

SOLUTION

To construct the transfer characteristic, we sweep v_{OUT} across the voltage range of interest and find the corresponding drain current I_{D2} of the p-channel load. We then determine the gate voltages of M_1 that make I_{D1} equal to I_{D2}. In the code below, I_{D2} is a vector, and I_{D1} a matrix.

```
VDS1 = .05: .01: 1.15;
ID2 = Wp*lookup(pch,'ID_W','VGS',VGS2,'VDS',VDD-VDS1,'L',L2);
ID1 = Wn*lookup(nch,'ID_W','VGS',nch.VGS,'VDS',VDS1,'L',L1)';
for m = 1:length(VDS1),
  VGS1(:,m) = interp1(ID1(m,:),nch.VGS,ID2(m));
end
```

The resulting transfer characteristic is shown in Figure 3.26. The voltage gain extracted from the slope of the transfer characteristic at the quiescent point is equal

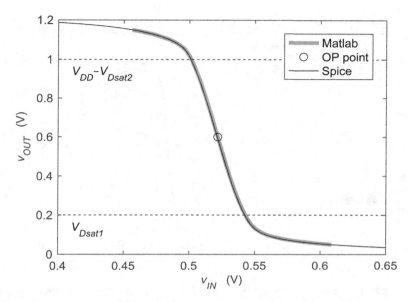

Figure 3.26 Transfer characteristic of the p-channel loaded IGS.

to −22.60, the same value as in Table 3.13. The horizontal lines in the plot mark the (approximate) output voltages for which the transistors leave saturation. The saturation voltages are found using (2.34) ($V_{Dsat} = 2/(g_m/I_D)$), which yields $2/(10 \text{ S/A}) = 200$ mV for both transistors in this example. Hence, the maximum output voltage swing that can be handled by this circuit is approximately ±400 mV. As the circuit nears these peak excitations, it becomes progressively less linear. We will quantify these non-linear effects in Chapter 4.

The thin solid line in Figure 3.26 was obtained from a SPICE simulation. We observe a good agreement with the curve computed using the lookup data in Matlab.

3.2.2 Resistive Load

For the resistively loaded CS stage in Figure 3.23(b), the expression for the LF voltage gain is obtained by replacing g_{ds2} of the p-channel load in (3.12) with $1/R$. This yields:

$$A_{v0} = -\frac{g_{m1}}{g_{ds1} + \dfrac{1}{R}} = -\frac{\left(\dfrac{g_m}{I_D}\right)_1}{\left(\dfrac{g_{ds}}{I_D}\right)_1 + \dfrac{1}{I_D R}} = -\frac{\left(\dfrac{g_m}{I_D}\right)_1}{\dfrac{1}{V_{EA1}} + \dfrac{1}{I_D R}}. \tag{3.13}$$

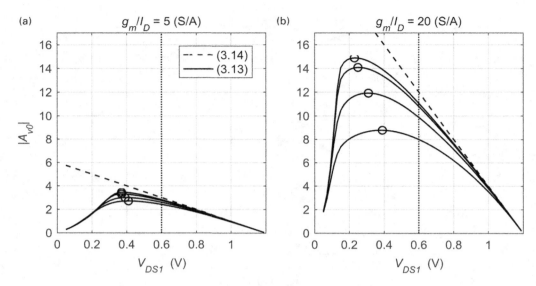

Figure 3.27 LF gain of the resistively loaded CS stage versus drain-to-source voltage. (a) $g_m/I_D = 5$ S/A and (b) $g_m/I_D = 20$ S/A. $V_{DD} = 1.2$ V and $L_1 = 0.1, 0.2, 0.5,$ and 1.0 μm (bottom up).

We see from this expression that the DC voltage drop across the load resistor ($I_D R$) plays a key role in setting the gain. In fact, for scenarios where the $I_D R \ll V_{EA1}$, the expression simplifies to:

$$A_{v0} \cong -\left(\frac{g_m}{I_D}\right)_1 I_D R. \tag{3.14}$$

To maximize the gain, one should make $I_D R$ as large as possible, but this reduces the available voltage swing, and additionally lowers V_{EA1}, due to the decrease in V_{DS1}. The latter effect gives rise to a maximum in the achievable gain.

Figure 3.27 plots (3.13) and (3.14) when V_{DS1} is swept over almost the entire supply voltage range, considering various gate lengths and two inversion levels: (a) strong inversion (5 S/A) and (b) moderate inversion (20 S/A). Circles mark the maxima, which are well below the gains obtained with the IGS and actively loaded CS stage. Note also that the corresponding drain-to-source voltages are well below $V_{DD}/2$ (dotted vertical line), which is non-ideal if large signal swing is desired. In the case of strong inversion, it would be impractical to operate the circuit at the maximum gain values, since V_{DS1} is very close to V_{DSat1} (400 mV) allowing essentially no signal swing. The situation is much improved in moderate inversion. Lastly, we see that for large L_1 (top curves) and large V_{DS1}, which leads to large V_{EA1}, the true voltage gain given by (3.13) is close to the simplified expression of (3.14).

Example 3.10 Sizing a CS Stage with Resistive Load

Consider a resistively loaded CS stage with $C_L = 1$ pF, $f_u = 1$ GHz, $V_{DD} = 1.2$ V and $FO = 10$. Find the values of V_{DS} and L that maximize the LF voltage gain and compute the corresponding load resistor value. Account for self-loading during sizing and validate the design using a SPICE simulation.

SOLUTION

We begin by defining a sweep range for the gate length and drain-to-source voltage (LL and UDS). Next, we find all g_m/I_D and g_{ds}/I_D values within this space for which f_T is equal to the required 10 GHz.

```
UDS = .1*(2:6);        %horizontal
LL  = .06: .01: .2;    %vertical
for k = 1:length(UDS)
  gmID(:,k)=lookup(nch,'GM_ID','GM_CGG',2*pi*fT,...
  'VDS',UDS(k),'L',LL);
  gdsID(:,k)=lookup(nch,'GDS_ID','GM_CGG',2*pi*fT,...
  'VDS',UDS(k),'L',LL);
end
```

Now we can compute the LF gain using (3.13) and determine the gate length, drain-to-source voltage and transconductance efficiency that maximize the expression:

```
AvoR = gmID./(gdsID + 1./(VDD-UDS(ones(length(LL),1),:)))
[a b] = max(AvoR);
[c d] = max(a);
AvoRmax = c
L    = LL(b(d))
VDS  = UDS(d)
gm_ID = gmID(b(d),d)
```

The maximum gain magnitude is found to be 8.49, along with $L = 0.11$ µm, $V_{DS} = 0.4$ V and $g_m/I_D = 18.04$. The remaining task is to de-normalize as usual, and iterate to accommodate the self-loading:

```
JD   = lookup(nch,'ID_W','GM_ID',gm_ID,'VDS',VDS,'L',L);
Cdd_W = lookup(nch,'CDD_W','GM_ID',gm_ID,'VDS',VDS,'L',L);
Cdd = 0;
for k = 1:5,
   gm = 2*pi*fT/10*(C+Cdd);
   ID = gm/gm_ID;
   W  = ID/JD;
   Cdd = W*Cdd_W;
end
```

We find $I_D = 368.4$ µA and $W = 87.92$ µm. Since we know the drain current and the voltage drop across the load resistor, we can readily compute $R = (V_{DD} - V_{DS})$

Table 3.14 Result summary.

	Design Values	SPICE
L_1 (nm)	110	–
W (μm)	87.92	–
V_{DS} (V)	0.400	–
g_m/I_D (S/A)	18.04	18.03
I_D (μA)	368.4	368.3
$|A_{vo}|$	8.487	8.493
V_{GS} (V)	0.4646	–
f_u (MHz)	1000	1000

divided by $I_D = 2.172$ kΩ. Table 3.14 summarizes the design values along with the SPICE-simulated numbers, which are in close agreement.

3.3   Differential Amplifier Stages

Differential amplifiers are key building blocks in analog design. At the core of a differential amplifier, we typical find a differential pair, as shown in Figure 3.28. The basic operation of this circuit is well described and analyzed in classical circuit design textbooks, such as [3], [4]. However, these treatments are typically based on the square-law, and have therefore become inaccurate.

The goal of this section is twofold. First, we want to analyze the large-signal behavior of the differential pair using the basic EKV model introduced in Chapter 2. This analysis sets the stage for the distortion analysis that will follow in Chapter 4. Secondly, we want to review the sizing of amplifiers that are based on the differential pair. A key difference compared to the IGS and CS stages discussed previously is that the transconductance is set by the tail current source I_0, rather than a voltage applied at the gate.

We begin with the large-signal analysis of the differential pair. To derive EKV-based expressions, we first rewrite (2.19) for M_1 and M_2:

$$v_{P1} - v_S = U_T \left(2(q_1 - 1) + log(q_1) \right)$$
$$v_{P2} - v_S = U_T \left(2(q_2 - 1) + log(q_2) \right). \tag{3.15}$$

We then subtract the two equations to isolate the incremental differential input, using (2.21) to relate the gate and pinch-off voltages:

$$v_{id} = v_{i1} - v_{i2} = nU_T \left[2(q_1 - q_2) + log\left(\frac{q_1}{q_2}\right) \right]. \tag{3.16}$$

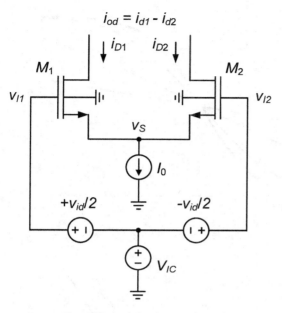

Figure 3.28 The differential pair.

To find the tail node voltage excursion v_S, we add the two instances of (3.15) and replace the quiescent point term $(V_{P1} + V_{P2})$ by:

$$2\frac{V_{IC} - V_T}{n} = 2V_P = 2(U_T(2(q_0 - 1) + log(q_0)) - V_S). \tag{3.17}$$

Here, q_0 designates the normalized charge density at the quiescent point. We then solve for v_S from the sum, and find the incremental tail voltage:

$$v_s = v_S - V_S = U_T\left[2q_0 - (q_1 + q_2) + \log\left(\frac{q_0}{\sqrt{q_1 q_2}}\right)\right]. \tag{3.18}$$

So far, the differential input voltage v_{id} and the incremental tail voltage v_s are functions of q_1 and q_2. To switch to the normalized EKV drain currents i_1 and i_2, we invert (2.22):

$$q_1 = 0.5\left[\sqrt{1 + 4i_1} - 1\right] \quad and \quad q_2 = 0.5\left[\sqrt{1 + 4i_2} - 1\right]. \tag{3.19}$$

We can now express the differential input voltage as a function of the normalized quantities i_1/i_0 or i_2/i_0, calling i_0 the sum of i_1 and i_2. Substituting i_{D1}/I_0 and i_{D2}/I_0 for i_1/i_0 or i_2/i_0 lets us numerically plot the transfer and source voltage curves of Figure 3.29, using the Matlab code below:

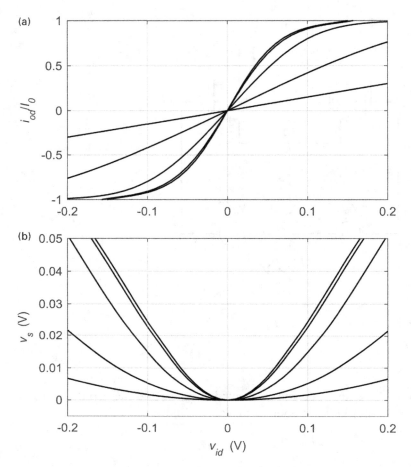

Figure 3.29 Large-signal characteristic of the differential pair as a function of differential input voltage considering five tail currents, (a) normalized differential output current and (b) Incremental component of the source node voltage. The five shown curves correspond to tenfold multiples of the normalized tail current i_o, from 0.01 (weak inversion, steepest curves) to 100 (strong inversion).

```
% data ====================
n   = 1.2;                      % subthreshold slope
io = 2*logspace(-2,2,5);        % normalized tail current I0/IS
vid = .01*(-20:20);             % input diff voltage range (V)
% compute ==================
UT = .026;
m   = (.05:.05:1.95); b = find(m==1);
for k = 1:length(io),
    i2   = .5*io(k)*m;
    q2   = .5*(sqrt(1+4*i2)-1);
    q1   = .5*(-1 + sqrt(1 + 4*(io(k)-q2.^2-q2)));
    vg = n*UT*(2*(q2-q1) + log(q2./q1));
    IoD_I0(k,:) = 2*interp1(vg,i2,vid,'spline')/io(k) - 1 ;
    q   = q2(b);
    vs = UT*(2*q - (q1+q2) + log(q./sqrt(q1.*q2)));
```

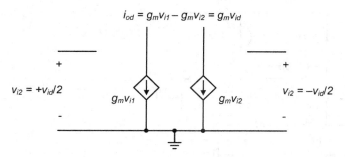

$$i_{od} = g_m v_{i1} - g_m v_{i2} = g_m v_{id}$$

$$v_{i2} = +v_{id}/2$$ $g_m v_{i1}$ $g_m v_{i2}$ $$v_{i2} = -v_{id}/2$$

Figure 3.30 Simplified small-signal model of the differential pair. (All parasitic elements are omitted.)

```
   VS(k,:) = interp1(vg,vs,vid,'spline');
end
```

The five shown curves correspond to tenfold multiples of the normalized tail current i_0, from 0.01 (weak inversion, steepest curves) to 100 (strong inversion). They confirm the well-known fact that the tail node voltage does not move significantly if the differential input is small. Lastly, note that the shown plots required only one parameter, the subthreshold slope factor n. We will see in Section 4.2 that this greatly simplifies the treatment of nonlinear distortion.

Given that the tail node does not move for small inputs allows us to turn it into an AC ground for small-signal modeling purposes, leading to the model shown Figure 3.30. Each half circuit sees half of the differential input, but after taking the difference at the output, the transconductance that links the (differential) input voltage and (differential) output current is simply g_m, just like in an IGS.

Given the similarity between the small-signal model of the IGS and the differential pair, everything that we learned in the previous sections is also applicable to amplifiers that are based on the differential pair. As mentioned initially, the main difference lies in how the bias point is established. The following example will repeat Example 3.1 for a differential pair to illustrate this.

Example 3.11 Sizing a Differential Pair with Ideal Current Source Loads

Size the differential amplifier shown in Figure 3.31 to achieve $f_u = 1$ GHz with a load capacitance $C_L = 1$ pF. Assume $g_m/I_D = 15$ S/A and $L = 60$ nm. The input common mode is $V_{IC} = 0.7$ V and the shown common-mode feedback (CMFB) circuit forces $V_{OC} = (V_{O1} + V_{O2})/2 = 1$ V at the operating point. Plot the large-signal transfer characteristic from the differential input to the differential drain current. Verify the results using SPICE simulations.

SOLUTION

We first compute the transconductance and drain currents of M_1 and M_2 as in Example 3.1, using (3.4) and (3.8):

```
gm = 2*pi*fu*CL;
ID = gm/gm_ID;
```

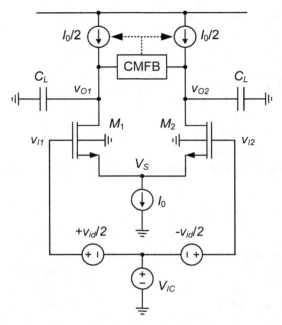

Figure 3.31 Differential pair amplifier. The common mode feedback (CMFB) block forces the output common-mode voltage to the desired value.

To find the width, we divide I_D by the current density J_D keeping in mind that the tail node (V_S) of the amplifier is not grounded. To find the actual source-to-bulk voltage, called V_{SB} in the lookup function calls below, we compute the gate-to-source voltage V_{GS} using the lookupVGS function:

```
VGS = lookupVGS(nch, 'GM_ID',gmID,'VDB',VDB,'VGB',VGB);
```

and derive V_{SB} from the difference $V_{GB} - V_{GS}$. We then find the current density using:

```
JD = lookup(nch,'ID_W','GM_ID',gmID,'VSB',VSB,'VDS',VDB-VSB)
```

Since f_u and L are the same as in Example 3.1, the drain currents of M_1 and M_2 again equal 419 μA. The device widths differ slightly, however; they are equal to 41.06 μm instead of 41.72 μm. The decrease in W is the result of (1) a larger V_{GS} (0.475 instead of 0.468 V) due to backgate bias and (2) a smaller V_{DS} (0.5254 instead of 0.6 V). The latter difference impacts the result due to DIBL, which is a strong effect with $L = 60$ nm. The source-to-bulk voltage V_{SB} (equal to the tail node voltage V_S) is 225.4 mV.

The circuit in Figure 3.31 is now simulated in SPICE with the given parameters ($V_{OC} = 1$ V, $V_{IC} = 0.7$ V, $W = 41.06$ μm, $L = 60$ nm, $I_0/2 = 419$ μA). The DC operating point simulation yields: $g_m/I_D = 6.31$ mS/419 μA $= 15.06$ S/A and $V_S = 226$ mV. The frequency response is shown below, indicating a good match with the targeted unity gain frequency of 1 GHz.

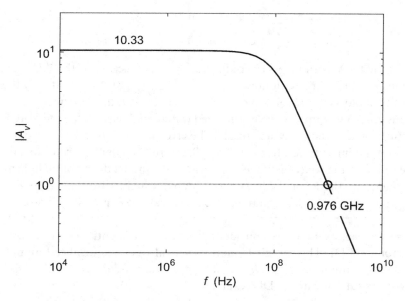

Figure 3.32 Simulated frequency response of the differential pair amplifier.

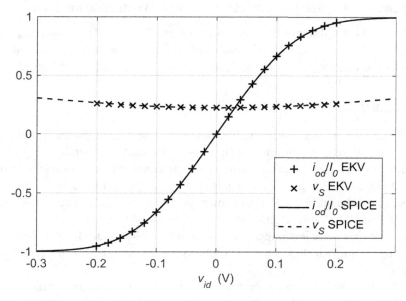

Figure 3.33 Normalized differential output current and tail node voltage (v_s) versus differential input voltage (v_{id}).

Lastly, we are to compare the transfer characteristic obtained by SPICE to the characteristic predicted by the basic EKV model. Therefore, we extract the basic EKV parameters considering the source and drain voltages used in the amplifier. This is done by running:

```
jd = lookup(nch,'ID_W','VGS',nch.VGS,'VSB',VSB, ...
'VDS', VDB-VSB, 'L',0.06);
y = XTRACT2(nch.VGS,jd);
```

Once the EKV parameters are obtained, the construction of the i_{od}/I_o and v_S versus v_{id} characteristics follows from (3.16) and (3.18). Note that only the subthreshold slope factor n and the specific current I_S are required.

Figure 3.33 compares the normalized differential current for the SPICE simulation and the analytical result. The error is less than 0.3% over the differential input range, from –0.2 to +0.2 V. In the same diagram, we also show the plots for the source voltage V_S. Because the model does not include mobility degradation, the V_S of the EKV model departs slightly from the SPICE result as v_{id} grows. The error between the two curves is about 8 mV as v_{id} approaches ±0.2 V.

As another example, we consider the classical differential amplifier with current-mirror load [3]. This circuit is more practical than the idealized amplifier in the previous example, and it is also used within larger circuits covered in this book (see low-dropout regulator (LDO) example in Chapter 5).

Example 3.12 Sizing a Differential Amplifier with Current-Mirror Load

Design the amplifier in Figure 3.34 to achieve f_u = 100 MHz, assuming C_L = 1 pF, V_{IC} = 0.7 V and V_{DD} = 1.2 V. To simplify this task, you may re-use results obtained earlier in this chapter (consider Example 3.6, Section 3.2.1 and Figure 3.25).

SOLUTION

The first step is to choose suitable transconductance efficiencies and gate lengths for the differential pair and the current-mirror load. To simplify this task, we can look at Example 3.6, Table 3.9(b), where we considered a CS stage with the same f_u, FO = 21.2, L = 0.34 μm and g_m/I_D = 17 S/A. Furthermore, in Section 3.2.1 we examined the addition of a p-channel load and concluded that its g_m/I_D should be about 10 S/A due to gain, V_{Dsat} and noise considerations. Finally, Figure 3.25 showed that there is diminishing return in the voltage gain for gate lengths larger than 0.5 μm. Our decision is therefore to use this length for our current mirror load.

Next, we proceed to the evaluation of the source and drain voltages of the input pair transistors. For $M_{1a,b}$, V_{SB1} is equal to zero due to the source-bulk tie shown in the schematic. V_{DS1} is unknown and we must find the source and drain voltages of $M_{1a,b}$ to compute it. Since we know the gate voltages of $M_{1a,b}$ (equal to the given V_{IC}), we can find the source voltage by subtracting V_{GS1} from V_{IC}. To find V_{GS1}, we use:

```
VGS1 = lookupVGS(nch,'GM_ID',gmID1,'L',L1);
```

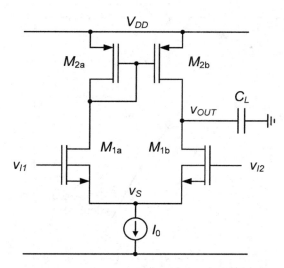

Figure 3.34 Differential amplifier with current-mirror load.

Here, we assumed the default value for V_{DS1} (0.6 V). This is fine, since the transistor is saturated and the gate length won't be minimum. We find that V_{GS1} is equal to 0.4661 V. Note that we can re-run this calculation later once an improved estimate of V_{DS1} is available. To find the drain voltage, we subtract V_{GS2} from V_{DD}.

```
VGS2 = lookupVGS(pch,'GM_ID',gmID2,'L',L2);
```

taking again the default value for V_{DS2}. We refine the result running the command below, where V_{GS2} serves as the new estimate for V_{DS2}.

```
VGS2 = lookupVGS(pch,'GM_ID',gmID2,'VDS',VGS2,'L',L2);
```

This yields $V_{GS2} = 0.5858$ V and thus $V_{DS1} = 0.3812$ V, which indicates that M_{1b} is saturated. Rerunning the commands a second time, we'll see that V_{GS1} increased slightly from 0.4661 to 0.4670 V. We also find that $V_S = 0.2330$ V.

To compute the LF voltage gain of the amplifier, we make use of the result from Figure 3.13. This is valid since the voltage gain of a differential amplifier with current-mirror load is the same as that of an actively loaded CS stage [4]. Hence:

```
gdsID1=lookup(nch,'GDS_ID','GM_ID',gmID1,'VDS',VDS1,'L',L1);
gdsID2=lookup(pch,'GDS_ID','GM_ID',gmID2,'VDS',VGS2,'L',L2);
Av0 = gmID1/(gdsID1 + gdsID2)
```

We find $A_{v0} = 31.06$. Finally, to determine the drain currents and the transistor widths, we repeat the same steps taken for the CS stage:

```
gm1 = 2*pi*fu*CL;
ID  = gm1/gmID1;
JD1 = lookup(nch,'ID_W','GM_ID',gmID1,'VDS',VDS1,'L',L1);
```

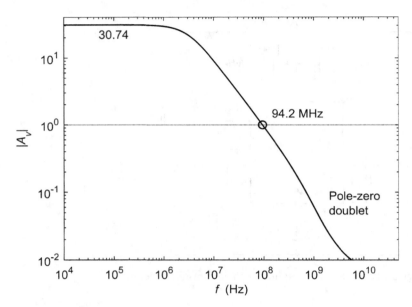

Figure 3.35 SPICE-simulated frequency response of the differential pair with current mirror load.

```
W1  = ID/JD1;
JD2 = lookup(pch,'ID_W','GM_ID',gmID2,'VDS',VGS2,'L',L2);
W2  = ID/JD2;
```

We find that the drain currents of $M_{1a,b}$ and $M_{2a,b}$ are 36.96 µA, W_1 is 18.07 µm and $W_2 = 12.13$ µm. The design is now completely specified and we can simulate it in SPICE. Figure 3.35 shows the obtained frequency response, which matches well with the computed numbers.

The simulation showed a tail node voltage of 0.2329 V and an output voltage of 0.6142 V, which agrees with the predicted values (0.2330 V and 0.6142 V). For the voltage gain, we measure 30.74, which is also close to the predicted value of 31.06. The unity gain frequency f_u is a little less than 100 MHz, because we neglected self-loading. Scaling the current and all widths by the factor S in (3.10) would correct this.

Even though the simulated unity gain frequency matches our expectations, it is worth paying attention to the pole-zero doublet caused by the current mirror [3] (see Figure 3.35). The pole of the doublet lies at approximately $f_T/2$ of the p-channel load devices,[5] which amounts to 580 MHz. This has little impact on our design with $f_u = 100$ MHz, but becomes a problem in faster circuits. One way to push the doublet to higher frequencies is to reduce L_2.

---

[5] The pole location is defined by the current mirror node resistance $1/g_{m2}$ and two times C_{gg2}, since two devices are loading the node. The result is $\omega_T/2$.

As a final example, we consider a differential amplifier with a resistive input driver. This is a very common scenario, for example in cascaded gain stages. A key aspect in this problem is the inclusion of the Miller effect in the bandwidth calculation, which we have so far not considered.

Example 3.13 Sizing a Differential Amplifier with Resistive Input Driver and Resistive Loads

Design the amplifier in Figure 3.36 to achieve a differential voltage gain of $|A_{vo}| = 4$ assuming $C_L = 50$ fF, $R_D = 1$ kΩ, $R_S = 10$ kΩ, $V_{SC} = 0.7$ V and $V_{DD} = 1.2$ V. Size the transistors assuming $L = 100$ nm and $g_m/I_D = 15$ S/A. Estimate the dominant and non-dominant pole frequency of the circuit and validate the design using SPICE simulations.

SOLUTION

The first step is to compute g_m from the voltage gain requirement. For this purpose, we first need to find the intrinsic gain of the transistors:

```
gm_gds = lookup(nch, 'GM_GDS', 'GM_ID', gm_ID, 'L', L);
```

Using this information, we can readily compute g_m using:

```
gm = 1/RL*(1/Av0 - 1./gm_gds).^-1;
```

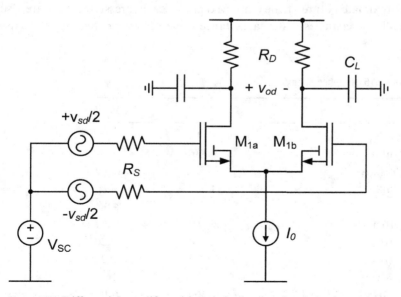

Figure 3.36 Differential amplifier with resistive load and resistive input driver.

This yields g_m = 4.93 mS, I_0 = 2 x 328 µA and W = 36.46 µm (using the usual de-normalization). To calculate the circuit's pole frequencies, we can invoke the following textbook expressions, which are based on the dominant pole approximation [3]:

$$\omega_{p1} = \frac{1}{b_1} \quad \omega_{p2} = \frac{b_1}{b_2},$$

(3.20)

where:

$$b_1 = R_s \left[C_{gs} + C_{gd} \left(1 + |A_{v0}| \right) \right] + R_L \left(C_{Ltot} + C_{gd} \right)$$
$$b_2 = R_s R_L \left[C_{gs} C_{Ltot} + C_{gs} C_{gd} + C_{Ltot} C_{gd} \right]$$

(3.21)

and $C_{Ltot} = C_L + C_{db}$. Note the multiplication of C_{gd} by the term $(1 + |A_{v0}|)$, which is due to the Miller effect.

Since we have already sized the transistors, we can readily compute the pole frequencies using the following capacitance estimates:

```
Cgs = W.*lookup(nch,  'CGS_W',  'GM_ID',  gm_ID,  'L',  L)
Cgd = W.*lookup(nch,  'CGD_W',  'GM_ID',  gm_ID,  'L',  L)
Cdd = W.*lookup(nch,  'CDD_W',  'GM_ID',  gm_ID,  'L',  L)
Cdb = Cdd-Cgd
CLtot = CL+Cdb;
```

Table 3.15 summarizes the obtained design values along with their SPICE validation (using an operating point analysis, followed by a pole-zero analysis). We observe close agreement between the calculated and simulated numbers. The largest discrepancy is in the non-dominant pole frequency, caused mainly by the approximate nature of the employed analytical expression. There are also some small discrepancies in the capacitance values. These are mainly due to assuming

Table 3.15 Result summary.

	Design values	SPICE
L (nm)	100	–
W (µm)	36.46	–
g_m/I_D (S/A)	15	15.09
g_m (mS)	4.93	4.95
$|A_{v0}|$	4	4.06
C_{gs} (fF)	27.94	28.01
C_{gd} (fF)	12.13	12.20
C_{db} (fF)	10.76	9.2
f_{p1} (MHz)	166	167
f_{p2} (GHz)	5.50	6.10

$V_{DS} = 0.6$ V in the Matlab calculation, while the SPICE simulation indicates $V_{DS} = 0.66$ V. These errors can be minimized by re-computing the capacitances with a better estimate of V_{DS}, but it would not be worth the effort in this example.

3.4 Summary

This chapter established a methodology for the systematic sizing of simple gain stages. The starting point for this development was the intrinsic gain stage (IGS), a single-transistor circuit with an ideal current source and capacitive load. To find the current I_D, we divide the required transconductance g_m by g_m/I_D. To find the device width, we "de-normalize" by dividing I_D with the drain current density J_D obtained from lookup tables.

Several examples served to illustrate how to pick appropriate gate lengths and transconductance efficiencies to match a given set of objectives, for example, maximum voltage gain, minimal current consumption, etc.

For circuits that are not constrained by speed requirements, weak inversion is a natural choice. In weak inversion, g_m/I_D is nearly constant and thus cannot serve as a distinguishing design knob. In this case, the current density J_D can be used as a substitute.

In high-speed circuits, extrinsic capacitances at the transistors' drain nodes can lead to noticeable self-loading and bandwidth reduction. We showed two iterative sizing methods that address this issue.

After establishing these basics, we extended the scope of the discussion and considered the sizing of common-source stages with active and resistive loads, as well as differential pair stages. We showed that the basic flows and considerations established for the IGS still apply.

3.5 References

[1] B. Murmann, *Analysis and Design of Elementary MOS Amplifier Stages*. NTS Press, 2013.
[2] P. Harpe, H. Gao, R. van Dommele, E. Cantatore, and A. van Roermund, "A 3nW Signal-Acquisition IC Integrating an Amplifier with 2.1 NEF and a 1.5fJ/conv-step ADC," in *ISSCC Dig. Tech. Papers*, 2015, pp. 382–383.
[3] P. R. Gray, P. Hurst, S. H. Lewis, and R. G. Meyer, *Analysis and Design of Analog Integrated Circuits*, 5th ed. Wiley, 2009.
[4] T. Chan Caruosone, D. A. Johns, and K. W. Martin, *Analog Integrated Circuit Design*, 2nd ed. Wiley, 2011.

4   Noise, Distortion and Mismatch

4.1   Electronic Noise

Electronic noise is a significant and fundamental performance limitation in electronic circuits. In MOS transistors, there are two effects to consider: thermal noise and flicker noise. Thermal noise is closely tied to thermodynamics, whereas flicker noise is due to material imperfections. Contrary to "man-made" noise, caused for example by unwanted fluctuations in the supply voltage, these noise sources are inherently present in transistors and are often times difficult to minimize. Therefore, it is important to explore the calculation of circuit noise within the g_m/I_D design framework presented in this book.

4.1.1   Thermal Noise Modeling

We begin by considering the effect of thermal noise in the intrinsic gain stage (IGS), as shown in Figure 4.1. The thermal noise of the transistor can be modeled using a current source with a power spectral density of

$$\frac{\overline{i_d^2}}{\Delta f} = 4kT\gamma_n g_m,   (4.1)$$

where k is the Boltzmann constant, T is the absolute temperature in Kelvin, and γ_n is a model parameter, ideally equal to 2/3 in a saturated transistor in strong inversion [1]. Due to a variety of second-order effects in short-channel transistors, γ_n is often somewhat larger, and takes on bias-dependent values in the range of 0.7...1.5 [2], [3] in strong inversion. In weak inversion, the parameter approaches the shot noise[1] limit with $\gamma_n = 0.5n$, where n is the subthreshold slope factor.

Figure 4.2 plots the γ parameters for the devices in our 65-nm technology. These data were computed by rearranging (4.1):

$$\gamma = \frac{1}{4kT}\frac{\overline{i_d^2}}{g_m} = \frac{1}{4kT}\frac{STH}{g_m},   (4.2)$$

---

[1] Shot noise is due to the discreteness of charge and stems from the randomness by which these charges cross a potential barrier (pn junction). Shot noise requires a DC current, while thermal noise (in strong inversion) exists without a DC current [1].

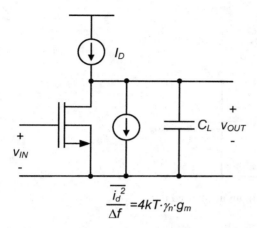

Figure 4.1 Intrinsic gain stage with thermal noise source.

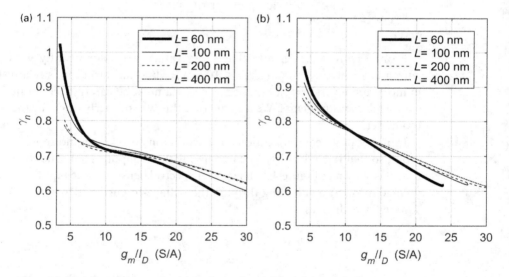

Figure 4.2 Noise parameters for (a) n-channel and (b) p-channel devices.

Here, STH is the variable name used in our lookup tables to represent the simulated thermal noise spectra density (constant across all frequencies). The Matlab code that implements the computation per (4.2) is given below:

```
kB = 1.3806488e-23;
L = [0.06, 0.1, 0.2, 0.4];
vgs = 0.2:25e-3:0.9;
for i=1:length(L)
    gm_id_n(:,i) = lookup(nch, 'GM_ID', 'VGS', vgs, 'L', L(i));
    gm_id_p(:,i) = lookup(pch, 'GM_ID', 'VGS', vgs, 'L', L(i));
```

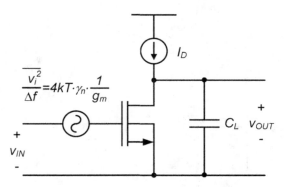

Figure 4.3 Intrinsic gain stage with input-referred noise.

```
gamma_n(:,i) = lookup(nch,'STH_GM','VGS',vgs,'L', ...
L(i))/4/kB/nch.TEMP;
gamma_p(:,i) = lookup(pch,'STH_GM','VGS',vgs,'L', ...
L(i))/4/kB/pch.TEMP;
end
```

From Figure 4.2, we note that the γ values vary significantly with channel length in strong inversion ($g_m/I_D < 10$ S/A), while the variation is relatively small in moderate inversion. In practice, such data must be carefully calibrated against actual lab measurements. We assume that this was properly done for the provided model set.

In many practical applications, it is common to refer the noise to the circuit's input, so that it can be directly compared with the input signal. The corresponding model is shown in Figure 4.3. The power spectral density of the input-referred noise voltage is found by dividing through the voltage-to-current conversion factor (g_m) squared, which gives

$$\frac{\overline{v_i^2}}{\Delta f} = 4kT\gamma_n \frac{1}{g_m}. \tag{4.3}$$

From this result, we see that lowering the input-referred noise requires an increase in the device's transconductance, which can be achieved in several ways. One way to increase g_m is to leave the device size unchanged and increase the drain current, which means higher power consumption. Another way to increase g_m is to keep I_D constant and widen the transistor. However, this option increases g_m/I_D, which comes at the price of lower transit frequency. Now, since the gain bandwidth product is equal to $\omega_u = \omega_T/FO$ (see Section 3.1), this means that the bandwidth will reduce for fixed fan-out ($FO = C_L/C_{gg}$). For a fair comparison, it is therefore useful to establish a figure of merit that considers both the noise and gain-bandwidth product simultaneously.

4.1.2 Tradeoff between Thermal Noise, GBW and Supply Current

As motivated by the above discussion, we define a figure of merit using the following ratio, which measures how efficiently the circuit trades the consumed supply current with noise and gain-bandwidth product:

$$\frac{\text{Gain} \cdot \text{Bandwidth}}{\text{Input-Referred Noise} \cdot \text{Supply Current}} = \frac{\dfrac{f_T}{FO}}{4kT\gamma_n \dfrac{1}{g_m} \cdot I_D} \propto f_T \cdot \frac{g_m}{I_D}. \tag{4.4}$$

Interestingly, we see from this result that once the FO is set (and assuming γ_n is approximately constant), the tradeoff is fully determined by the product of transit frequency ($g_m/2\pi C_{gg}$) and g_m/I_D. Intuitively, we want the transistor to generate large g_m with little current, and at the same time present a small capacitance.

It is now interesting to ask whether there is an optimum bias point where (4.4) is maximized. To investigate, we take advantage of the basic EKV model and analyze the impact of g_m/I_D on the product on the right-hand side of (4.4). We consider the transit frequency first:

$$\omega_T = \frac{g_m}{C_{gg}} = \frac{\dfrac{g_m}{I_D}}{\dfrac{C_{gg}}{I_D}}. \tag{4.5}$$

The denominator can be written as follows, if the transistor is saturated and edge effects (due to small geometries) are negligible:

$$\frac{C_{gg}}{I_D} \approx \frac{\frac{2}{3} WLC_{ox}}{2nU_T^2 \mu C_{ox} \frac{W}{L} i} = \frac{1}{3} \frac{L^2}{nU_T^2 \mu} \frac{1}{i} = \frac{1}{3} \frac{L^2}{nU_T^2 \mu} \frac{1}{q^2 + q}. \tag{4.6}$$

Hence, per (2.29) we find:

$$\omega_T \approx 3 \frac{\mu U_T}{L^2} q. \tag{4.7}$$

The figure of merit of (4.4) now becomes:

$$\omega_T \frac{g_m}{I_D} \approx 3 \frac{\mu}{nL^2} \frac{q}{1+q} = 3 \frac{\mu}{nL^2} \left(1 - nU_T \frac{g_m}{I_D} \right) = 3 \frac{\mu}{nL^2} (1 - \rho), \tag{4.8}$$

where ρ is the normalized transconductance efficiency as defined in (2.31). We see that the product ω_T times g_m/I_D varies linearly from zero (deep in weak inversion) to a constant value of $3\mu/nL^2$ deep in strong inversion.[2] This is confirmed by the plot of Figure 4.4, which represents the $f_T \times g_m/I_D$ curves versus g_m/I_D for real transistors with various channel lengths. If g_m/I_D remains larger than 10 S/A, the figure of merit varies linearly with the transconductance efficiency. However, once we enter strong inversion, the curve departs from the straight line and bends as we go deep into strong inversion. This is due to mobility degradation effects, which were discussed in Chapter 2.

We can draw several conclusions from the data of Figure 4.4: (1) the curves show optima in the moderate-to-strong inversion region, near $g_m/I_D = 7$ S/A. (2) The peak of the curve is highest for the shortest channel length, simply because the transit frequency is maximized for short channels. This confirms that short-channel transistors can be operated at lower current levels for a given bandwidth and noise specification.

As discussed in more detail in [4], [5], the observed optimum can be leveraged to minimize the power consumption of a variety of circuits, such as RF low-noise amplifiers (LNAs). We will now consider a basic example to outline this idea in its simplest form.

Example 4.1 Sizing of a Low-Noise IGS

Size the circuit of Figure 4.3 such that the input-referred thermal noise is equal to 1 nV/rt-Hz. Choose the inversion level such that the optimum tradeoff between noise, gain-bandwidth product and current consumption is achieved.

SOLUTION

For operation near the optimum tradeoff, we choose $g_m/I_D = 7$ S/A (see Figure 4.4). Since no constraint on low-frequency gain is given, we chose the shortest channel and let $L = 60$ nm. The value of γ_n is approximately 0.8 for this length and inversion level (see Figure 4.2). From the noise specification, we can now compute the required transconductance using (4.3):

$$g_m = \frac{4kT\gamma_n}{\overline{v_i^2}} = \frac{4kT \cdot 0.8}{\left(\dfrac{1\,\text{nV}}{\sqrt{\text{Hz}}}\right)^2} = 13.2\,\text{mS}.$$

This fixes the current, $I_D = g_m/(g_m/I_D) = 1.9$ mA. The width of the transistor can now be determined after finding the current density for a 60-nm n-channel at $g_m/I_D = 7$ S/A. Using the lookup function, we find $I_D/W = 84.4$ μA/μm, and thus $W = I_D/(I_D/W) = 22.4$ μm. The circuit is now fully specified.

---

[2] Notice that the constant slope, which is equal to $3\mu U_T/L^2$ in moderate and weak inversion, offers a simple method to assess the magnitude of the low-field mobility. The expression of the slope doesn't involve any other physical quantity than the gate length.

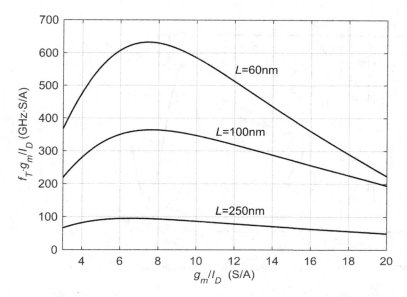

Figure 4.4 Product of transit frequency and transconductance efficiency for an n-channel transistor. $V_{DS} = 0.6$ V.

As a final remark, it is important to realize that the tradeoffs in more complex circuits are not as simple as in the previous example. For instance, it may be the case that low noise is valued higher than large bandwidth, in which case the figure of merit of (4.4) becomes inappropriate. In this case, one will find that pushing the circuit toward moderate or weak inversion is preferred (see examples in Chapters 5 and 6). In any case, the take-home that remains is that noise-constrained circuit optimization can be carried out systematically using a g_m/I_D-centric approach.

4.1.3 Thermal Noise from Active Loads

Another basic problem in thermal noise optimization concerns the excess noise from active loads. Figure 4.5 shows the circuit considered previously, but now the ideal current source is replaced with a p-channel device, which contributes additional noise. The input-referred noise now becomes

$$\frac{\overline{v_i^2}}{\Delta f} = 4kT\gamma_n \frac{1}{g_{m1}} + 4kT\gamma_p \frac{g_{m2}}{g_{m1}^2} = 4kT\gamma_n \frac{1}{g_{m1}}\left(1 + \frac{\gamma_p}{\gamma_n}\frac{g_{m2}}{g_{m1}}\right), \qquad (4.9)$$

where the bracketed term on the right-hand side is an excess noise factor due to the p-channel device. To minimize the excess noise, we must minimize g_{m2}/g_{m1}, which is equivalent to minimizing $(g_m/I_D)_2/(g_m/I_D)_1$, since both transistors carry the same bias current. Once $(g_m/I_D)_1$ has been chosen through circuit optimization (for example, as discussed above), this means that $(g_m/I_D)_2$ should be made as small as possible.

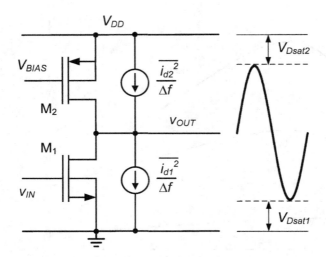

Figure 4.5 Common-source stage with noise from active load included.

However, since $V_{Dsat2} \cong 2/(g_m/I_D)_2$, this will force us to sacrifice output swing (see Figure 4.5).

In a circuit where dynamic range is the key metric of interest, there exists an optimum value for $(g_m/I_D)_2$. The dynamic range of a circuit is defined as the ratio of maximum signal power to the noise power. Regardless of how this circuit is used (within a feedback configuration or open-loop), maximizing the dynamic range will require maximizing the following term:

$$K = \frac{(V_{DD} - V_{Dsat1} - V_{Dsat2})^2}{1 + \dfrac{\gamma_p}{\gamma_n}\dfrac{g_{m2}}{g_{m1}}} \cong \frac{\left(V_{DD} - \dfrac{2}{(g_m/I_D)_1} - \dfrac{2}{(g_m/I_D)_2}\right)^2}{1 + \dfrac{\gamma_p}{\gamma_n}\dfrac{(g_m/I_D)_2}{(g_m/I_D)_1}}. \qquad (4.10)$$

The numerator of this expression scales the signal power, and the denominator scales the noise power. The dynamic range of the circuit is therefore proportional to K. We will now inspect the optimum of K through an example.

Example 4.2 Choosing g_m/I_D of a p-Channel Load for Maximum Dynamic Range

Evaluate (4.10) versus $(g_m/I_D)_2$ assuming the following parameters: $V_{DD} = 1.2$ V, $\gamma_p = \gamma_n$, and $(g_m/I_D)_1 = 7$ S/A (optimum from Figure 4.4). Repeat for $V_{DD} = 0.9$ V and again for $V_{DD} = 1.2$ V, but now with $(g_m/I_D)_1$ ranging from 5...25 S/A in steps of 5 S/A.

SOLUTION

The result for the first part is shown in Figure 4.6(a), which plots K normalized to its maximum value along the sweep. We see that the maximum is relatively

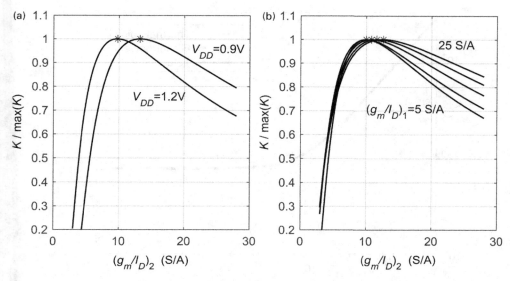

Figure 4.6 Normalized dynamic range scaling factor K as a function of $(g_m/I_D)_2$. (a) for $V_{DD} = 1.2$ and 0.9 V with $(g_m/I_D)_1 = 7$ (A/S). (b) $(g_m/I_D)_1 = 5...25$ S/A and $V_{DD} = 1.2$ V.

shallow, but choosing a $(g_m/I_D)_2$ that is too small can be costly, and the roll-off to the left of the optimum is steep. For reduced V_{DD}, the optimum shifts to the right, since now there is a larger premium on signal headroom, and this creates a bias toward larger $(g_m/I_D)_2$ (smaller V_{Dsat2}). Similar conclusions can be drawn from Figure 4.6(b). The location of the optimum is a relatively weak function of $(g_m/I_D)_1$. In general, we see that biasing the load device in moderate inversion is a good choice for the shown parameter ranges.

4.1.4 Flicker Noise (1/f Noise)

Unlike thermal noise, the power spectral density of flicker noise is not constant, but decreases inversely proportional to frequency. With flicker noise included in the model, the total input referred noise of the IGS (as considered in the previous section) behaves as shown in Figure 4.7. At low frequencies, the flicker noise component dominates, whereas at high frequencies the thermal noise dominates. The frequency at which the thermal and flicker power spectral densities are equal is called the flicker noise corner frequency (f_{co}).

There are two known physical mechanisms causing flicker noise: carrier number fluctuations due to trapping (McWorther model) and mobility fluctuations (Hooge model). Most modern device models, such as the PSP model [6] used for the circuit simulations in this book, take both effects into account. However, it is known that carrier number fluctuation effects dominate for today's technologies and for the biasing conditions found in typical analog circuits [7], [8].

The most widely accepted model for flicker noise was first proposed in [9]. Over the years, this model has been refined for compatibility with the physical variables

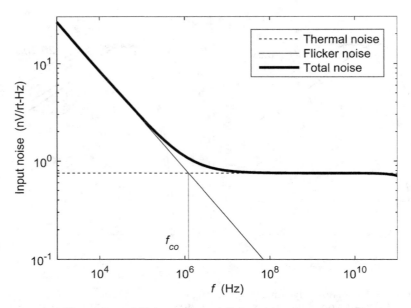

Figure 4.7 Illustration of flicker noise and flicker noise corner (f_{co}).

used in modern device models, for instance the surface potential in the PSP model [6]. Using g_m/I_D as the variable of interest, flicker noise due to number fluctuations, referred to the gate of the transistor (as in Figure 4.3), can be expressed as [7]

$$\frac{\overline{v_i^2}}{\Delta f} = \left(1 + \alpha\mu C_{ox}\frac{I_D}{g_m}\right)^2 \left(\frac{q_e}{C_{ox}}\right)^2 \cdot \frac{kT\lambda N_t}{WL} \cdot \frac{1}{f} = \frac{K_f}{WL} \cdot \frac{1}{f}. \tag{4.11}$$

In this expression, q_e stands for electron charge and the modeling parameters λ and N_t describe the so-called tunnel attenuation distance and charge trap density, respectively. The parameter α controls the bias dependence. Per this model, the input-referred flicker noise is minimized in weak inversion (largest g_m/I_D). However, the effect is relatively weak, as can be seen from Figure 4.8. While the thermal noise floor changes significantly with gate bias, the flicker noise component changes only by a small amount. Figure 4.9 takes a closer look at the flicker noise component versus g_m/I_D at a fixed frequency. This plot confirms that the simple model of (4.11) captures the trend of the actual device behavior (modeled though a complex equation in the underlying PSP model) reasonably well.

Since the flicker noise is relatively independent of gate bias, the primary way to reduce it is to increase the device area, WL, as seen from (4.11). All other parameters are outside the control of the designer. If increasing the device size does not bring the noise to the desired level, circuit-level techniques, such as chopping and correlated double sampling can be employed [10].

From an application perspective, it is important to note that flicker noise tends to be significant only in low-bandwidth circuits. This is illustrated in Figure 4.10,

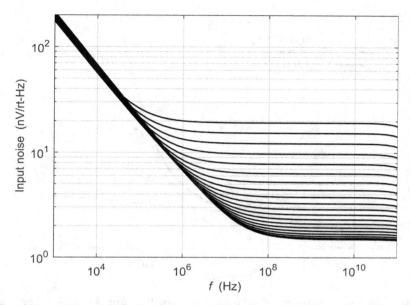

Figure 4.8 Simulated input-referred noise density of an NMOS device with $W = 10$ μm, $L = 100$ nm and $V_{DS} = 0.6$ V. V_{GS} is swept from 0.3 V (top curve) to 0.8 V (bottom curve), with corresponding g_m/I_D values between 29 and 5 S/A. We see that the input-referred flicker noise is only weakly bias dependent.

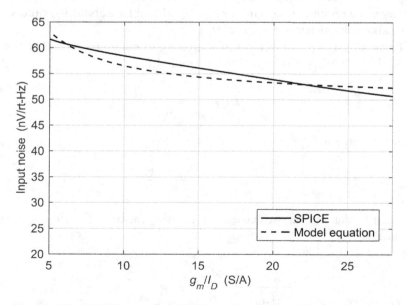

Figure 4.9 Simulated input referred flicker noise density of an NMOS device with $W = 10$ μm and $L = 100$ nm, measured at $f = 12$ kHz (solid line). Also shown is the simple model of (4.11), with $\alpha\mu C_{ox} = 1.3$ V^{-1} (dashed line).

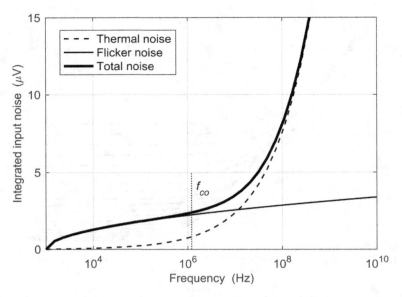

Figure 4.10 Running integral of the total input noise PSD from Figure 4.7. The noise integrated up to the flicker noise corner f_{co} is typically negligible in a wideband system.

which shows the running integral of the noise PSDs from Figure 4.7. For wideband systems, whose bandwidth is much larger than the flicker noise corner, the total integrated noise is dominated by the thermal component. As a rule of thumb, the flicker noise becomes irrelevant for bandwidths that are about 1–2 orders of magnitude above the flicker noise corner (f_{co}).

From a design perspective, it is interesting to compute the flicker noise corner analytically. We begin by equating (4.3) and (4.11):

$$4kT\gamma_n \frac{1}{g_m} = \frac{K_f}{WL} \cdot \frac{1}{f_{co}}. \tag{4.12}$$

Now, solving for f_{co} gives:

$$f_{co} = \frac{K_f}{4kT\gamma_n} \cdot \frac{g_m}{WL}. \tag{4.13}$$

To gain further insight, we can express g_m using the basic EKV model, using (2.28) and (2.16):

$$f_{co} = \frac{K_f}{4kT\gamma_n} \cdot \frac{2U_T\mu C_{ox} \frac{W}{L} q}{WL} = \frac{K_f}{4kT\gamma_n} \cdot \frac{2U_T\mu C_{ox} q}{L^2}. \tag{4.14}$$

Here, q is the normalized mobile charge density introduced in Chapter 2. From this result, we see that f_{co} becomes small for small q (pushing the device toward weak inversion); this is aligned with the trend seen in Figure 4.8. The only other option for making f_{co} small is to work with long channels. Either one of these options is incompatible with high-speed design, and hence the flicker noise corner tends to be rather high (tens of megahertz) in devices that were sized for wideband operation. This aside, the most important take-home from (4.14) is that the flicker noise corner is fully defined once L and g_m/I_D (and hence q) have been chosen.

Example 4.3 Estimation of the Flicker Noise Corner Frequency

Plot the flicker noise corner frequency versus g_m/I_D for an n-channel device with $L = 60$, 100, 200 and 40 nm.

SOLUTION

One way to solve this problem is to obtain estimates of K_f and γ_n, and use (4.13) to compute f_{co}. However, there exists a more straightforward path using the flicker noise drain current PSD (measured at $f = 1$ Hz) that is stored in our lookup tables:

$$SFL = \left.\frac{i^2_{d,flicker}}{\Delta f}\right|_{f=1\,Hz}. \tag{4.15}$$

Assuming an exact $1/f$ roll-off for the flicker noise, f_{co} is simply:

$$f_{co} = \frac{SFL}{STH}, \tag{4.16}$$

where STH is the stored thermal noise current density that we already introduced in (4.2). The Matlab code for the corresponding computation is given below:

```
L = [0.06, 0.1, 0.2, 0.4];
vgs = 0.2:25e-3:0.9;
for i=1:length(L)
    gm_id_n(:, i) = lookup(nch, 'GM_ID', 'VGS', vgs,...
    'L', L(i));
    fco(:, i) = lookup(nch, 'SFL_STH', 'VGS', vgs, 'L', L(i));
end
```

Figure 4.11 shows the resulting curves, which confirm our earlier observations. The flicker noise corner frequency reduces for larger g_m/I_D or larger L. For the shortest channel in strong inversion, f_{co} approaches 50 MHz.

A final aspect to consider is the behavior of the flicker noise with technology scaling. If the trap density N_t stays constant and W, L and C_{ox} are scaled by the same factor (as in classical Dennard scaling [11]), then there is to first order no change

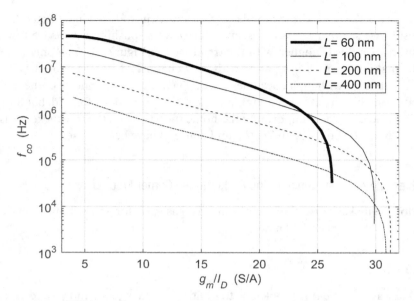

Figure 4.11 Flicker noise corner frequency of an n-channel device versus g_m/I_D and for various channel lengths. $V_{DS} = 0.6$ V and $V_{SB} = 0$ V.

in f_{co}. In reality, the situation is more complex and depends on the inversion level and the relative impact of mobility fluctuations [12] (which were largely ignored in our discussion). The general tendency is that the flicker noise corner frequency increases in scaled transistors.

4.2   Nonlinear Distortion

While noise determines the smallest signals that a circuit can process, nonlinear distortion sets a bound on the largest signal that can be handled. MOS devices suffer from several nonlinear effects. We will limit our discussion here to nonlinearity in the transistor's transconductance (g_m) and output conductance (g_{ds}).

4.2.1   Nonlinearity of the MOS Transconductance

It is customary to analyze nonlinear distortion by expanding the drain current as a Taylor series around the quiescent point (V_{GS}, I_D) [12]. The variables used to form the Taylor series are the incremental variables i_d and v_{gs} that add to the quiescent point variables to represent the total quantities:

$$i_D = I_D + i_d \qquad v_{GS} = V_{GS} + v_{gs} \; . \tag{4.17}$$

Using this convention, we can write:

$$i_d = a_1 v_{gs} + a_2 v_{gs}^2 + \dots + a_m v_{gs}^m, \tag{4.18}$$

where the coefficients a_k represent the kth order derivatives of the current i_d with respect to v_{gs}, divided by $k!$

$$a_k = \frac{1}{k!} \frac{d^k (i_d)}{dv_{gs}^k}.$$

(4.19)

Since the first derivative with respect to v_{gs} is simply the transconductance, we can also write:

$$a_1 = g_{m1} \quad a_2 = \frac{1}{2} g_{m2} \quad a_3 = \frac{1}{6} g_{m3} \quad etc.,$$

(4.20)

where the g_{mk} terms are the kth derivatives of the current.

Figure 4.12 plots g_{mk} versus V_{GS}, considering the common-source n-channel transistor with $L = 100$ nm and a constant drain voltage $V_{DS} = 0.6$ V. We see that g_{m1} saturates for large V_{GS} due to mobility degradation in strong inversion. Deep in strong inversion, the transconductance even starts decaying, as evident from the negative g_{m2} for $V_{GS} > 1$ V.

Knowing the coefficients of the Taylor expansion, we can reconstruct the original drain current. In Figure 4.13, we compare four reconstructed drain currents when taking one, two, three and four coefficients of the Taylor expansion into account. The vertical line marks the quiescent point ($V_{GS} = 0.45$ V). The bold black lines mark the ranges within which the reconstructed currents are within 1% of the original curve. Outside these ranges, the differences become large. We see the accuracy of the Taylor expansion increase steadily as the number of derivatives grows.

Instead of evaluating the derivatives of the Taylor expansion numerically from actual drain currents, we can also take advantage of the basic EKV model and derive analytical expressions from the differentials of (2.28), (2.23), and (2.21). The first four g_{mk} are given by:

$$g_{m1} = I_S \left(\frac{1}{nU_T} \right) q,$$

$$g_{m2} = I_S \left(\frac{1}{nU_T} \right)^2 \frac{q}{2q+1},$$

$$g_{m3} = I_S \left(\frac{1}{nU_T} \right)^3 \frac{q}{(2q+1)^3},$$

$$g_{m4} = I_S \left(\frac{1}{nU_T} \right)^4 \frac{q(1-4q)}{(2q+1)^5}.$$

(4.21)

All the above terms are proportional to I_S and the kth power of $1/nU_T$, and they are multiplied by factors that depend on the inversion level via the normalized mobile

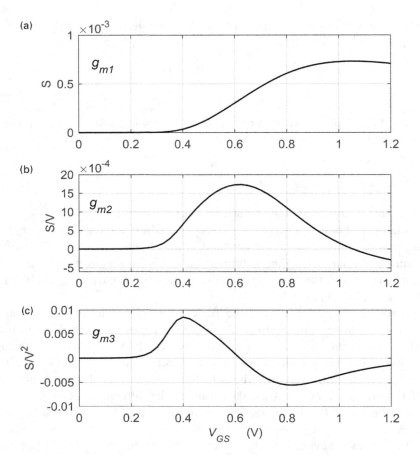

Figure 4.12 (a) g_{m1}, (b) g_{m2}, and (c) g_{m3} versus V_{GS} of a common-source n-channel device with $L = 100$ nm and $V_{DS} = 0.6$ V. The plotted values are normalized to a width of 1 μm.

charge density q. Recall that q and g_m/I_D are related per (2.29), which is repeated here for convenience:

$$\frac{g_m}{I_D} = \frac{1}{nU_T}\frac{1}{q+1}.$$ (4.22)

Figure 4.14 compares the first three EKV derivatives to the "real" g_{mk} from Figure 4.12, plotted against V_{GS} (left) and g_m/I_D (right). As we go from weak to strong inversion, we see that the curves diverge once mobility degradation enters the picture.

A more circuit-oriented way to quantify nonlinear distortion is through harmonic distortion analysis [13], using a sinusoid as the input signal:

$$i_d = a_1 v_{gs,pk}\cos(\omega t) + a_2 \left(v_{gs,pk}\cos(\omega t)\right)^2 + \dots$$ (4.23)

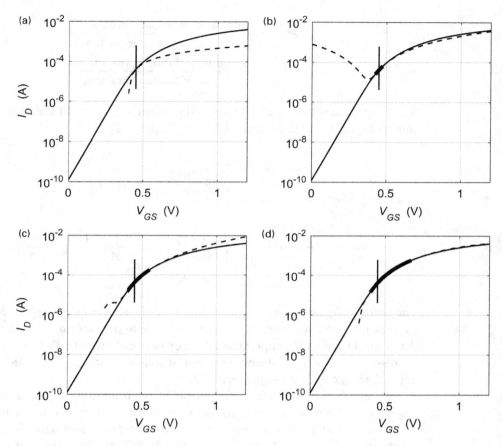

Figure 4.13 n-channel MOS transistor drain currents versus V_{GS} (solid lines). The dashed lines represent the reconstructed drain currents using Taylor expansions of (a) 1st order up (d) 4th order. The bold black lines around the quiescent points (vertical line) represent ranges where the reconstructed and original currents differ by less than 1%.

The powers of the applied sinusoid expand as follows with (4.20):

$$
\begin{aligned}
i_d \quad &= \frac{1}{4} g_{m2} v_{gs,pk}^2 \\
&+ \left(g_{m1} v_{gs,pk} + \frac{3}{24} g_{m3} v_{gs,pk}^3 \right) \cos(\omega t) \\
&+ \frac{1}{4} g_{m2} v_{gs,pk}^2 \cos(2\omega t) + \frac{1}{24} g_{m3} v_{gs,pk}^3 \cos(3\omega t) + \ldots
\end{aligned}
\tag{4.24}
$$

The first line represents the DC component, the second the fundamental, and the third the harmonics. The harmonics are typically compared to the fundamental, which gives rise to the following fractional harmonic distortion metrics:

$$HD_2 = \frac{\text{Amplitude of second harmonic}}{\text{Amplitude of fundamental tone}}$$

$$HD_3 = \frac{\text{Amplitude of third harmonic}}{\text{Amplitude of fundamental tone}}. \qquad (4.25)$$

Approximate expressions of HD_2 and HD_3 (when $v_{gs,pk}$ is small) are derived below and are expressed in terms of the EKV model per (4.21):

$$HD_2 = \left| \frac{\frac{1}{4} g_{m2} v_{gs,pk}^2}{g_{m1} v_{gs,pk} + \frac{3}{24} g_{m3} v_{gs,pk}^3} \right| \approx \frac{1}{4} \left| \frac{g_{m2}}{g_{m1}} \right| v_{gs,pk} = \frac{1}{4} \left(\frac{1}{nU_T} \right) \frac{1}{1+2q} v_{gs,pk},$$

$$HD_3 = \left| \frac{\frac{1}{24} g_{m3} v_{gs,pk}^3}{g_{m1} v_{gs,pk} + \frac{3}{24} g_{m3} v_{gs,pk}^3} \right| \approx \frac{1}{24} \left| \frac{g_{m3}}{g_{m1}} \right| v_{gs,pk}^2 = \frac{1}{24} \left(\frac{1}{nU_T} \right)^2 \frac{1}{(1+2q)^3} v_{gs,pk}^2.$$

$$(4.26)$$

The ratios g_{mk}/g_{m1} boil down to functions of q only, after normalizing by $(g_m/I_D)^{(k-1)}$. In weak inversion, all normalized terms converge to one since $q \ll 1$. In this case, the expansion of (4.20) approximates the exponential given by the weak inversion expression of (2.27). In strong inversion, it follows that g_{m2}/g_{m1} divided by g_m/I_D approaches 1/2, whereas higher order ratios tend to zero.

Between weak and strong inversion, the normalized ratios evolve as shown in Figure 4.15. When we compare the data for the "real" transistor to the EKV predictions, we see that these compare well in weak inversion and partly in moderate inversion. But, when g_m/I_D gets smaller than 10 S/A, the magnitude of the "real" ratios decreases rapidly before changing sign. The sign change comes from the zero crossings in g_{m2} and g_{m3}, as already seen in Figure 4.14.

Figure 4.16 shows the fractional harmonic distortion per (4.26) versus g_m/I_D and for three different signal amplitudes. We see that the third-order distortion is a stronger function of g_m/I_D and also the signal amplitude (due to the square dependence seen in (4.26)).

Figure 4.17 compares HD_2 and HD_3 of the real transistor and the EKV prediction for a signal amplitude of 10 mV. To obtain the "real" data, we performed a time domain simulation of the device and evaluated the harmonic distortion products in the drain current (AC grounded). The results match closely with the analytical prediction in weak and moderate inversion. In strong inversion, the differences are due to mobility degradation. Most notably is the null in HD_3, which coincides with the zero crossing of g_{m3} in Figure 4.14 for $V_{GS} = 0.617$ V, or $g_m/I_D = 9.03$ S/A. This null has been recognized in the literature [14], but it is difficult to leverage in practice due to its sensitivity to process parameters and temperature.

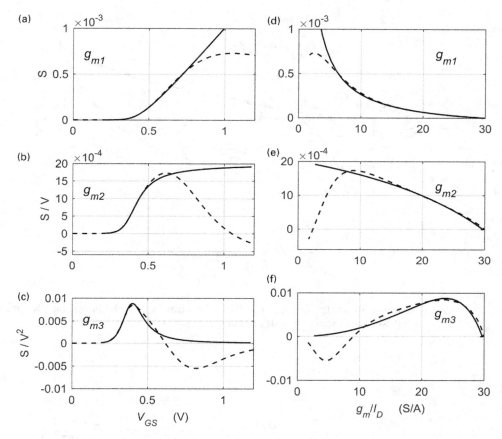

Figure 4.14 (a), (b), (c) EKV (solid lines) and "real" (dashed lines) g_{mk} versus V_{GS} and (d), (e), (f) versus g_m/I_D. (Same transistor data as in Figure 4.12.)

4.2.2   Nonlinearity of the MOS Differential Pair

We will now study the nonlinearities in the differential pair, as shown in Figure 3.28, considering AC-grounded drain terminals. As in the previous section, we take advantage of the basic EKV model to obtain analytical derivatives, similar to (4.21). Mathematically, we start with the pinch-off expression (2.23) and: (1) evaluate the differentials dV_{P1} and dV_{P2} with respect to the normalized mobile charge densities q_1 and q_2, (2) evaluate the differentials di_1 and di_2 of the normalized saturated drain currents i_1 and i_2 given by (2.22), and (3) eliminate dq_1 and dq_2 using (2.21). Since the sum of the normalized drain currents must equal zero, we substitute di with di_1 and $-di_2$ and arrive at the expression below for the input voltage differential:

$$dv_{id} = nU_T \left[\frac{di_1}{q_1} - \frac{di_2}{q_2} \right] = nU_T \left[\frac{1}{q_1} + \frac{1}{q_2} \right] di. \tag{4.27}$$

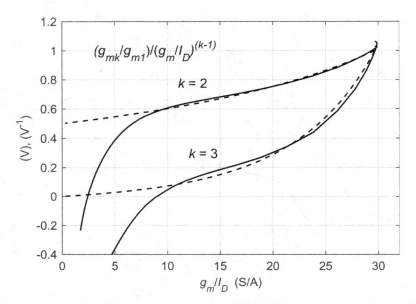

Figure 4.15 Ratios $(g_{mk}/g_{m1})/(g_m/I_D)^{(k-1)}$ for a real transistor (solid lines) and EKV model (dashed lines). (Same transistor data as in Figure 4.12.)

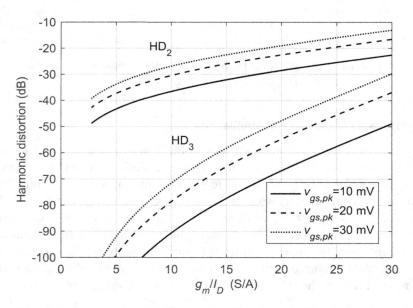

Figure 4.16 Fractional second- and third-order harmonic distortion metrics derived from the basic EKV model (4.26), considering various signal amplitudes. (Same transistor data as in Figure 4.12.)

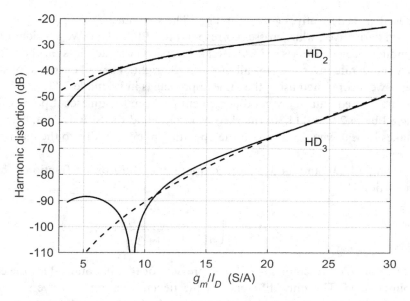

Figure 4.17 Fractional second- and third-order harmonic distortion of the n-channel drain current for a 10-mV peak input sine wave. Solid lines: simulation data, dashed lines: analytical prediction ($L = 100$ nm, $V_{DS} = 0.6$ V).

Next, we compute the derivative of the differential output current with respect to the differential input:

$$\frac{di_{od}}{dv_{id}} = 2\frac{I_S}{nU_T} \cdot \frac{q_1 q_2}{q_1 + q_2}.$$
(4.28)

When v_{id} equals zero, q_1 and q_2 are equal. Replacing the latter by q, we obtain the following expression of the differential transconductance at the operating point:

$$g_{m1} = \frac{I_S}{nU_T} q.$$
(4.29)

Note that this is the same result as in (4.21) (for the common-source configuration). The higher-order derivatives differ significantly. The second-order term turns out to be zero at the quiescent point, which is explained by the symmetry of the transfer function (we will illustrate this below). For the third-order term, we obtain:

$$g_{m3} = -I_S \left(\frac{1}{nU_T}\right)^3 \frac{q(1+3q)}{2(1+2q)^3}.$$
(4.30)

Notice the similarity with the third order expression in (4.21), namely the occurrence of I_S and $1/(nU_T)^3$ followed by a function of q that captures the inversion level.

    Figure 4.18 examines the differential pair's transfer characteristic and its derivatives graphically. The plots show the normalized differential drain current i_{od}/I_0

as well as its derivatives g_{m1}, g_{m2} and g_{m3}. The solid lines in the plot represent EKV-based derivatives, while crosses correspond to SPICE data. We consider two transconductance efficiencies: 27 S/A (weak inversion) for the plots on the left, and 15 S/A (moderate inversion) for all plots on the right. The tail currents for these two cases were computed using the same approach as in Example 3.11.

The plotted data show good agreement between the quiescent point values predicted by (4.29) and (4.30) (marked by circles) and the SPICE data. The plot also makes it clear why g_{m2} is zero at the operating point; it is due to the maximum seen in g_{m1}.

Using (4.29) and (4.30), we compute the third-order fractional harmonic distortion HD_3:

$$HD_3 = \frac{1}{24}\left(\frac{1}{nU_T}\right)^2 \frac{(1+3q)}{2(1+2q)^3} v_{id,pk}^2 .$$
(4.31)

This expression resembles the HD_3 expression of (4.26), obtained for the common-source stage. The only difference is the dependence on the inversion level (q). Comparing the two expressions, we see that the HD_3 of the differential pair is better than that of the CS stage. The difference is 6 dB in weak inversion, and 2.5 dB in strong inversion.

Note that for both the common-source stage and the differential pair, we only need to know the subthreshold factor n and the inversion level to compute HD_3. The specific current I_S and the threshold voltage V_T play no role. Furthermore, the subthreshold factor n is known to lie in a relatively narrow range, and is fully defined by the maximum value of g_m/I_D in weak inversion.

In Figure 4.19(a), we compare the HD_3 predicted by (4.31) to a SPICE simulation (bold line) for $v_{id,pk}$ = 10 mV. The thin lines enclosing the SPICE curve correspond to typical boundaries of 1.0 and 1.5 for the subthreshold factor. This result makes it clear that even when we don't have perfect knowledge of n, it is still possible to predict the distortion with good accuracy.

It is also interesting to compare (4.31) with the classical results found in the literature [13] for ideal square-law devices:

$$HD_3 = \frac{1}{32}\left(\frac{v_{id,pk}}{V_{GS}-V_T}\right)^2$$
(4.32)

and for bipolar transistors:

$$HD_3 = \frac{1}{48}\left(\frac{v_{id,pk}}{U_T}\right)^2 .$$
(4.33)

Equation (4.32) is plotted in Figure 4.19(b) using V_T data from SPICE (0.5...0.38 V along the given sweep). We see that the square-law result does not fit the circuit behavior satisfactorily. One issue is that it does not account for mobility degradation in strong inversion. Another issue is that the square-law expression does not

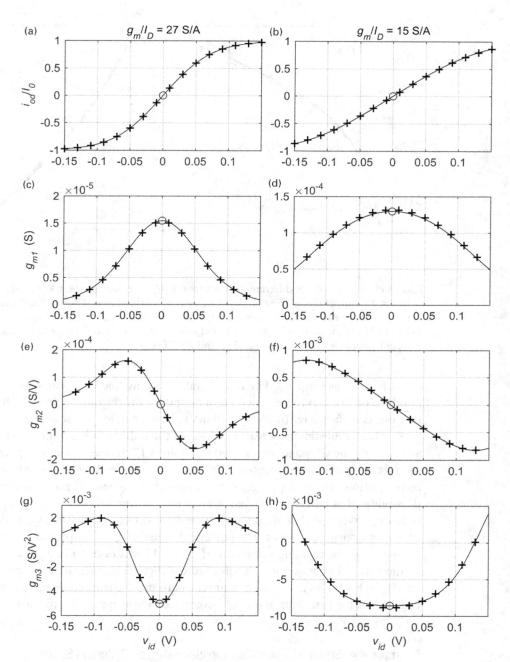

Figure 4.18 (a), (b) Transfer characteristics of the differential amplifier, followed by (c), (d) the first derivative, (e), (f) second derivative and (g), (h) third derivative. The plots on the left are for weak inversion ($g_m/I_D = 27$ S/A), while the plots on the right correspond to moderate inversion ($g_m/I_D = 15$ S/A). The solid lines and circles correspond to the basic EKV model, while crosses mark SPICE data. The plotted derivatives are normalized to a width of 1 µm and assume $V_{GB} = V_{DB} = 0.8$ V, $L = 100$ nm.

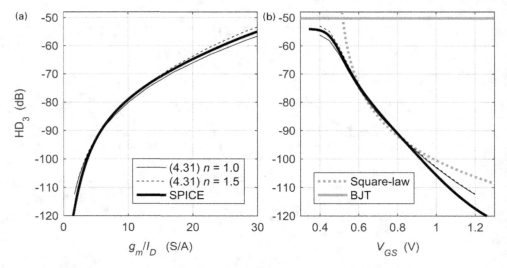

Figure 4.19 (a) HD$_3$ of the differential pair versus g_m/I_D, considering the EKV model with $n = 1.5$ (upper dashed line) and $n = 1$ (lower solid thin line).[3] The bold curve is the corresponding SPICE simulation result ($L = 100$ nm, $V_{GB} = 1.2$ V, $V_{DB} = 1.2$ V, $v_{id,pk} = 10$ mV). (b) The same data as in (a) are plotted against V_{GS}, and showing the two classical textbook results for the ideal square law and for BJTs.

apply for V_{GS} nearing V_T. The horizontal bold-gray line in Figure 4.19(b) corresponds to (4.33). It is merely useful as a bound for the maximum possible distortion that one can encounter (for $v_{in,pk} = 10$ mV) given a perfectly exponential device.

Another interesting observation concerns the null that we saw in the HD$_3$ characteristic of the CS stage represented in Figure 4.17. It does not show up here since the symmetry of the differential pair "rejects" terms leading to the null in the CS configuration. The fact that g_{m2} crosses zero at the quiescent point eliminates the possibility of a null in g_{m3}. What is even more surprising is the good agreement in strong inversion that we see in Figure 4.19(a) but not in (b). In (a), the effect of mobility degradation on the numerical value of g_{m3} is captured to first order through a corresponding reduction of g_m/I_D. This is essentially a distortion/shift of the x-axis that does not occur in Figure 4.19(b), where the data are plotted versus V_{GS}. This serves as an important reason to use the transconductance efficiency rather than the gate voltage (or gate overdrive) when computing the distortion.

Example 4.4 Sizing a Differential Amplifier Based on Distortion Specs

Consider the differential amplifier shown in Figure 4.20. We want to design this amplifier so that it achieves a small-signal differential voltage gain of $A_v = 2$, while maintaining an HD$_3$ of -60 dB. Plot the required tail current as a function

---

[3] The HD$_3$ curve for $n = 1.5$ in Figure 4.19 lies above the $n = 1$ curve since the q dependent factor in (4.31) grows faster than the $(1/nU_T)^2$ factor (recall that n and q are related through g_m/I_D via (4.22)).

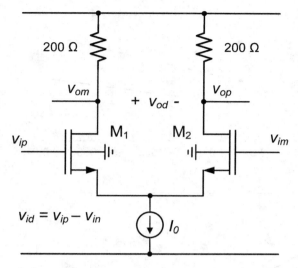

Figure 4.20 Differential amplifier.

of input amplitude v_{id_pk} ranging from 10 to 40 mV. Also, plot the corresponding device width. Repeat for $HD_3 = -70$ dB. Assume $L = 100$ nm and validate the results using SPICE simulations.

SOLUTION

We begin by obtaining an approximate estimate for n using

```
UT = .026;
L  = 0.1;
n  = 1/UT/max(lookup(nch,'GM_ID', 'L', L))
```

This yields $n = 1.29$. Note that this value was found using the default terminal voltages assumed in the lookup function, and the actual terminal voltages will be different. However, as we have seen from Figure 4.19, it is not necessary to know n exactly.

Now we can use (4.31) to find the values of q, g_m/I_D and I_D that give the desired HD_3 for a given input amplitude:

```
vid  = 10e-3;
HD3o = -60;
q  = logspace(-2,1,100)';
HD3 = 20*log10(1/24*1/(n*UT)^2*(1+3*q)./(2*(1+2*q).^3)*vid^2);
qo  = interp1(HD3,q,HD3o);
gmIDo = 1./(n*UT*(1+qo))
```

To compute the transconductance, we assume $g_m = A_v/R_L = 10$ mS, which neglects the output conductance of the transistors. This is acceptable due to the low target voltage gain. The current then follows via:

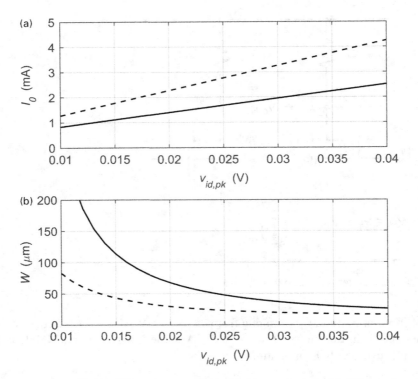

Figure 4.21 (a) Amplifier tail current (I_0) and (b) transistor width (W) as a function of differential input amplitude ($v_{id,pk}$) for two distortion specifications: $HD_3 = -60$dB (solid line) and $HD_3 = -70$dB (dashed line). Channel length $L = 100$ nm.

```
ID = gm/gmID
```

We can now also determine the width of the transistors using

```
W = ID/lookup(nch, 'ID_W', 'GM_ID', gmID, 'L', L)
```

Note that we are ignoring here the voltage dependence of the current density and simply evaluate at the default values of V_{DS} and V_{SB}. Repeating the above calculations for the given range of input amplitudes and for $HD_3 = -70$ dB yields the plots shown in Figure 4.21.

To validate these results, we run SPICE simulations for the extreme points in Figure 4.21. The simulation uses a fixed common-mode input voltage of 1 V. The results are summarized in Table 4.1.

As we can see, there is close agreement between the predicted and simulated distortion performance, despite the large number of approximations made in the analysis and sizing procedure. The largest discrepancy occurs for the highest inversion level (lowest g_m/I_D). At this point, the assumption that the drain swing has no influence on distortion starts to become invalid.

Table 4.1 Result comparison.

I_0 (mA)	W (μm)	$v_{id,pk}$ (mV)	**Theory** g_m/I_D (S/A)	**SPICE** g_m/I_D (S/A)	**Theory** HD_3 (dB)	**SPICE** HD_3 (dB)
0.826	321	10	24.2	25.7	−60	−61.0
2.51	25.7	40	8.0	8.7	−60	−60.4
1.26	82.7	10	15.8	17.2	−70	−70.6
4.25	16.1	40	4.7	5.53	−70	−67.5

One of the design-related take-homes from the above example is that improving the distortion performance or handling a large signal typically costs extra current. To achieve smaller distortion, we must increase q (see (4.31)), and this leads to lower g_m/I_D. With g_m fixed, this means that more current must be invested.

The situation is similar for larger signals, which require lower q when the same distortion level must be maintained. From the data of Table 4.1, we see that maintaining the same HD_3 for a fourfold increase in signal amplitude requires current scaling by 2.51/0.826 = 3.0x and 4.25/1.26 = 3.37x, respectively. We can compare this outcome to a prediction from the simplistic square-law expression of (4.32), by which $V_{GS} - V_T$ must increase by a factor of four to accommodate a 4x larger signal. Furthermore, using also the square-law approximation of $g_m = 2I_D/(V_{GS} - V_T)$, this would translate into a 4x penalty in I_D to maintain the same g_m. We see that the g_m/I_D-based expressions lead to more realistic numbers.

In summary, the strength of the distortion equations presented in this section is that they are valid at all inversion levels, and yield estimates that are very close to the behavior seen in SPICE-level simulations. This allows the designer to study the design space of distortion-limited circuits using simple, yet accurate Matlab scripts.

4.2.3 Inclusion of the Output Conductance

All results so far assume that the drain voltage of the transistor does not change (i.e. it is an AC ground) or at least does not change significantly (as in Example 4.4). However, in circuits with large voltage gain this condition is often not met, and the drain current changes caused by v_{gs} may lead to substantial amplitudes in v_{ds}. This has little effect on distortion with long-channel transistors. However, the situation is different for short channel devices, where effects like DIBL create a significant link between the drain voltage and the drain current. To consider the dependence on both v_{gs} and v_{ds}, a two-dimensional Taylor expansion is required [15]:

$$
\begin{aligned}
i_d &= g_{m1}v_{gs} + g_{ds1}v_{ds} \\
&\quad + \frac{1}{2}g_{m2}v_{gs}^2 + x_{11}v_{gs}v_{ds} + \frac{1}{2}g_{ds2}v_{ds}^2 \\
&\quad + \frac{1}{6}g_{m3}v_{gs}^3 + \dots
\end{aligned}
\tag{4.34}
$$

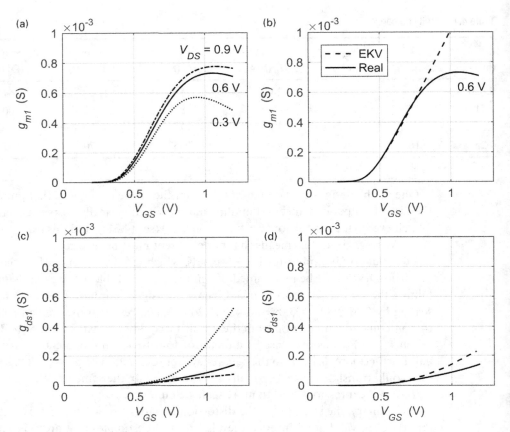

Figure 4.22 First-order derivatives of the drain current versus V_{GS}, for an n-channel with $L = 100$ nm. (a) Illustration of the V_{DS} dependence of g_{m1} and (c) g_{ds1}. Plots (b) and (d) show comparisons with the basic EKV model for $V_{DS} = 0.6$ V.

where:

$$g_{mk} = \frac{d^k i_d}{dv_{gs}^k}; \quad x_{pq} = \frac{d^{p+q} i_d}{dv_{gs}^p dv_{ds}^q}; \quad g_{dsk} = \frac{d^k i_d}{dv_{ds}^k}. \tag{4.35}$$

In the previous section, only the derivatives of i_d with respect to the gate voltage were considered (g_{m1}, g_{m2}, etc.). These terms resurface in the 2D expression above, but we must add derivatives with respect to the drain voltage (leading to g_{ds1}, g_{ds2}, ...) as well as cross derivatives (x_{11}, ...).

To get a feel for what the derivatives look like, we make use of a function (blkm.m)[4] that computes the derivatives from our lookup table data. The results for the first-order derivatives are plotted in Figure 4.22(a) and (c), considering three values of V_{DS}

---

[4] The blkm.m function evaluates the first and second derivatives of the drain current given the transistor type, L, V_{GS} and V_{DS}. The first derivatives are identical to the lookup table data for GM and GDS. The second derivatives are obtained by numerically differentiating this lookup table data (leading to g_{m2} and g_{ds2}, and the cross term x_{11}).

(0.3, 0.6 and 0.9 V). In Figure 4.22(b) and (d), we compare the "real" derivatives for $V_{DS} = 0.6$ V to the basic EKV counterparts (g_{m1} from (4.21), and g_{ds1} given by (2.42)). In these plots, we can clearly see the impact of mobility degradation for large values of V_{GS}. In both Figure 4.22(b) and (d), we see the "real" g_{m1} and g_{ds1} diverge rapidly from the EKV prediction once V_{GS} exceeds 0.7 V. In terms of normalized mobile charge densities, this corresponds to approximately 3, or a g_m/I_D of about 7 S/A.

To create a similar comparison for the second-order derivatives, we use g_{m2} from (4.21) and the following analytical expressions for g_{ds2} and x_{11}:

$$x_{11} = g_{m1}S_{IS} - g_{m2}S_{VT} \tag{4.36}$$

$$g_{ds2} = g_{ds1}S_{IS} - x_{11}S_{VT} + I_D dS_{IS} - g_{m1}dS_{VT}. \tag{4.37}$$

Here, dS_{IS} and dS_{VT} are the derivatives with respect to the drain voltage of S_{IS} and S_{VT} defined in (2.37). The results are illustrated in Figure 4.23, again for three values of V_{DS} (0.3, 0.6 and 0.9 V) and comparing the "real" derivatives for $V_{DS} = 0.6$ V to the basic EKV counterparts discussed above.

We observe that the impact of mobility degradation on the second-order derivatives is even stronger than for the first-order terms in Figure 4.22. The effects are already visible when $V_{GS} = 0.5$ V ($q = 1$ or g_m/I_D somewhat less than 15 S/A). Since the evaluation of higher-order derivatives of the EKV model gets quite complex (and potentially inaccurate due to taking multiple derivatives of I_S and V_T), we restrict the comparison to the second-order derivatives only.

As a final step, to create a direct connection with g_m/I_D, we re-plot the data from Figure 4.23 in Figure 4.24, but now with the transconductance efficiency on the x-axes. An interesting observation here is that the g_{ds2} curves for the real transistor and the basic EKV expression in Figure 4.24(d) match almost exactly. This is not the case in Figure 4.23(d), where we plot against V_{GS}. The explanation for this is the same as for Figure 4.19. The effect of mobility degradation on the numerical value of g_{ds2} is captured to first order through a corresponding reduction of g_m/I_D, which distorts the x-axis to compensate for the discrepancy seen in the V_{GS}-based plot.

Now that we have a handle on the 2D Taylor expansion, we re-analyze the harmonic distortion. We consider a common-source stage loaded by a resistor R (with conductance $Y = 1/R$). The objective is to set up a power series that directly links v_{gs} to v_{ds} per the following equation:

$$v_{ds} = a_1 v_{gs} + a_2 v_{gs}^2 + a_3 v_{gs}^3 + \ldots \tag{4.38}$$

To identify the coefficients, we first note that the incremental drain current i_d must be equal to $-Y v_{ds}$. Setting this product equal to the right-hand side of (4.34) and performing a coefficient comparison for the first-order term yields:

$$a_1 = -\frac{g_{m1}}{Y + g_{ds1}}. \tag{4.39}$$

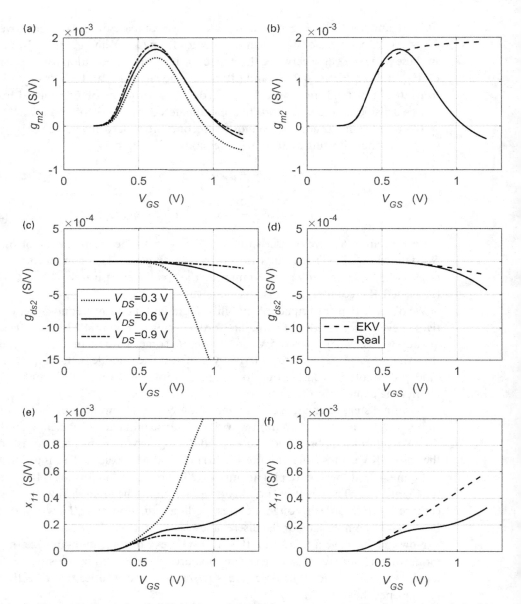

Figure 4.23 Second-order derivatives of the drain current versus V_{GS}, for an n-channel with $L = 100$ nm. (a) Illustration of the V_{DS} dependence of g_{m2}, (c) g_{ds2} and (e) the cross term x_{11}. Plots (b), (d) and (f) show the respective comparisons with the basic EKV model for $V_{DS} = 0.6$ V.

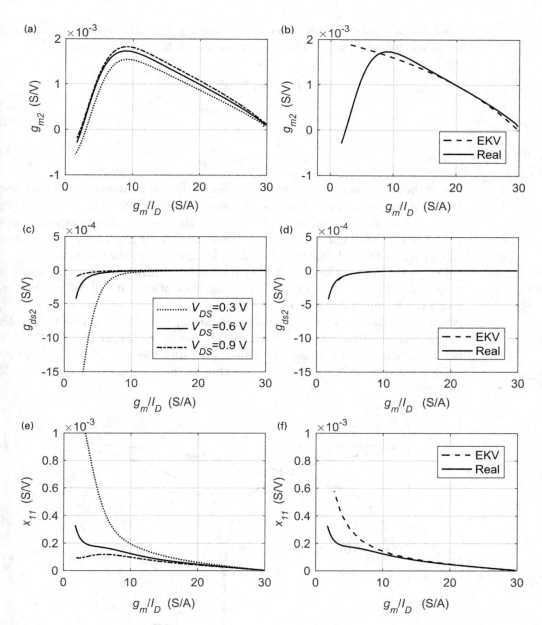

Figure 4.24 Second-order derivatives of the drain current versus g_m/I_D, for an n-channel with $L = 100$ nm. (a) Illustration of the V_{DS} dependence of g_{m2}, (c) g_{ds2} and (e) the cross term x_{11}. Plots (b), (d) and (f) show the respective comparisons with the basic EKV model for $V_{DS} = 0.6$ V.

Note that a_1 is simply the low-frequency voltage gain of the circuit. Higher order terms also follow from a coefficient comparison for the respective order:

$$a_2 = -\frac{\frac{1}{2}g_{m2} + x_{11}a_1 + \frac{1}{2}g_{ds2}a_1^2}{Y + g_{ds1}},$$

$$a_3 = -\frac{\frac{1}{6}g_{m3} + \cdots}{Y + g_{ds1}}.$$

$$(4.40)$$

Notice the occurrence of g_{ds2} and x_{11} in addition to g_{m2} in the numerator of a_2. These terms capture the impact of the drain voltage on the second-order distortion, as well as the drain-gate interaction. It is foreseeable from the numerical values seen in Figure 4.23 and Figure 4.24 that we cannot neglect the cross-derivative x_{11}, even though it is multiplied only by a_1, while g_{ds2} is multiplied by the square of a_1.

Due to the opposite signs of some terms in the numerator of a_2 (note that a_1 is negative), and the fact that Y controls the magnitude of each, it is likely that arrangements can be found that null a_2 and thus HD_2. We explore this idea in Figure 4.25, which shows HD_2 versus g_m/I_D for various R ranging from 100 Ω to 1 $M\Omega$. Consider for instance the curve for $R = 100$ kΩ. We see that for g_m/I_D near 15 S/A, HD_2 falls below –90 dB, which is much better than the –40 dB seen in Figure 4.17 for an AC-grounded drain. Unfortunately, operating deep within the HD_2 notch would

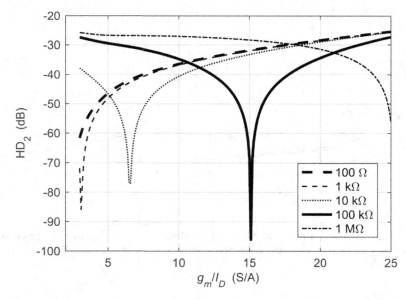

Figure 4.25 HD_2 of a resistively loaded common-source stage versus g_m/I_D. The different curves correspond to R values listed in the legend. ($V_{DS} = 0.6$ V, $v_{id\_pk} = 10$ mV, $W = 1$ μm and $L = 100$ nm).

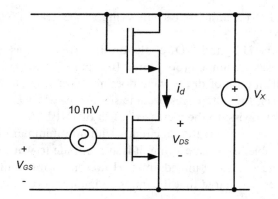

Figure 4.26 Simulation circuit to test the impact of a (small) amount of drain swing on the harmonic distortion of the drain current (i_d). The voltage V_X is adjusted such that $V_{DS} = 0.6$ V at the operating point of the circuit (for direct comparison with the simulation result of Figure 4.17).

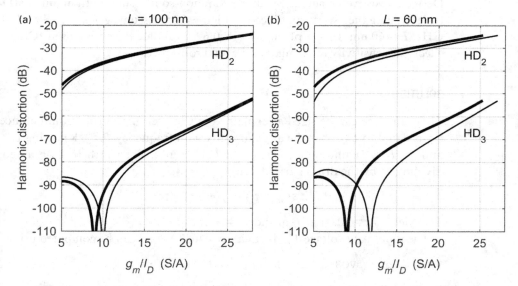

Figure 4.27 Fractional second and third order harmonic distortion of the drain current for (a) $L = 100$ nm and (b) $L = 60$ nm. Bold lines: SPICE simulation data from Figure 4.17, thin lines: SPICE simulation data for the circuit of Figure 4.31 (n-channel, $V_{DS} = 0.6$ V).

require very tight tolerances on R. However, a reduction of 20 dB would already be welcome, and may not necessitate tight control. We will investigate this further in the examples that follow.

Notice that for $R = 100$ Ω, the shape of the HD₂ curve approaches the plot in Figure 4.16 (for an AC-grounded drain). This is because drain-induced effects become negligible when R, and thus the voltage gain, is small. It is interesting to investigate one such special case that is commonly encountered. Figure 4.26 shows a simplified model of a common-source circuit with a same-size cascode transistor stacked on top. This leads to a finite, but small swing at the drain node of the common-source transistor.

Figure 4.27 compares the HD_2 and HD_3 of the drain current in this circuit to the case where the drain is ac-grounded (bold line). In Figure 4.27(a), we see that for $L = 100$ nm, the small amount of drain swing does not have a significant effect on HD_2 and HD_3. However, the differences become significant when L is reduced to 60 nm in Figure 4.27(b), owing to the increased effect of DIBL (larger S_{IS} and S_{VT} in (4.36) and (4.37)). We conclude that in circuits with non-minimum L, where the drain swing is comparable to the gate swing, it is appropriate to work with distortion models that assume an AC-grounded drain. However, for minimum-length devices, the drain influence is so large that it cannot be ignored, even if the drain swing is as small as the gate swing.

Example 4.5 Sizing of a Resistively Loaded CS Stage with Low HD_2

Design a resistively loaded, n-channel common-source stage with minimal HD_2 by leveraging the "sweet spot" seen in Figure 4.25. Assume $g_m/I_D = 15$ S/A, $f_u = 1$ GHz, $L = 60$ nm, $C_L = 1$ pF, and $V_{DS} = 0.6$ V. Validate the design in SPICE and check the sensitivity to changes in the load resistance.

SOLUTION

The transistor can be sized as explained in Chapter 3. The transconductance g_{m1} is equal to C_L times the angular unity gain frequency, $2\pi f_u$. Knowing g_{m1} and $(g_m/I_D)_1$, we find the drain current I_D (419 µA). The width W is obtained dividing I_D by the drain current density J_D:

```
JD = lookup(nch,'ID_W','GM_ID',gm_ID,'L',L);
```

This yields $J_D = 10.04$ µA/µm and $W = 41.72$ µm.

The input gate voltage V_{GS} is equal to 0.4683 V, and can be computed using:

```
VGS = lookupVGS(nch,'GM_ID',gm_ID,'L',L);
```

We make the implicit assumption that the load resistor is perfectly linear (if it were a nonlinear device, for example a p-channel device, additional nonlinear terms must be taken into consideration). Since the numerator of a_2 in (4.40) is a quadratic expression in a_1, the gain that nulls HD_2 becomes:

$$a_1 = \frac{-x_{11} + \sqrt{x_{11}^2 - g_{m2}g_{ds2}}}{g_{ds2}}. \tag{4.41}$$

Having the gain a_1, all we need to do is invert (4.39) to get the load conductance that nulls HD_2:

$$Y = 1/R = -\frac{g_{m1}}{a_1} - g_{ds1}. \tag{4.42}$$

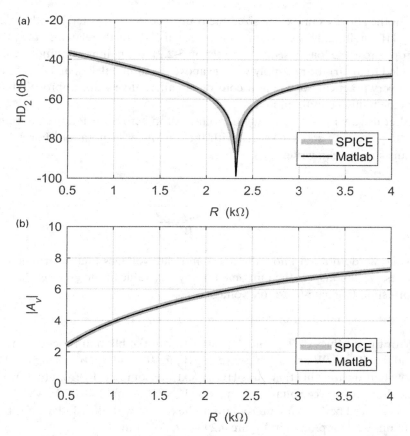

Figure 4.28 Impact of the actual load resistance on (a) HD_2 and (b) voltage gain. Solid black lines correspond to the 2D Taylor expansion derived from the lookup data, the thick gray lines represent SPICE simulation data.

To proceed with these expressions, we must know the values of g_{m1}, g_{ds1}, g_{m2}, g_{ds2} and x_{11}, five parameters that are obtained running the blkm.m function described earlier. We find $a_1 = -6.024$ and $1/Y = 2.321$ kΩ. Since R is a linear resistor and we know the drain current, we know also the voltage drop $I_D R$, which equals 0.972 V. Because V_{DS} was fixed a priori to 0.6 V, the power supply voltage must equal 1.57 V.

To get a feel for the accuracy requirements of the load resistor, we sweep Y over a wide range, keep all other parameters constant, and re-evaluate a_1, a_2 and the fractional second harmonic distortion. The result is shown in Figure 4.28 and compared to a SPICE simulation for $v_{gs,pk} = 10$ mV. We see that pushing HD_2 below -80 dB is possible, but requires rather strict tolerances on the load resistor.

In the above example, we sized the transistor as in Chapter 3, without considering HD$_2$ nulling. However, since the stage gain is directly related to the DC voltage drop across the load resistor (see Section 3.2.2), the nulling condition (via the gain a_1) led to an arbitrary supply voltage above the nominal V_{DD} of 1.2 V. Since this is not very practical, we will now consider an alternative sizing approach that considers the nulling condition first.

Let us call $V_R = V_{DD} - V_{DS}$ the voltage drop across the load resistor and a_1 the voltage gain, to stay consistent with the Taylor series notation. For a 1 µm wide transistor, we can write:

$$a_1 = -\frac{\dfrac{g_{m1}}{W}}{\dfrac{g_{ds1}}{W} + \dfrac{J_D}{V_R}}.$$

(4.43)

While we do not yet know the drain and gate voltages that determine the Taylor series, we can inspect the outcome for any V_{DS} value at the given g_m/I_D and L. The corresponding gate-to-source voltage is:

```
VGS = lookupVGS(nch,'GM_ID',gm_ID, 'L',L);
```

With these values of V_{GS} and V_{DS}, we can run the blkm.m function and not only find g_{m1}/W, g_{ds1}/W and J_D, but also g_{m2}/W, g_{ds2}/W and x_{11}/W. Using (4.41), we can now find the a_1 value that nulls HD$_2$. Next, we compute V_R from (4.43), and to get the drain-to-source voltage that makes $(V_R + V_{DS})$ equal to 1.2 V, we turn V_{DS} into a vector and perform an interpolation. We end up with the Matlab code below, and obtain the data plotted in Figure 4.29 for $L = 60$ nm.

```
% data ===================
VDD = 1.2;
L = .06;
UDS = .2: .02: .64;
gm_ID = (5:20);
% compute ================
for k = 1:length(gm_ID),
    UGS = lookupVGS(nch,'GM_ID',gm_ID(k),'VDS',UDS,'L',L);
    y  = blkm(nch,L,UDS,UGS);
    gm1_W  = y(:,:,1);
    gds1_W = y(:,:,2);
    Jd1    = y(:,:,3);
    gm2_W  = y(:,:,4);
    gds2_W = y(:,:,5);
    x11_W  = y(:,:,6);
    A1   = (x11_W - sqrt(x11_W.^2 - gm2_W.*gds2_W))./gds2_W;
    UR   = diag(Jd1./(gm1_W./A1 - gds1_W));
    z(k,:) = interp1(VDD-UDS'-UR, [UGS  (VDD-UR) diag(A1) ...
    diag(gm1_W)],0);
end
VDS = z(:,2);
```

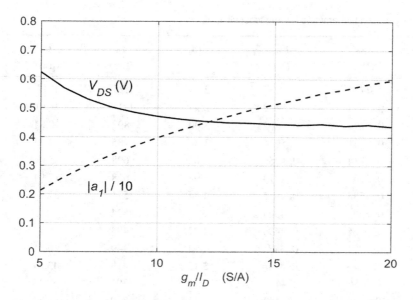

Figure 4.29 Voltage gain (divided by ten) and drain-to-source voltage that null HD$_2$ in a resistively loaded CS stage. Parameters: V_{DD} = 1.2 V, L = 60 nm.

```
VGS = z(:,1);
a1  = z(:,3);
```

Interestingly, the parameter values plotted in Figure 4.29 apply to any resistively-loaded CS stage with the given gate length. We will now illustrate how this data can be used for sizing in the following example.

Example 4.6 Sizing of a Resistively Loaded CS Stage with Low HD$_2$ and V_{DD} = 1.2 V

Repeat Example 4.5, leveraging the data of Figure 4.29 for V_{DD} = 1.2 V and L = 60 nm. Size the stage for g_m/I_D values ranging from 5 to 20 S/A. Validate the design in SPICE for g_m/I_D = 15 S/A and find the tolerances on V_{DS} and R that keep HD$_2$ within –70 and –60 dB.

SOLUTION

To find I_D, we divide the desired transconductance ($2\pi f_u C_L$) by g_m/I_D as usual and to find W, we divide I_D by J_D. The resulting drain currents, widths and load resistances are listed in Table 4.2, which also incorporates the data from Figure 4.29.

We now perform a SPICE simulation using the data for g_m/I_D = 15 S/A in Table 4.2. The simulation circuit is similar to Figure 3.5, which allows us to set the DC drain voltage to the desired 444.6 mV. We observe that g_m/I_D = 14.99 S/A, I_D = 419.7 μA, V_{GS} = 0.4815 V, and $|a_1|$ = 5.15, which are all close to the pre-computed

Table 4.2 Summary of sizing parameters leading to a null in HD$_2$.

g_m/I_D (S/A)	V_{GS} (V)	V_{DS} (V)	$|a_1|$	I_D (µA)	W (µm)	R (kΩ)
5	0. 7726	0. 6233	2.13	1,257	8.14	459
6	0. 7218	0. 5682	2.59	1,047	9.30	603
7	0. 6794	0. 5302	3.00	898	10.81	746
8	0. 6431	0. 5034	3.36	785	12.71	887
9	0. 6114	0. 4850	3.68	698	15.06	1,024
10	0. 5836	0. 4714	3.98	628	17.91	1,160
11	0. 5588	0. 4615	4.24	571	21.36	1,293
12	0. 5366	0. 4547	4.49	524	25.49	1,424
13	0. 5165	0. 4499	4.72	483	30.43	1,552
14	0. 4981	0. 4484	4.94	449	36.35	1,675
15	**0. 4813**	**0. 4446**	**5.13**	**419**	**43.51**	**1.803**
16	0.4656	0.4424	5.31	397	52.18	1.929
17	0.4505	0.4453	5.51	370	62.73	2.042
18	0.4368	0.4396	5.65	349	76.05	2.178
19	0.4228	0.4430	5.84	331	92.78	2.289
20	0.4099	0.4365	5.96	314	114.9	2.430

values. However, the unity gain frequency is only 978 MHz (instead of 1 GHz), since we did not consider the effect of self-loading.

Per the SPICE output depicted in Figure 4.30, the lowest HD$_2$ that we can achieve lies near -78 dB, which is much better than the -33 dB reached for the same g_m/I_D when the output is AC-grounded (see Figure 4.17). However, to achieve this performance, the DC drain voltage V_{DS} requires tight tolerances. To stay within -70 dB and -60 dB, the tolerances on V_{DS} are ±8 mV and ±27 mV, respectively. Interestingly, the tolerances on the load resistance are much less demanding as illustrated by the ±10% curves in Figure 4.30. This is because the optimal V_{DS} does not change much in moderate and weak inversion, as seen from Figure 4.29. The changes in R affect the current, and thus also g_m/I_D, but the V_{DS} curve is flat, thus not requiring a change in V_{DS} to maintain low distortion.

4.3   Random Mismatch

One of the most important features of integrated circuit technology is that the electrical parameters of identically drawn (and closely spaced) components match with good accuracy. This motivates the design of "ratiometric" and symmetrical circuits that exploit this feature. Examples include fully differential amplifier circuits, current mirrors, and a variety of resistive and capacitive networks used for feedback and voltage or current division.

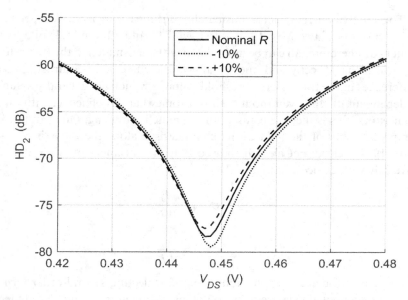

Figure 4.30 SPICE simulation of HD_2 versus V_{DS} for an input amplitude of 10 mV and the parameters listed in the bold row of Table 4.2 ($g_m/I_D = 15$ S/A). Three curves are shown: one for the nominal resistor value and two additional ones representing $\pm10\%$ variations in R.

Unfortunately, even though the matching of integrated components is very good, it is never perfect and in many cases the IC designer must pay attention to mismatch-induced non-idealities. Consequently, there exists a significant body of literature devoted to this topic (see e.g. [16]–[18] and the contained reference lists). The purpose of this section is to review how mismatch may affect the choice of g_m/I_D.

When speaking of mismatch, it is important to distinguish systematic and random effects. Systematic mismatch between two transistors may be caused by a non-symmetric layout, or different layer densities in the surroundings of the devices in question. Such issues are relatively well understood and can be mitigated through appropriate layout techniques [18]. Once all significant sources of systematic mismatch have been eliminated, we are left with random mismatch, caused for example by line edge roughness and random doping fluctuations.

The focus of this section is on random mismatch. We will first review models for random mismatch, and subsequently consider the effects of random mismatch in a few circuit primitives of general interest, from a g_m/I_D-centric perspective.

4.3.1 Modeling of Random Mismatch

By far the most common way to quantify random mismatch is through deviations in the transistors' threshold voltage (V_T) and current factor (β). The choice of these parameters was originally inspired by the square-law model ($\beta = \mu C_{ox} W/L$), but is applicable regardless of the transistor's I-V law.

The relative current factor mismatch $\Delta\beta/\beta$ measures by which percentage the drain currents of two MOSFETs differ, with V_T and all terminal voltages perfectly matched. Therefore, $\Delta\beta$ can be assimilated with a mismatch in the transistors' channel width that directly and linearly affects the current. In contrast, ΔV_T captures differences in the transistors' threshold voltages, which are generally assumed to be independent of $\Delta\beta$. Even though V_T is a somewhat ill-defined quantity in modern transistors, this is inconsequential for mismatch modeling. One can think of ΔV_T simply as a shift of the I-V transfer characteristic along the V_{GS} axis.

In 1989, Pelgrom et al. [19] showed that the variances of $\Delta\beta/\beta$ and ΔV_T scale inversely with device area:

$$\sigma^2_{\Delta\beta/\beta} = \frac{A^2_\beta}{WL},$$ (4.44)

$$\sigma^2_{\Delta V_T} = \frac{A^2_{VT}}{WL}.$$ (4.45)

Note that in the above expressions, we are neglecting second-order terms related to the spacing of the devices. The matching parameters A_β and A_{VT} are technology dependent. Typical values for A_{VT} and A_β for a 65-nm technology are 3.5 mV-μm and 1 %-μm, respectively. As an example, for an n-channel device with a gate area of 1 μm², the standard deviation of the threshold voltage is 3.5 mV. As technology scales, A_{VT} tends to improve proportional to the effective gate oxide thickness [20], while A_β is almost unaffected by scaling, being in the range of 1–2% for many recent process generations [17].

In recent years, several improvements to Pelgrom's basic mismatch models have been proposed, mainly to incorporate second-order effects in the fabrication of short-channel devices. The most significant issue that has surfaced recently is that the threshold variance no longer scales directly with $1/L$ [21]. This is because the pocket implants near the drain and source invalidate the uniform doping assumption that underpins (4.45). In a transistor with pocket implants, the center region of the channel does not play a significant role in defining the threshold, and extending it therefore does not have an impact on its variability. To account for this issue, a new model has been proposed [17]:

$$\tilde{A}^2_{VT} = \frac{A^2_{VT}}{WL} + \frac{B^2_{VT}}{f(WL)},$$ (4.46)

where $f(WL)$ is still to be determined, but will likely be close to W itself, and independent of L. The take-home for analog designers is that they should no longer take better matching in longer transistors for granted. A conservative guideline is to use (4.44) and (4.45) with $L = L_{min}$, regardless of which channel length is used.

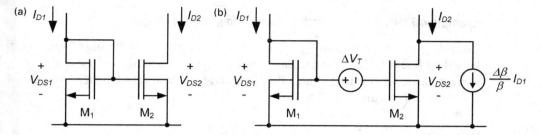

Figure 4.31 Current mirror. (a) Ideal circuit with matched transistors. (b) Circuit model with mismatch.

4.3.2   Effect of Mismatch in a Current Mirror

Consider the basic current mirror circuit shown in Figure 4.31(a). If both transistors are identical and $V_{DS1} = V_{DS2}$, the drain currents must also be exactly equal ($I_{D1} = I_{D2}$). Any deviation from this behavior must be due to mismatch in the transistors' electrical parameters.

Figure 4.31(b) shows a model of the current mirror circuit with both threshold and current factor mismatch included. The input current I_{D1} is assumed to be constant, and the error introduced by mismatch lies in the difference $I_{D2} - I_{D1}$. Assuming equal drain-source voltages, this difference is

$$\Delta I_D = I_{D2} - I_{D1} \cong g_{m2}\Delta V_T + \frac{\Delta \beta}{\beta} I_{D1}. \tag{4.47}$$

It is assumed here that ΔV_T is small, so that it can be referred from gate to drain using the transistor's small-signal transconductance (g_{m2}). The error in the output current I_{D2}, relative to the fixed input current I_{D1} is therefore

$$\frac{\Delta I_D}{I_{D1}} \cong \frac{g_{m2}}{I_{D1}}\Delta V_T + \frac{\Delta \beta}{\beta} \cong \frac{g_{m1}}{I_{D1}}\Delta V_T + \frac{\Delta \beta}{\beta}, \tag{4.48}$$

where the last simplification is justifiable because the difference in the devices' transconductance is small. As a final step, we express the mismatch in terms of variances, assuming that the threshold and current factor terms are statistically independent:

$$\sigma^2_{\Delta I_D / I_{D1}} \cong \left(\frac{g_{m1}}{I_{D1}}\right)^2 \sigma^2_{\Delta V_T} + \sigma^2_{\Delta \beta / \beta}. \tag{4.49}$$

As we can see from this result, g_m/I_D plays a role in the matching performance, and this deserves further investigation. Another important fact is that the $\Delta \beta / \beta$ term is usually negligible. We will illustrate this point using an example.

Example 4.7 Random Mismatch Estimation in a Current Mirror

Consider a current mirror using n-channel devices with $W/L = 50$ μm/ 60 nm, biased such that $g_m/I_D = 10$ S/A. Calculate the standard deviation of the drain current mismatch due to $\Delta\beta$ and ΔV_T alone, and also the total mismatch. Assume $A_{VT} = 3.5$ mV-μm and $A_\beta = 1$ %-μm.

SOLUTION

The first mismatch component due to ΔV_T follows by evaluating the first term of (4.49), and using (4.45) with the given numbers:

$$\sigma_1 \cong 10\frac{S}{A} \cdot \frac{3.5mV}{\sqrt{50 \cdot 0.06}} = 2.02\%.$$

Similarly, the second mismatch component due to $\Delta\beta$ is (using (4.44)):

$$\sigma_2 \cong \frac{1\%}{\sqrt{50 \cdot 0.06}} = 0.58\%.$$

The total mismatch is:

$$\sigma_{\Delta I_D/I_{D1}} \cong \sqrt{\sigma_1^2 + \sigma_2^2} = 2.10\%.$$

Therefore, we conclude that $\Delta\beta$ mismatch plays no significant role in a typical current mirror. As we can see from the above example, the $\Delta\beta$ mismatch would become comparable to the ΔV_T component only if g_m/I_D was reduced significantly. From the above numbers, we see that the mismatch components would be comparable for $g_m/I_D = 2.5$ S/A, which is impractically small.

With the insight gained from the previous example, we can approximate the standard deviation of the current mismatch as given in (4.50), which neglects $\Delta\beta$ mismatch:

$$\sigma_{\Delta I_D/I_{D1}} \cong \frac{g_{m1}}{I_{D1}} \frac{A_{VT}}{\sqrt{WL}}. \tag{4.50}$$

A natural question to ask at this point is which value of g_m/I_D should be used in the practical realization of a current mirror. In some cases, the answer to this question follows from constraints unrelated to matching. For example, as we have seen in Example 4.2, there exists an optimum value for g_m/I_D that maximizes the dynamic range of a CS stage with an active load device (which is usually the output device of a current mirror). In situations where mismatch is a bigger concern than noise performance, we can distinguish two scenarios: (1) We have a fixed area (WL) available for the current mirror, or (2) the area is unconstrained, but the current is fixed. We begin by investigating the first scenario.

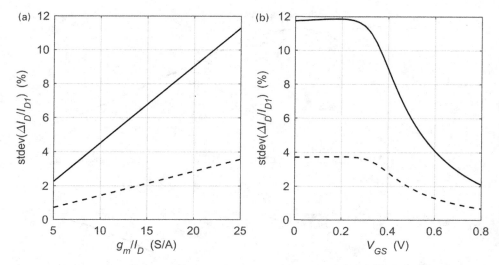

Figure 4.32 Standard deviation of current mirror mismatch for two different device sizes: $W/L =$ 10 μm/ 60 nm (solid line) and $W/L = 100$ μm/ 60 nm (dashed line), assuming $A_{VT} = 3.5$ mV-μm. (a) Mismatch as a function of g_m/I_D. (b) Mismatch as a function of V_{GS} ($V_{DS} = 0.6$ V).

When the device area is fixed, it is clear from (4.50) that we must make g_m/I_D as small as possible. Since we assumed that WL is fixed and A_{VT} is a constant process parameter, this is the only degree of freedom available. The ultimate lower bound for g_m/I_D then usually follows from voltage headroom constraints, since V_{Dsat} increases as g_m/I_D is lowered (see Chapter 2, (2.34)). As an example, Figure 4.32(a) plots the standard deviation of the current mismatch versus g_m/I_D for two device sizes (and using (4.50)). In Figure 4.32(b), the variable on the x-axis is changed to V_{GS}, a representation often seen in the literature [22]. As expected, we observe from this plot that the mismatch versus V_{GS} curve has the same shape as the g_m/I_D versus V_{GS} curve that we have already seen in earlier chapters.

The conclusion from Figure 4.32 is that the current mirror transistors should be operated at the highest possible V_{GS}, or, in other words, the smallest possible g_m/I_D for good matching. While this conclusion is engrained in the thinking of most IC designers, it is important to remember that it assumes that the device area is fixed. We will now look at the second case where the current is fixed, which corresponds to a more typical scenario. We often want to design a current mirror for a given current, regardless of area (within reasonable limits).

We can gain some insight about the tradeoffs for the fixed current scenario by expressing g_m/I_D using (2.32), which is repeated here for convenience:

$$\frac{g_m}{I_D} = \frac{1}{nU_T} \frac{2}{\sqrt{1 + 4\dfrac{I_D}{I_{Ssq}}\dfrac{L}{W} + 1}} . \tag{4.51}$$

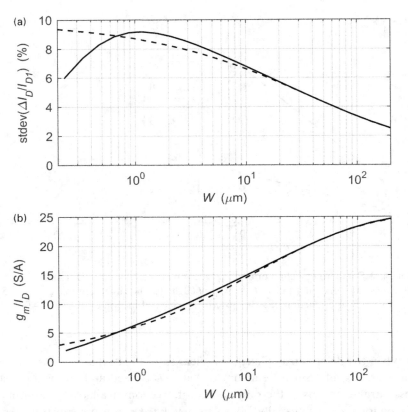

Figure 4.33 Standard deviation of n-channel current mirror mismatch versus device width for $I_D = 100\ \mu\text{A}$, $A_{VT} = 3.5\ \text{mV-}\mu\text{m}$ and $L = L_{min} = 60\ \text{nm}$. (b) Corresponding g_m/I_D. Solid lines: real transistor. Dashed lines: Basic EKV model.

We now substitute this expression into (4.50), which gives

$$\sigma_{\Delta I_D/I_{D1}} \cong \frac{1}{nU_T} \frac{2}{\sqrt{1 + 4\dfrac{I_D}{I_{Ssq}}\dfrac{L}{W} + 1}} \frac{A_{VT}}{\sqrt{WL}}. \tag{4.52}$$

This closed-form expression allows us to estimate the mismatch at arbitrary inversion levels. For additional insight, let us consider the special cases of weak and strong inversion. In weak inversion, $I_D \ll I_{Ssq}W/L$ and (4.52) simplifies to

$$\sigma_{\Delta I_D/I_{D1}} \cong \frac{1}{nU_T} \frac{A_{VT}}{\sqrt{WL}}. \tag{4.53}$$

On the other hand, in strong inversion $I_D \gg I_{Ssq}W/L$ and we obtain

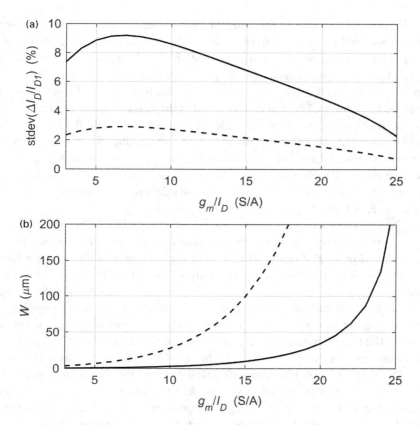

Figure 4.34 (a) Standard deviation of n-channel current mirror mismatch versus g_m/I_D for two different drain currents: 100 µA (solid line) and 1 mA (dashed line), assuming $A_{VT} = 3.5$ mV-µm and $L = L_{min} = 60$ nm. (b) Corresponding device width.

$$\sigma_{\Delta I_D / I_{D1}} \cong \frac{1}{nU_T} \sqrt{\frac{I_D}{I_{Ssq}}} \frac{A_{VT}}{L}. \tag{4.54}$$

The surprising result from (4.54) is that the mismatch is independent of transistor width in strong inversion. The only way to improve the matching is to employ a longer channel. That, however, has also become somewhat ineffective in modern technologies due to the above-discussed impact of pocket implants. Per the conservative approximation proposed in Section 4.3.1, we can refine the above result as follows:

$$\sigma_{\Delta I_D / I_{D1}} \cong \frac{1}{nU_T} \sqrt{\frac{I_D}{I_{Ssq}}} \frac{A_{VT}}{\sqrt{L \cdot L_{min}}}. \tag{4.55}$$

where L_{min} is the minimum channel length and also the channel length at which A_{VT} was measured.

Figure 4.33(a) plots the current mirror mismatch for a minimum-length n-channel and $I_D = 100\ \mu A$ (dashed line). The dashed line corresponds to (4.52) and the solid line represents real transistor data. The latter curve was generated using (4.50), and g_m/I_D was computed using the lookup function for each value of W. For reference, Figure 4.33(b) shows the corresponding g_m/I_D (the dashed line is the EKV prediction from (4.50), and the solid line is the real transistor data).

We see that the curves in Figure 4.33(a) match well in weak inversion (large W), indicating that the matching improves with increasing W, as seen from (4.53). In strong inversion (small W), the dashed curve approaches a constant value (as predicted by (4.54)). However, the solid curve (real transistor) bends down in strong inversion. This can be explained by mobility degradation, which can be assimilated with a reduction of I_{Ssq} in (4.52).

The real transistor data from Figure 4.33 is plotted again in Figure 4.34, but with g_m/I_D placed on the x-axis and for a second current value. We now see the tradeoffs as a function of g_m/I_D more directly. The take-home from this chart is that, unlike in the area-constrained scenario (Figure 4.32), it may be acceptable to operate a current mirror in moderate or weak inversion, if the area overhead is acceptable to the designer. In any case, this tends to be necessary in modern low-voltage designs to manage headroom constraints (small V_{Dsat} requires large g_m/I_D). On the other hand, the large junction capacitances associated with the large widths may become an issue and the designer must pick the suitable tradeoff on a case-by-case basis.

As a final remark, it is worth pointing out that there remains one second-order argument in favor of strong inversion. As shown in Figure 4.35(a), careless layout can cause substantial voltage drop (ΔV) between the source terminals of the mirror devices. The current I_{WIRE} is not necessarily just the mirror current, but can also include the current of other blocks. Figure 4.35(b) shows an equivalent model for this situation, making it clear that the voltage drop is referred to the mirror output via g_{m2} (just like threshold voltage mismatch, analyzed above). To attenuate the error due to ΔV, g_{m2} should be made as small as possible. For fixed current level, this means that g_m/I_D must be minimized as well, calling for strong inversion. Operating

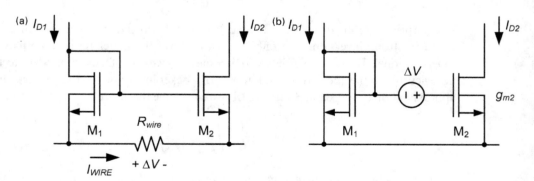

Figure 4.35 Current mirror. (a) Circuit with IR drop in the source connection. (b) Equivalent model.

the transistor in weak or strong inversion is still possible, but the designer is advised to think about potential IR drop issues carefully.

4.3.3 Effect of Mismatch in a Differential Amplifier

In a differential amplifier, the most important mismatch-related issue is input referred (random) offset and its temperature dependence. To analyze these effects, we consider the basic fully-differential gain stage shown in Figure 4.36(a). The presented analyses can be extended for more complex structures (such as folded-cascode amplifiers, two-stage amplifiers, etc.); the discussed principles remain the same.

The circuit has a differential input and output, with a common-mode feed-back circuit (CMFB) defining the common-mode operating point of the output nodes. Both the n-channel and p-channel pairs in this circuit will exhibit mismatch, shown as $\Delta\beta/\beta$ and ΔV_T in the schematic. To achieve a symmetric representation, it is common to split the mismatch and assign half of the errors to each half circuit.

To capture the net effect of all mismatch errors, one commonly defines the so-called input-referred offset voltage of the amplifier. The offset voltage V_{OS}, shown in Figure 4.36(b), is obtained by reflecting all mismatches to the input of the amplifier. Just as in the case of the current mirror example treated above, it is straightforward to show that the contribution of the $\Delta\beta/\beta$ mismatch to V_{OS} is usually overshadowed by the ΔV_T terms. Hence, one can show that a good approximation for the input offset is given by:

$$V_{OS} \cong \Delta V_{T1} + \frac{\Delta V_{T2}}{2}\frac{g_{m2a}}{g_{m1a}} + \frac{\Delta V_{T2}}{2}\frac{g_{m2b}}{g_{m1b}} \cong \Delta V_{T1} + \Delta V_{T2}\frac{g_{m2}}{g_{m1}}, \qquad (4.56)$$

where g_{m1} and g_{m2} are the average transconductances of $M_{1a,b}$ and $M_{2a,b}$, respectively. Since the (average) drain currents I_{D1} and I_{D2} are equal, we can re-write this result as shown below.

$$V_{OS} \cong \Delta V_{T1} + \Delta V_{T2}\frac{\left(\frac{g_m}{I_D}\right)_2}{\left(\frac{g_m}{I_D}\right)_1}. \qquad (4.57)$$

As expected, we see that the threshold mismatch of the input pair directly contributes to the equivalent input offset. More importantly, we see that the mismatch contribution of the active load devices ($M_{2a,b}$) can be minimized via the ratio of the transconductance efficiencies of $M_{2a,b}$ and $M_{1a,b}$. In a typical realization, the designer will try to make this ratio as small as possible, usually dictating small $(g_m/I_D)_2$. This is once again a situation where small g_m/I_D can be beneficial. However, in practice, voltage headroom constraints may impose limits to making $(g_m/I_D)_2$ small, due to the associated increase in V_{Dsat2}.

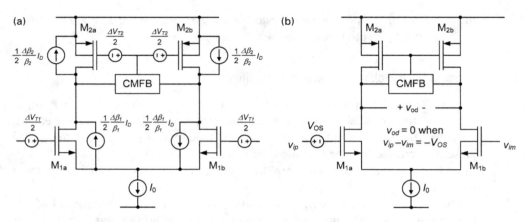

Figure 4.36 Differential amplifier with active loads. (a) Circuit with all mismatch sources. (b) Equivalent circuit with a single input referred offset voltage source (V_{OS}).

Let us now turn our attention to the temperature dependence of the offset voltage, which is a significant concern in precision instrumentation. Interestingly, and as pointed out in the literature [17], [23], the main contribution to the input-referred offset drift comes from the $\Delta\beta$ mismatch terms. Even though the input offset contribution due to ΔV_T is larger, this component does not drift, since it is primarily caused by a mismatch in doping, which does not depend on temperature. This was confirmed in [23], which shows that both the ΔV_T and $\Delta\beta/\beta$ mismatch are essentially temperature independent. However, while the ΔV_T mismatch appears directly at the input,[5] the $\Delta\beta/\beta$ terms refer to the input via the device transconductances, which can vary significantly with temperature. To investigate, we can refer the current factor mismatch terms to the input of the amplifier and find

$$V_{OS,\Delta\beta} = \left(\frac{\Delta\beta_1}{\beta_1} + \frac{\Delta\beta_2}{\beta_2} \right) \frac{I_D}{g_{m1}} = \frac{\Delta\beta_{tot}}{\beta_{tot}} \frac{I_D}{g_{m1}}. \tag{4.58}$$

The offset drift is therefore

$$V_{OS,drift} = \frac{dV_{OS,\Delta\beta}}{dT} = \frac{\Delta\beta_{tot}}{\beta_{tot}} \cdot \frac{d}{dT} \frac{I_D}{g_{m1}}. \tag{4.59}$$

That is, the drift is governed by the temperature dependence of the $(g_m/I_D)^{-1}$ of the input devices. Figure 4.37 shows a SPICE simulation of $d/dT(I_D/g_m)_1$, assuming that the differential amplifier is operated with a temperature-independent bias current (I_0). We observe that the temperature dependence decreases in weak inversion

---

[5] Strictly speaking, this is true only for the input pair. The threshold mismatch of $M_{2a,b}$ is referred to the input via a ratio of g_m/I_D terms (see (4.57)). If the input pair and the load devices are operated at different inversion levels, the temperature coefficients of the g_m/I_D terms may differ and lead to an additional drift component.

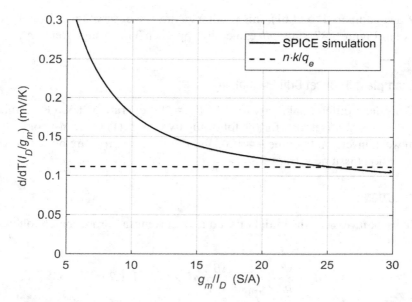

Figure 4.37 Temperature coefficient of $(g_m/I_D)^{-1}$ for an n-channel device with $L = 100$ nm and constant bias current at $T = 300$ K.

and approaches a constant value. This asymptote in weak inversion can be easily computed:

$$V_{OS,drift,W.I.} = \frac{\Delta\beta_{tot}}{\beta_{tot}} \cdot \frac{d}{dT} n U_T = \frac{\Delta\beta_{tot}}{\beta_{tot}} n \frac{k}{q_e}. \tag{4.60}$$

Here, k is the Boltzmann constant and q_e stands for electron charge. Furthermore, note that we can express this result in terms of the initial offset per (4.58):

$$V_{OS,drift,W.I.} = \frac{\Delta\beta_{tot}}{\beta_{tot}} \cdot \frac{n U_T}{T} = \frac{V_{OS,\Delta\beta}}{T}. \tag{4.61}$$

This is a well-known result that also holds for the offset drift in bipolar amplifiers. The drift is equal to the initial offset due to $\Delta\beta$, divided by absolute temperature.

Let us now investigate why the temperature coefficient of $(g_m/I_D)^{-1}$ increases in strong inversion. Under the assumption of constant current, the drift is due to changes in g_m. The transconductance is proportional to mobility$^{1/b}$, where $b = 2$ for the square law, and $b \to 1$ for a velocity-saturated short-channel transistor. The mobility itself is proportional to T^a, where $a = -2...-3$. Using this information, we obtain:

$$V_{OS,drift,S.I.} = \frac{a}{b} \frac{V_{OS,\Delta\beta}}{T}. \tag{4.62}$$

When compared to (4.61), this result explains the observed increase of the derivative in Figure 4.37, since a/b can easily approach 3 in strong inversion.

Example 4.8 Offset Drift Estimation

Consider a differential pair with $\Delta\beta_{tot}/\beta_{tot} = 1\%$, operating at 300 K. Estimate the expected offset drift for the following two cases: (1) the transistors are biased in weak inversion (assume $n = 1.3$), and (2) the transistors are biased in strong inversion (assume $g_m/I_D = 5$ S/A and $a/b = 3$).

SOLUTION

In weak inversion, the initial offset due to current factor mismatch follows from (4.58):

$$V_{OS,\Delta\beta} = \frac{\Delta\beta_{tot}}{\beta_{tot}}\frac{I_D}{g_{m1}} = \frac{\Delta\beta_{tot}}{\beta_{tot}}nU_T = 1\% \cdot 1.3 \cdot 26\,\mathrm{mV} = 338\ \mu\mathrm{V}.$$

The offset drift is now easily computed using (4.61):

$$V_{OS,\mathrm{drift,W.I.}} = \frac{V_{OS,\Delta\beta}}{T} = 1.1\,\mu\mathrm{V/K}.$$

In strong inversion, the initial offset is:

$$V_{OS,\Delta\beta} = \frac{\Delta\beta_{tot}}{\beta_{tot}}\frac{I_D}{g_{m1}} = 1\% \cdot \frac{1}{5\,\mathrm{S/A}} = 2\,\mathrm{mV}.$$

Using (4.62), the offset drift is

$$V_{OS,\mathrm{drift,S.I.}} = 3\frac{V_{OS,\Delta\beta}}{T} = 20\,\mu\mathrm{V/K}.$$

The take-home from the above example is that the offset drift of a CMOS amplifier strongly depends on the inversion level of the constituent transistors. If low offset drift is desired, it helps to operate in weak inversion.

4.4   Summary

This chapter reviewed the fundamentals of noise, distortion and mismatch analysis using a g_m/I_D-centric framework. We showed that the inversion level, represented in our work by its proxy g_m/I_D, plays a key role in quantifying and minimizing these non-idealities.

In the discussion on thermal noise, we showed that the product of g_m/I_D and transit frequency is a useful figure of merit that defines the optimum inversion level in circuits that value low noise and wide bandwidth simultaneously. In addition, we saw that excess noise from biasing devices can be minimized by proper choice of g_m/I_D. In the modeling of flicker noise, g_m/I_D can be used to quantify the bias dependency of the flicker noise PSD, although this effect is weak for a typical CMOS process.

The treatment of nonlinear distortion leveraged the basic EKV equations defined in Chapter 2 to derive distortion metrics that hold for all levels of inversion. These expressions contain the normalized inversion charge q, and the subthreshold slope factor n, which have direct links to g_m/I_D. We showed through numerical examples that the harmonic distortion of a common-source stage and a differential pair can be accurately predicted using the proposed framework. As a result, the designer can explore the distortion tradeoff space across all inversion levels using Matlab scripts, instead of exhaustive SPICE-level simulations.

In our review of mismatch analysis, we showed that g_m/I_D emerges directly as an important parameter in the design expressions that quantify mismatch. Our investigation looked at current mirrors and differential amplifiers, and we formulated guidelines on how to pick the inversion level in these circuits to minimize the effect of mismatch. We also showed that the temperature drift of the input-referred offset in differential amplifiers is linked to the temperature behavior of $(g_m/I_D)^{-1}$.

4.5   References

[1] P. R. Gray, P. Hurst, S. H. Lewis, and R. G. Meyer, *Analysis and Design of Analog Integrated Circuits*, 5th ed. Wiley, 2009.

[2] A. J. Scholten, L. F. Tiemeijer, R. van Langevelde, R. J. Havens, A. T. A. Zegers-van Duijnhoven, and V. C. Venezia, "Noise Modeling for RF CMOS Circuit Simulation," *IEEE Trans. Electron Devices*, vol. 50, no. 3, pp. 618–632, Mar. 2003.

[3] G. D. J. Smit, A. J. Scholten, R. M. T. Pijper, R. van Langevelde, L. F. Tiemeijer, and D. B. M. Klaassen, "Experimental Demonstration and Modeling of Excess RF Noise in Sub-100-nm CMOS Technologies," *IEEE Electron Device Lett.*, vol. 31, no. 8, pp. 884–886, Aug. 2010.

[4] A. Shameli and P. Heydari, "A Novel Power Optimization Technique for Ultra-Low Power RFICs," *Proc. International Symposium on Low Power Electronics and Design (ISLPED)*, 2006, pp. 274–279.

[5] A. Mangla, C. C. Enz, and J.-M. Sallese, "Figure-of-Merit for Optimizing the Current-Efficiency of Low-Power RF Circuits," Proc. International Conference on Mixed Design of Integrated Circuits and Systems (MIXDES), 2011, pp. 85–89.

[6] G. Gildenblat, X. Li, W. Wu, H. Wang, A. Jha, R. Van Langevelde, G. D. J. Smit, A. J. Scholten, and D. B. M. Klaassen, "PSP: An Advanced Surface-Potential-Based MOSFET Model for Circuit Simulation," *IEEE Trans. Electron Devices*, vol. 53, no. 9, pp. 1979–1993, Sep. 2006.

[7] T. Boutchacha, G. Ghibaudo, and B. Belmekki, "Study of Low Frequency Noise in the 0.18 μm Silicon CMOS Transistors," in *Proc. International Conference on Microelectronic Test Structures*, 1999, pp. 84–88.

[8] C. C. Enz and E. A. Vittoz, *Charge-Based MOS Transistor Modeling: The EKV Model for Low-Power and RF IC Design*. John Wiley & Sons, 2006.

[9] K. K. Hung, P. K. Ko, C. Hu, and Y. C. Cheng, "A Unified Model for the Flicker Noise in Metal-Oxide-Semiconductor Field-Effect Transistors," *IEEE Trans. Electron Devices*, vol. 37, no. 3, pp. 654–665, Mar. 1990.

[10] C. C. Enz and G. C. Temes, "Circuit Techniques for Reducing the Effects of Op-amp Imperfections: Autozeroing, Correlated Double Sampling, and Chopper Stabilization," *Proc. IEEE*, vol. 84, no. 11, pp. 1584–1614, 1996.

[11] R. H. Dennard, F. H. Gaensslen, V. L. Rideout, E. Bassous, and A. R. LeBlanc, "Design of Ion-Implanted MOSFET's with Very Small Physical Dimensions," *IEEE J. Solid-State Circuits*, vol. 9, no. 5, pp. 256–268, Oct. 1974.

[12] M. J. Knitel, P. H. Woerlee, A. J. Scholten, and A. Zegers-Van Duijnhoven, "Impact of Process Scaling on 1/f Noise in Advanced CMOS Technologies," in *Proc. IEDM*, 2000, pp. 463–466.

[13] W. Sansen, "Distortion in Elementary Transistor Circuits," *IEEE Trans. Circuits Syst. II*, vol. 46, no. 3, pp. 315–325, Mar. 1999.

[14] B. Toole, C. Plett, and M. Cloutier, "RF Circuit Implications of Moderate Inversion Enhanced Linear Region in MOSFETs," *IEEE Trans. Circuits Syst. I*, vol. 51, no. 2, pp. 319–328, Feb. 2004.

[15] S. C. Blaakmeer, E. A. Klumpetink, D. M. W. Leenaerts, and B. Nauta, "Wideband Balun-LNA with Simultaneous Output Balancing, Noise-Cancelating and Distortion-Canceling," *IEEE J. Solid-State Circuits*, vol. 43, no. 6, pp. 1341–1350, 2008.

[16] P. R. Kinget, "Device Mismatch and Tradeoffs in the Design of Analog Circuits," *IEEE J. Solid-State Circuits*, vol. 40, no. 6, pp. 1212–1224, June 2005.

[17] M. Pelgrom, H. Tuinhout, and M. Vertregt, "A Designer's View on Mismatch," in *Nyquist AD Converters, Sensor Interfaces, and Robustness*, A. H. M. van Roermund, A. Baschirotto, and M. Steyaert, Eds. Springer, 2013, pp. 245–267.

[18] A. Hastings, *The Art of Analog Layout*, 2nd ed. Prentice Hall, 2005.

[19] M. J. M. Pelgrom, A. C. J. Duinmaijer, and A. P. G. Welbers, "Matching properties of MOS transistors," *IEEE J. Solid-State Circuits*, vol. 24, no. 5, pp. 1433–1439, Oct. 1989.

[20] M. J. M. Pelgrom, H. P. Tuinhout, and M. Vertregt, "Transistor Matching in Analog CMOS Applications," in *IEDM Tech. Digest*, 1998, pp. 915–918.

[21] C. M. Mezzomo, A. Bajolet, A. Cathignol, R. Di Frenza, and G. Ghibaudo, "Characterization and Modeling of Transistor Variability in Advanced CMOS Technologies," *IEEE Trans. Electron Devices*, vol. 58, no. 8, pp. 2235–2248, Aug. 2011.

[22] Tony Chan Carusone, D. A. Johns, and K. W. Martin, *Analog Integrated Circuit Design*, 2nd ed. Wiley, 2011.

[23] P. Andricciola and H. P. Tuinhout, "The Temperature Dependence of Mismatch in Deep-Submicrometer Bulk MOSFETs," *IEEE Electron Device Lett.*, vol. 30, no. 6, pp. 690–692, June 2009.

5 Practical Circuit Examples I

The purpose of this chapter is to show how we can use the concepts outlined in earlier chapters to design practical circuits in a systematic manner. We begin with the design of basic auxiliary circuits for bias current generation and distribution and then move to higher-complexity examples involving class-A amplifier stages. Specifically, we cover the design of a low-dropout voltage regulator (LDO), an RF low-noise amplifier (LNA) and a charge amplifier. At the end of the chapter, we address considerations and possible design flows for process-corner-aware design. Here, we re-visit the charge-amplifier example to illustrate the suggested approach.

5.1 Constant Transconductance Bias Circuit

The circuit shown in Figure 5.1 is commonly used as a bias current generator [1]. The prime merit of the circuit is that it can be designed to make the transconductance of M_2 approximately constant, regardless of process and temperature variations. This circuit is typically used to bias amplifier circuits, like the differential stage shown in the right part of Figure 5.1. If $M_{6a,b}$ obey the same "physics" as M_2 (similar inversion level and channel length), their transconductances will also be stable over process and temperature.

The core of the circuit consists of two self-biased current mirrors, $M_1 - M_2$ and $M_3 - M_4$. While the in- and output currents of the upper current mirror are linearly related, those of the lower current mirror are not (see Figure 5.2). During circuit start-up (small currents), the output current of the lower current mirror (I_{D1}) is larger than the input current, since the voltage drop across R is still negligible and W_1 is larger than W_2. The upper current mirror causes more current to enter the lower current mirror consequently, which in turn increases the output current, and so on. Meanwhile, since the voltage drop across R is getting larger, the current gain of the lower current mirror starts decreasing until I_{D1} and I_{D2} equilibrate (circle in Figure 5.2). A start-up circuit, discussed in [2], is needed to initiate this mechanism.

As we shall show below, the point where the currents equalize is set by R and the width ratio of M_1 and M_2. Once self-biased, the currents are essentially constant, except for minor changes caused by drain-to-source voltage variations.

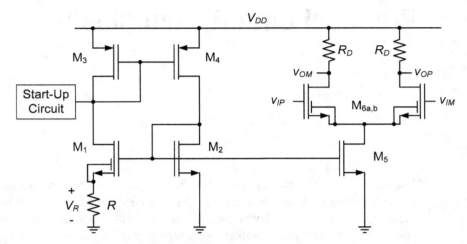

Figure 5.1 Self-biased, constant g_m current generator ($M_1 - M_4$). As an example, the circuit is used to bias a differential pair amplifier ($M_5 - M_{6a,b}$). Details of the required start-up circuit are omitted for simplicity.

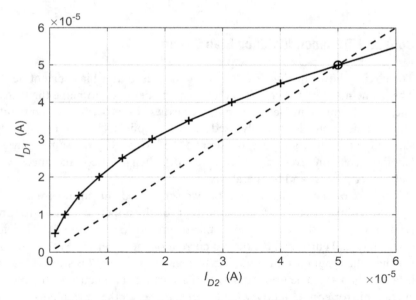

Figure 5.2 Illustration of the nonlinear relation between the lower current mirror input and output currents.

We will now use the basic EKV model of Chapter 2 to analyze the temperature dependence of the circuit. We start from (2.29), repeated here for convenience:

$$g_{m2} = \frac{1}{nU_T}\frac{I_{D2}}{q_2 + 1}.$$
(5.1)

Per (2.23), (2.21) and assuming I_{D2} equal to I_{D1}:

$$RI_{D2} = V_{GS2} - V_{GS1} = n(V_{p2} - V_{p1}) = nU_T\left[2(q_2 - q_1) + log\left(\frac{q_2}{q_1}\right)\right].$$

(5.2)

After eliminating I_{D2} between the above two equations, g_{m2} equals the reciprocal of R, times a factor that depends on the normalized mobile charge densities q_1 and q_2:

$$g_{m2} = \frac{1}{R}\frac{2(q_2 - q_1) + log\left(\frac{q_2}{q_1}\right)}{1 + q_2} = \frac{1}{R}F(q_1, q_2).$$

(5.3)

The temperature dependence of g_{m2} is therefore determined by R and by the transistor's inversion levels. R may be an external precision resistor, in which case g_{m2} can be very stable, as we will see below. If R is an on-chip resistor, the variations can still be relatively small, especially when a doped polysilicon resistor is used.

To get a closer view of the temperature dependence of the factor F, we first note that for $I_{D1} = I_{D2}$, we can write:

$$\frac{W_1}{W_2} = \frac{q_2^2 + q_2}{q_1^2 + q_1}.$$

(5.4)

This expression follows from (2.22) and (2.16) and assumes that the devices' threshold voltages are equal. In our circuit, this is ensured by placing M_2 in a separate well ($V_{SB2} = 0$). We now consider two extreme cases: weak and strong inversion. Deep in weak inversion, q_1 and q_2 are very small. Equations (5.3) and (5.4) then lead to:

$$g_{m2} = \frac{1}{R}log\left(\frac{q_2}{q_1}\right) = \frac{1}{R}log\left(\frac{W_1}{W_2}\right).$$

(5.5)

On the other hand, q_1 and q_2 are large in strong inversion and we find:

$$g_{m2} = \frac{1}{R}\frac{2(q_2 - q_1)}{q_2} = \frac{2}{R}\left(1 - \sqrt{\frac{W_2}{W_1}}\right).$$

(5.6)

Regardless of whether the transistor is in weak or strong inversion, we see that the two expressions depend only on the width ratio of M_1 and M_2 and the value of R. Between these two extremes, the behavior is controlled by the factor F. After eliminating q_1 between (5.3) and (5.4), and connecting F to the transconductance efficiency of M_2 via (5.1), one can construct the curves shown in Figure 5.3. These show the trend versus the normalized transconductance efficiency p for various width ratios. The asterisks on the left side of the plot correspond to (5.6), whereas the circles on the right side represent (5.5).

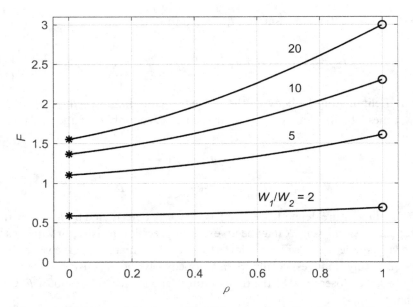

Figure 5.3 Evolution of the factor F in (5.3), plotted against the normalized transconductance efficiency ρ. W_1/W_2 are equal to 2, 5, 10 and 20, as annotated. The asterisks and circles correspond to (5.6) and (5.5), respectively.

If the load resistors R_D of the differential pair in Figure 5.1 are made from the same material as R, the voltage gain of the amplifier is stabilized, even if R and R_D vary in terms of their absolute value. In weak inversion, we have:

$$g_{m6}R_D = \frac{W_5}{W_2}\frac{R_D}{R}\ln\left(\frac{W_1}{W_2}\right).$$

(5.7)

In strong inversion, a similar result follows. It is important to note that $M_{6a,b}$ and M_2 need to operate at similar conditions, which implies that the channel lengths of $M_{6a,b}$ and M_2 must be identical for this result to hold. In addition, since M_5 and M_2 form a current mirror, the gate length of M_5 must also be equal to that of M_2.

Example 5.1 Sizing of a Constant Transconductance Bias Circuit

Design a constant-g_m bias circuit with currents I_{D1} and I_{D2} equal to 50 µA. Assume $V_{DD} = 1.2$ V, $V_R = 0.1$ V, and all $L = 0.5$ µm. Evaluate (via SPICE simulations) the design's sensitivity to temperature and V_{DD} variations. Note: The choice of the gate length corresponds to a low-speed design.

SOLUTION

To make the circuit as balanced as possible, we assume that the p-channel current mirror transistors not only have the same widths $W_3 = W_4$, but also the same

5.1 Constant Transconductance Bias Circuit 169

drain-to-source voltages. The sum of the voltage drops across the diode-connected transistors M_2 and M_3 is therefore equal to the supply voltage. This turns out to be acceptable since V_{DD} is equal to 1.2 V in our process.

Consider first the n-channel current mirror, and let us sweep the transconductance efficiency of the diode-connected transistor M_2 from strong to weak inversion. We then calculate the corresponding gate-to-source voltages V_{GS2} and evaluate the current densities J_{D2}, J_{D1} and J_{D3} of M_2, M_1 and M_3:

```
JD2 = diag(lookup(nch,'ID_W','VGS',VGS2,'VDS',VGS2,'L',L));
JD1 = diag(lookup(nch,'ID_W','VGS',VGS2-VR,'VDS',VGS2-VR,'L',L));
JD3 = diag(lookup(pch,'ID_W','VGS',VDD-VGS2,'VDS',...
VDD-VGS2,'L',L));
```

Next, we find the widths by dividing I_D (50 μA) through the current densities. The results are plotted in Figure 5.4(a) versus $(g_m/I_D)_2$, along with the other transconductance efficiencies in Figure 5.4(b).

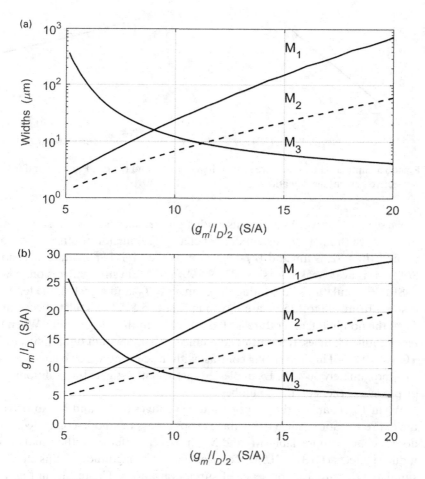

Figure 5.4 Widths and transconductance efficiencies of the current reference circuit versus the g_m/I_D of M_2. The nominal current is set to 50 μA.

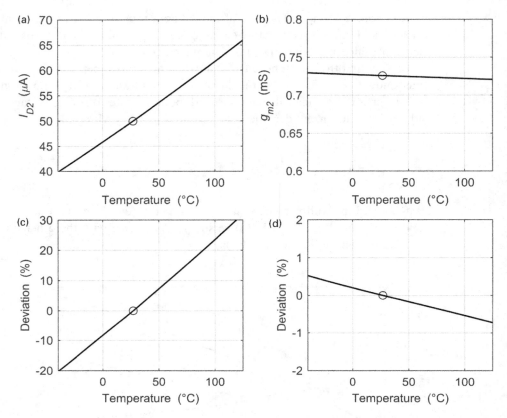

Figure 5.5 Simulated temperature sensitivity of (a) I_{D2} and (b) g_{m2}. Plots (c) and (d) show the respective percentage deviations.

We see that the widths and transconductance efficiencies span relatively wide ranges, even though the current doesn't change. Taking for instance $W_2 = 15$ μm, we find $W_1 = 82.59$ μm and $W_3 = 6.99$ μm. While M_2 is in moderate inversion (13.29 S/A), M_1 is near weak inversion (21.59 S/A) and M_3 in strong inversion (6.86 S/A).

SPICE simulations illustrating the changes in I_{D2} and g_{m2} when the temperature is swept from –40 to 125 °C are shown in Figure 5.5. The simulated current I_{D2} at the nominal temperature is 50.6 μA, close to the design value. When the temperature changes, the transconductance g_{m2} stays within tight limits, between +0.5 to –0.8%. This is in strong contrast with the current, which changes from –20% to approximately 30%. In Example 5.10, we will investigate the additional influence of process corners on these numbers.

When V_{DD} changes, the drain-to-source voltages of M_1 and M_4 absorb almost the entire variation, since the voltage drops across the diode-connected transistors do not change appreciably. M_1 and M_4, however, suffer from the usual drain-induced effects (DIBL, CLM). To investigate the magnitude of this effect, we simulated the circuit in presence of supply variations. The results in Figure 5.6

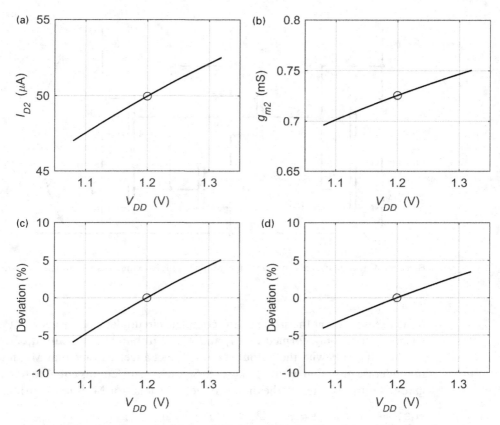

Figure 5.6 Supply voltage sensitivity of (a) I_{D2} and (b) g_{m2}. Plots (c) and (d) show the respective percentage deviations.

indicate changes in the current and the transconductance of about $\pm 10\%$. Cascoding M_1 and M_4 is one way to reduce the V_{DD} dependence.

5.2 High-Swing Cascoded Current Mirror

Figure 5.7 shows a current mirror that achieves both high output resistance and large output swing [3]. The core of the circuit is the current mirror consisting of M_1 and M_3. The input current I_{in} feeds the diode-connected transistor M_1, while M_3 delivers the output current I_{out}, nominally identical to I_{in}. M_4 serves to reduce the impact of output voltage changes on the current, while the purpose of M_2 is to equalize V_{DS3} and V_{DS4} to minimize the systematic mismatch between I_{out} and I_{in}. Lastly, M_6 and M_7 "compute" the cascode bias voltage such that it tracks process and temperature variations.

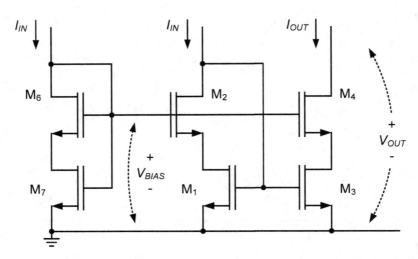

Figure 5.7 Cascode current mirror ($M_1 - M_4$) with high-swing bias circuit ($M_6 - M_7$).

Before we look at the sizing of the complete circuit, let us have a look at the core current mirror only, formed by M_1 and M_3. Without M_4, the output current will vary significantly with the output voltage. To get a feel, assume that M_1 and M_3 are in moderate inversion with $(g_m/I_D)_1 = 20$ S/A with gate lengths equal to 0.5 μm. The gate-to-source voltage of the diode-connected transistor M_1 is found using:

```
VGS1 = lookupVGS(nch,'GM_ID',gmID1,'L',L);
VGS1 = lookupVGS(nch,'GM_ID',gmID1,'VDS',VGS1,'L',L);
```

Note that the evaluation of V_{GS1} is done in two steps, since we do not know the drain voltage a priori. We begin with the default value 0.6 V and then make use of the obtained estimate to perform an iteration. We don't need additional iterations since the gate-to-source voltage is only a weak function of V_{DS}. We find that the gate-to-source voltage of M_1 is equal to 0.4380 V. We now compute the drain current density J_{D1}:

```
JD1 = lookup(nch,'ID_W','VGS',VGS1,'VDS',VDS1,'L',L);
```

Let's assume that the input current I_{in} is equal to 100 μA. Dividing I_{in} by J_{D1}, we then obtain $W_1 = 121$ μm). We take the same width for M_3 and now look at I_{D3} as a function of the output voltage $V_{OUT} = V_{DS3}$:

```
ID3 = W1*lookup(nch,'ID_W','VGS',VGS1,'VDS',VDS3,'L',L);
```

Figure 5.8(a) shows the current as the output voltage is swept from zero to 1.2 V. We see that below the drain saturation voltage V_{Dsat1} (equal to 0.1 V per (2.34)) the current drops rapidly, whereas beyond, it varies by about ±10% with respect to the reference value. When the output voltage is equal to V_{GS1}, the output current equals I_{in} (marked by a circle).

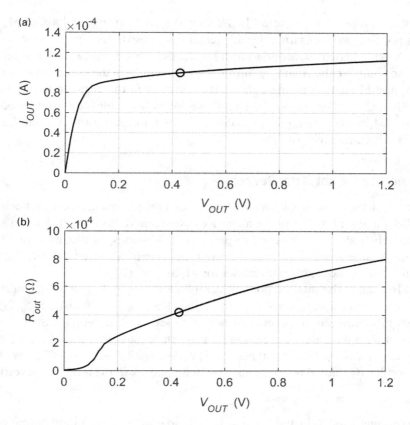

Figure 5.8 (a) Drain current and (b) output resistance of the two-transistor current mirror M_1 and M_3 in the absence of the cascodes. When V_{OUT} is equal to V_{GS1} (circle), the in- and output currents are equal to 100 μA.

In Figure 5.8(b), we see the output resistance increasing from 20 to 80 kΩ as the voltage changes from 0.2 to 1.2 V. If we divide the current density J_{D1} by the normalized output conductance g_{ds1}/W, we obtain the Early voltage, V_{EA1}, as defined in Section 2.3.6.

```
gds1 = lookup(nch,'GDS_W','VGS',VGS1,'VDS',VDS1,'L',L);
VEA1 = JD1./gds1
```

We find that $V_{EA1} = 4.19$ V. For many practical circuits, this value is not large enough. Adding the cascode device M_4, as shown in Figure 5.7 is the most obvious way to increase the output resistance. The price to pay, however, is an increase of the compliance voltage,[1] owing to the series stacking of M_3 and M_4. To achieve the largest possible output swing (smallest possible compliance voltage), we must minimize the

---

[1] The compliance voltage is the lowest voltage that the output terminal can accept before the output resistance of the current mirror starts to decrease sharply.

voltage drops across M_3 and M_4. We therefore lower the gate voltage of M_4 as much as possible, while ensuring that M_3 remains saturated. The drain voltage excursions of M_3 caused by V_{OUT} are now much smaller than in the non-cascoded current mirror.

To complete the circuit, we introduce M_2 to equalize the drain-source voltages of M_1 and M_3. This minimizes the systematic error in the mirror ratio, as mentioned earlier. Given this augmented circuit, we will now establish the sizing procedure of $M_1 - M_4$ while looking for two objectives: high output resistance and the smallest possible compliance voltage.

5.2.1   Sizing the Current Mirror Devices

In the sizing example that will follow, we assume reasonably long channels for simplicity and set $L = 500$ nm for all devices. Note that the usable gate lengths are sometimes bounded by frequency response considerations, as discussed in Chapter 3. Furthermore, we consider the same output current as above (100 μA) and keep the same transconductance efficiencies for M_1 and M_3 (20 S/A).

To achieve the smallest possible compliance voltage, V_{DS3} and V_{DS1} must be close to the drain saturation voltages. With the chosen g_m/I_D (20 S/A), V_{Dsat1} is equal to 100 mV. However, in practice, it is best to set the drain voltage of M_1 somewhat larger than this value. The reason for this can be gleaned from Figure 5.8(b), which shows that the output resistance of M_3 varies rapidly around its V_{Dsat}. We therefore consider V_{DS1} equal to V_{Dsat} plus a small margin V_X in the following investigation.

```
VDS1 = 2./gmID1 + Vx;
```

Assuming $V_X = 0$, 50 and 100 mV, we will now study the resulting output resistance of the mirror core (M_1 and M_3) and the overall output resistance of the complete circuit. We begin by computing the Early voltage (V_{EA1}) of the mirror core:

```
VGS1 = lookupVGS(nch,'GM_ID',gmID1,'VDS',VDS1,'L',L);
VEA1 = lookup(nch,'ID_GDS','VGS',VGS1,'VDS',VDS1,'L',L);
```

When V_X is equal to zero, the Early voltage is only 0.556 V and the output resistance is 0.556 V/100 μA = 5.56 kΩ. Setting V_X to 50 mV or 100 mV boosts the Early voltage to 1.733 or 2.285 V, respectively. Even though the large Early voltage for $V_X = 100$ mV seems attractive at first glance, we will conclude that $V_X = 50$ mV offers a better compromise with only 150 mV voltage drop across M_3.

The required gate bias voltage V_{BIAS} (see Figure 5.7) is the sum of V_{GS2} and V_{DS1}. To find V_{GS2}, we interpolate to find a design point that makes J_{D1} equal to J_{D2} (since $I_{D1} = I_{D2}$ and we want and $W_1 = W_2$). The interpolation is done within a small search range (S) added to V_{GS1}, which is known to be somewhat smaller than V_{GS2}.

```
JD1  = lookup(nch,'ID_W','VGS',VGS1,'VDS',VDS1,'L',L);
S    = .001*(0:50);
JD2  = lookup(nch,'ID_W','VGS',UGS1+S,'VDS',...
VGS1-VDS1, 'VSB',VDS1,'L',L);
VGS2 = interp1(JD2/JD1,VGS1+S,1);
VBIAS = VDS1 + VGS2;
```

Table 5.1 Current mirror parameters as a function of the margin V_X.

V_X (mV)	0	50	100
V_{DSI} (mV)	100	150	200
V_{GSI} (mV)	427	430	430
V_{BIAS} (mV)	538	602	662
W (µm)	146.2	131.4	128.0

Table 5.2 Early voltage of the two-transistor core (V_{EA1}) and the four-transistor circuit (V_{EA}) for the three considered margins V_X.

V_X (mV)	0	50	100
V_{EA1} (V)	0.5559	1.7286	2.2846
A_4	87.25	71.31	59.95
V_{EA} (V)	48.51	123.26	136.97
R_{out} (MΩ)	0.49	1.23	1.37

Finally, we divide I_{in} by J_{D1} to find the width of the four transistors:

```
W  = Iin/JD1;
```

A summary of the resulting design parameters is give in Table 5.1, considering the three possible margins.

We now compute the Early voltage V_{EA} of the four-transistor circuit. When the input and output voltages of the current mirror are equal, all we need to do is to multiply the Early voltage of the mirror core (V_{EA1}) by the gain (A_4) of the cascode stage:

$$A_4 = \frac{\dfrac{g_{m4}}{I_D} + \dfrac{g_{mb4}}{I_D} + \dfrac{g_{ds4}}{I_D} + \dfrac{1}{V_{EA1}}}{\dfrac{g_{ds4}}{I_D}}. \tag{5.8}$$

From here, we can find the overall output resistance R_{out} by dividing the global Early voltage (V_{EA}) by the current, leading to the data listed in Table 5.2. The zero-margin implementation clearly represents the worst option, given the small output resistance of the core current mirror. The choice of $V_X = 50$ mV represents a good compromise between compliance voltage and output resistance.

5.2.2 Sizing the Cascode Bias Circuit

The bias voltage V_{BIAS}, which sets the gate potential of M_2 and M_4, is defined by the diode-connected stack formed by M_6 and M_7 (see Figure 5.7). Though a single transistor would suffice to fulfill the function, partitioning into two devices in series

Table 5.3 Width of M_7 versus the margin V_X.

V_X (V)	0	50	100
W_7 (µm)	29.02	13.78	8.13

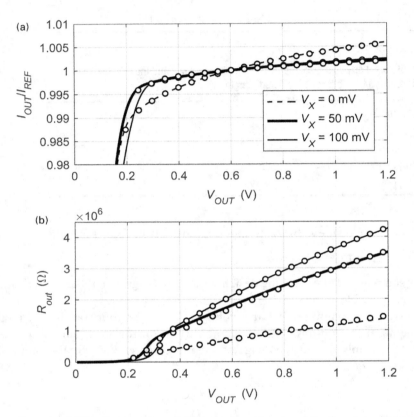

Figure 5.9 (a) Drain currents and (b) output resistances of the four-transistor current mirror, considering margins of $V_X = 0$, 50 and 100 mV. The output current is normalized to the value at $V_{OUT} = 0.6$ V (I_{REF}). Circles represent analytical predictions; the solid lines are from SPICE simulations.

is preferable, owing to the improved matching obtained when M_6 is a replica of M_2 [2]. The fact that the drain voltages of M_6 and M_2 are not identical leads to a small (and negligible) source voltage difference. The final parameter to compute is the width of M_7, which operates in the triode region. To find W_7, we evaluate the current density:

```
JD7 = lookup(nch,'ID_W','VGS',Vbias,'VDS',VDS1,'L',L);
```

Next, we can compute the width from the ratio I_{In}/J_{D7}. Table 5.3 lists W_7 for the three margins considered.

The solid lines in Figure 5.9(a) illustrate SPICE simulations of the output current versus V_{OUT}. The currents are normalized to the value at 0.6 V output voltage. When the margin V_X is 50 mV or larger, the output current shows less than 0.2% variation for V_{OUT} ranging from 0.3 to 1.2 V. The fact that the zero-margin curve departs considerably from the two other curves confirms the typical guideline that V_{DS1} should not be smaller than V_{Dsat1} + 50 mV [3]. The improvement that the 100 mV margin offers is not worth pursuing, given the associated increase of the compliance voltage.

Figure 5.9(b) shows the simulated output resistances, showing similar trends for all margins. When V_x is 50 mV or larger, R_{out} increases by a factor of three to almost four compared to the zero-margin case. When V_{OUT} is equal to V_{GS1} (0.43 V), the results agree with the resistances predicted in Table 5.2.

The circles in Figure 5.9 represent the currents and output resistances obtained analytically. To find these points, we evaluate the impact of the output voltage V_{OUT} on the drain voltage of M_3 and compute the actual change in the output current. The calculation is done by means of an interpolation equalizing the current densities of M_3 and M_4 as done above.

A key take-home from the above example is that the transconductance efficiency plus margin (V_X) controls the entire sizing procedure. Regardless of the current, the compliance voltage is set by choosing g_m/I_D. The widths follow from the desired current and the device current density at the chosen g_m/I_D.

5.3   Low-Dropout Voltage Regulator

Voltage regulators are used whenever a device or a sub-circuit requires a well-controlled power supply derived from the main supply. The voltage is lowered by means of a series device and possibly a shunt transistor in parallel with the load. The objective of the regulator is not only to keep the output voltage at a well-defined value, but also to shield the output against any noise present on the main supply. The challenge is to make the dropout voltage (the voltage drop across the series transistor) as small as possible, leading to the notion of a low-dropout regulator (LDO).

The common-drain configuration in Figure 5.10(a), is typically not suitable for LDOs, because the dropout voltage is of the order of the threshold voltage, which is large compared to the supply voltage of a modern CMOS technology. The common-source configuration in Figure 5.10(b) is more appropriate, since the dropout voltage can be as small as the drain saturation voltage. Let's consider a concrete example with a supply voltage (V_{DD}) of 1.2 V and a load current of $I_D = 10$ mA. Figure 5.11 plots the required device widths versus the regulated output voltage (V_{OUT}) and assuming minimum-length ($L = 60$ nm) devices. We see that the CD configuration (dashed curve) is not practical, since dropout voltages below 0.4 V require unacceptably large transistors. For the common-source transistor (solid lines), the widths are more reasonable, even with transconductance efficiencies as

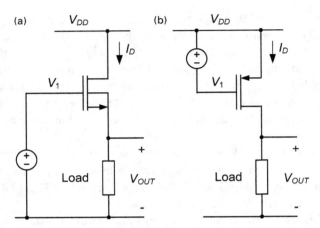

Figure 5.10 Options for series device: (a) common drain n-channel, and (b) common-source p-channel.

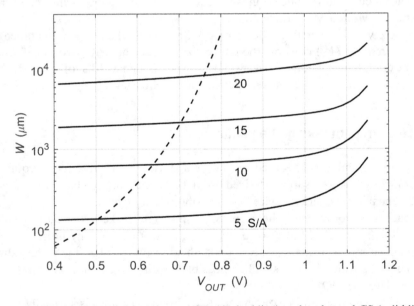

Figure 5.11 Widths of the n-channel CD (dashed line) and p-channel CS (solid lines) series transistors versus the desired regulated output voltage. The supply voltage is 1.2 V, the output current is 10 mA, and $L = 60$ nm. The g_m/I_D of the CS transistor varies from 5 to 20 S/A and the gate voltage of the CD device (V_1) is set to 1.2 V.

large as 15 S/A (moderate inversion). Note also that the CS device widths don't change much over a large range of dropout voltages. Based on this comparison, we consider only the CS configuration from now on.

Another important specification of LDOs is the rejection of power supply noise, caused for example by switching transients in surrounding digital blocks. The series

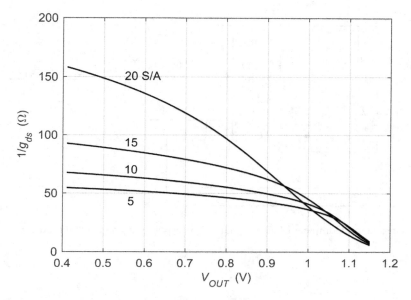

Figure 5.12 Resistance of the p-channel series transistor versus regulator output voltage and for various g_m/I_D, ranging from 5 to 20 S/A.

transistor and the output load of the LDO form a voltage divider that transfers a fraction of this noise to the output terminal. As shown in Figure 5.12, the resistance of the CS series transistor can be relatively small, and on the same order as the load resistance R_L (equal to 0.9 V/10 mA = 90 Ω in our example). For a dropout voltage of 0.3 V, the plot indicates series resistances between 40 and 70 Ω, depending on g_m/I_D.

To quantify the rejection of supply noise we introduce the concept of power supply rejection (PSR) [3]. The PSR is defined as the ratio of supply variation and output variation. For the circuit in Figure 5.10(b), we find:

$$PSR_{OL} = \frac{v_{dd}}{v_{out}} = \frac{Y_L + g_{ds}}{g_{ds}}, \tag{5.9}$$

where the subscript "OL" stands for open-loop, i.e. no regulation is applied to the series device. From the above numbers, it follows that the PSR of an unregulated circuit will be of the order of 2 to 3. A feedback loop is required to improve the performance.

Figure 5.13(a) shows a commonly used LDO topology containing a feedback loop and a differential amplifier that senses the difference between the desired voltage output (V_{REF}) and the actual V_{OUT}. Figure 5.13(b) shows the corresponding small-signal model. The differential amplifier boils down to a voltage controlled current source $g_{ma}(v_{out} - v_{ref})$ with an output conductance of g_{dsa}. Both are connected to the power supply line, as discussed in [4]. The circuit in the dashed box represents the series transistor M_1. The output load consists of the parallel combination of the conductance Y_L and the capacitance C_L.

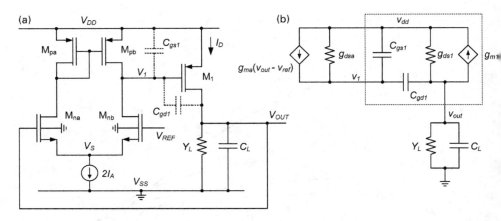

Figure 5.13 (a) Complete LDO circuit. (b) Corresponding small-signal model.

5.3.1   Low-Frequency Analysis

First, let us focus on the low-frequency behavior of the LDO, ignoring all capacitors in the small-signal model. The regulator can then be analyzed as a typical series-shunt feedback loop [3] with a feedback transfer function $f = 1$. The closed-loop transfer function of the LDO is:

$$\frac{v_{out}}{v_{ref}} = \frac{A_1 A_a}{1 + A_1 A_a},$$

(5.10)

where A_1 and A_a are the voltage gains of the CS series transistor and the differential amplifier, respectively, and their product is the loop gain. With large loop gain, the closed-loop transfer function approaches unity. Further analysis shows that the low-frequency PSR is:

$$PSR = \frac{Y_L + g_{ds1} + g_{m1} A_a}{g_{ds1}} = PSR_{OL} + \left(\frac{g_m}{g_{ds}}\right)_1 A_a \cong \left(\frac{g_m}{g_{ds}}\right)_1 A_a.$$

(5.11)

The next aspect that we will consider is the regulator's output resistance, which is given by:

$$R_{out} = \frac{v_{out}}{i_{load}} \cong \frac{1}{g_{m1} A_a}.$$

(5.12)

The output resistance also sets the so-called "load regulation," which characterizes the change in the output voltage as the LDO load current is varied ($\Delta V_{OUT}/\Delta I_{LOAD}$, often specified as a ratio of percentages). With these preliminaries in place, we will now look at a sizing example.

Example 5.2 Basic Sizing of an LDO

Consider the LDO of Figure 5.13, with a DC load current of 10 mA, $V_{OUT} = 0.9$ V and $V_{DD} = 1.2$ V. Size the series transistor and the differential amplifier to maximize the circuit's loop gain. Compute the LDO's PSR and output resistance. Validate the design in SPICE.

SOLUTION

We define two primary sizing parameters: $(g_m/I_D)_1$, the transconductance efficiency of the series transistor M_1 (gm_ID in the Matlab code below) and $(g_m/I_D)_n$, the transconductance efficiency of the differential amplifier (gm_IDn). We start with the sizing of the series transistor M_1. The low-frequency loop gain contributed by this device is:

$$A_1 = \frac{g_{m1}}{Y_L + g_{ds1}} = \frac{\left(\dfrac{g_m}{I_D}\right)_1}{\dfrac{1}{V_{OUT}} + \left(\dfrac{g_{ds}}{I_D}\right)_1}.$$

We can find g_{ds}/I_D via g_m/I_D through the lookup function command below:[2]

```
gds_ID = lookup(pch,'GDS_ID','GM_ID',gm_ID,'VDS',VDD-V,'L',L);
```

According to (2.34), $(g_m/I_D)_1$ must be larger than 6.6 S/A to keep the V_{Dsat} of M_1 smaller than the dropout voltage (0.3 V). To find the width W_1, we divide the drain current by the corresponding current density:

```
JD = lookup(pch,'ID_W','GM_ID',gm_ID,'VDS',VDD-V,'L',L);
```

Table 5.4 enumerates the gains that can be achieved for $(g_m/I_D)_1$ values between 7 S/A and 12 S/A. We don't consider larger transconductance efficiencies since the widths become impractical. The top entries in each cell are for $L_1 = 100$ nm, the bottom entries for $L_1 = 200$ nm. We also list the gate-to-source voltage V_{GS1}, the gate-to-source and gate-to-drain capacitances C_{gs1} and C_{gd1}, since we will need these later. With $L_1 = 100$ nm, the gain is around 5 for $W \cong 1000$ μm. Going to 200 nm improves the gain very little, and is very costly as far as layout area is concerned.
    We now consider the differential amplifier. Its voltage gain is given by:

$$A_a = \frac{\left(\dfrac{g_m}{I_D}\right)_n}{\left(\dfrac{g_{ds}}{I_D}\right)_n + \left(\dfrac{g_{ds}}{I_D}\right)_p}.$$

[2] The full sizing procedure (including code for Example 5.3) is available in the Matlab file Sizing_LDO1.m.

Table 5.4 Sizing data for the CS series transistor M_1. The top entries in each cell are for $L_1 = 100$ nm, the bottom entries for $L_1 = 200$ nm.

$(g_m/I_D)_1$ (S/A)	7	8	9	10	11	12
A_1	3.37	3.86	4.33	4.78	5.20	5.61
	3.75	4.31	4.84	5.36	5.88	6.39
W_1 (μm)	419	544	700	890	1122	1402
	721	943	1216	1543	1934	2396
V_{GS1} (V)	0.7266	0.6916	0.6620	0.6366	0.6146	0.5952
	0.6978	0.6619	0.6320	0.6069	0.5853	0.5667
C_{gs1} (pF)	0.393	0.497	0.622	0.770	0.942	1.143
	1.190	1.514	1.896	2.337	2.841	3.411
C_{gd1} (pF)	0.170	0.213	0.267	0.334	0.415	0.513
	0.330	0.408	0.506	0.625	0.768	0.938

The drain-to-source and the gate-to-source voltages of the p-channel mirror devices are fixed by V_{GS1} and hence the gate length is the only degree of freedom for their sizing. Since we value gain more than bandwidth in this sub-circuit, we opt for 0.5 μm channel length. This allows us to find $(g_{ds}/I_D)_p$:

```
gds_IDp = diag(lookup(pch,'GDS_ID','VGS',VGS,'VDS',VGS,'L',Lp));
```

Consider now the differential pair. Owing to the low V_{DD}, the pair will have a rather small drain-to-source voltage, somewhere between 0.1 and 0.3 V. The source voltage V_S (see Figure 5.13) is equal to $(V_{OUT} - V_{GSn})$ and the drain voltage V_D is $(V_{DD} - V_{GS1})$. We turn the source voltage V_S into a variable and run the Matlab code below to evaluate $(g_m/I_D)_n$ and $(g_{ds}/I_D)_n$ (gm_IDn and gds_IDn, respectively). For high gain, we assume $L_n = 0.5$ μm in these calculations.

```
VS = .2: .02: .5;
for k = 1: length(VS),
  US = VS(k);
  gm_IDn(:,k)  = lookup(nch,'GM_ID','VGS',V-US, ...
  'VDS',VD-US,'VSB',US,'L',Ln);
  gds_IDn(:,k) = lookup(nch,'GDS_ID','VGS',V-US, ...
  'VDS',VD-US,'VSB',US,'L',Ln);
end
```

The voltage gain A_a of the differential pair stage then follows via:

```
Aa = gm_IDn./(gds_IDn + gds_IDp(:,ones(1,length(VS))))
```

Here, A_a is a matrix with its columns defined by V_D, and the rows defined by V_S. V_D is a function of V_{GS1}, and thus $(g_m/I_D)_1$, whereas V_S is controlled by $(g_m/I_D)_n$. Finally, the overall loop gain is found through the product of A_1 and A_a:

```
gain = Aa.*A1(:,ones(1,length(gm_ID)));
```

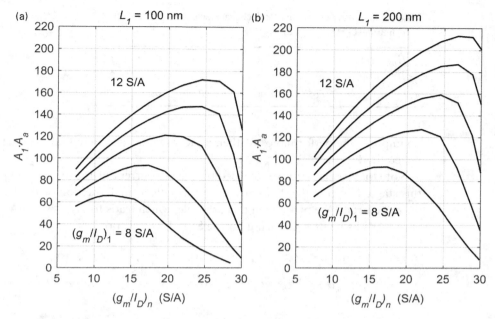

Figure 5.14 LDO loop gain versus the transconductance efficiency of the differential pair. (a) $L = 100$ nm, (b) $L = 200$ nm.

The results are plotted in Figure 5.14, with $(g_m/I_D)_n$ on the x-axis and $(g_m/I_D)_l$ as a parameter on the curves, and considering two choices for L_l. As expected, the loop gain increases with increasing transconductance efficiencies. However, a very rapid fall-off is visible when $(g_m/I_D)_n$ becomes large. This increases the tail node voltage V_S and brings V_{DSn} close to the saturation voltage. Consequently, there exists an optimal $(g_m/I_D)_n$ for every $(g_m/I_D)_l$. These values are summarized in Table 5.5 for $L_l = 100$ nm.

Based on the data in Table 5.4 and Table 5.5, we make the following design choices:

- Series transistor: $L_l = 100$ nm, $(g_m/I_D)_l = 10$ S/A.
- Differential amplifier: $L_p = L_n = 500$ nm with $(g_m/I_D)_n = 20$ S/A.

The loop gain is 120 ($A_1 = 4.78$ and $A_a = 25.2$), while V_S (obtained by means of an interpolation) equals 0.4040 V. For the power supply rejection, we find:

$$PSR = PSR_{OL} + \left(\frac{g_m}{g_{ds}}\right)_1 A_a = 259.$$

Note that this number is about 48 dB larger than the PSR of the series device without regulation, which is only 2.13 per (5.9). The output resistance is:

$$R_{out} = \frac{v_{out}}{i_{load}} \cong \frac{1}{\left(\frac{g_m}{I_D}\right)_1 I_{LOAD} A_a} = 0.4\,\Omega.$$

Table 5.5 Summary of the transconductance efficiencies and loop gains at the maxima of Figure 5.14(a).

$(g_m/I_D)_l$ (S/A)	8	9	10	11	12
$(g_m/I_D)_n$ (S/A)	12	16	20	22	25
$A_1 A_a$	65	90	120	145	170

To improve the performance further, a more elaborate differential amplifier would be required.

Note that we have not yet decided on the tail current of the differential pair at this stage. The amplifier bias current is normally a small fraction of the LDO's output current. We pick 2% in this example, which leads to $2I_{Dn} = 0.2$ mA. This choice lets us finalize the sizing for the SPICE verification:

```
JDp = lookup(pch,'ID_W','VGS',VGS,'VDS',VGS,'L',Lp);
Wp  = IDn/JDp;
JDn = lookup(nch,'ID_W','VGS',V-VS,'VDS', VD-VS,'VSB',VS,'L',Ln);
Wn  = IDn/JDn;
```

The resulting widths are $W_p = 20.43$ μm and $W_n = 127.1$ μm. The transconductance g_{ma} and the output conductance g_{dsa} of the differential pair are found via:

```
gma  = gm_ID_n*IDn
gdsa = (gds_IDp + gds_IDn)*IDn
```

We obtain $g_{ma} = 2$ mS and $g_{dsa} = 79.2$ μS (12.6 kΩ). Table 5.6 compares the obtained parameters with a SPICE simulation. The predicted and simulated numbers are in good agreement.

5.3.2 High-Frequency Analysis

We will now investigate the frequency behavior of the LDO. The LDO's loop amplifier resembles a two-stage Miller topology, having a dominant pole p_1, a non-dominant pole p_2, and a right half-plane zero z, due to the gate-to-drain capacitance C_{gdl} of the series transistor. Detailed analysis reveals the following expressions for the poles and the zero of the open-loop amplifier:

$$p_1 \cong -\frac{1}{R_1\left(C_{gs1}+C_{gd1}\left(1+A_1\right)\right)+R_2\left(C_{gd1}+C_L\right)}$$

$$p_2 \cong -\frac{R_1\left(C_{gs1}+C_{gd1}\left(1+A_1\right)\right)+R_2\left(C_{gd1}+C_L\right)}{R_1 R_2\left(C_{gs1}C_{gd1}+C_L\left(C_{gs1}+C_{gd1}\right)\right)} \qquad (5.13)$$

$$z = +\frac{g_{m1}}{C_{gd1}}.$$

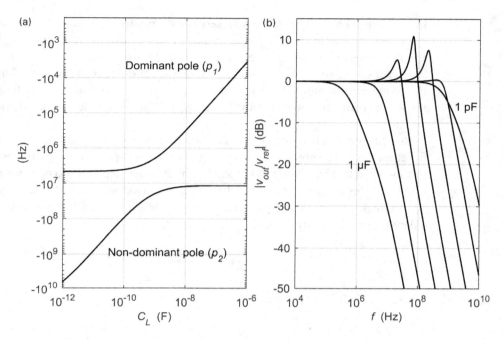

Figure 5.15 (a) Dominant and non-dominant pole frequencies of the LDO as C_L is varied from 1 pF to 1 μF. (b) Closed-loop frequency response considering the same range of C_L values (step size of 10x).

Table 5.6 SPICE verification data.

	V_S (mV)	V_{DS} (mV)	A_1	A_a	PSR	R_{out} (Ω)
Analysis	404	159.4	4.78	25.25	259	0.40
SPICE	403.4	159.4	4.78	25.7	257	0.39

In these expressions, R_1 is the output resistance $1/g_{dsa}$ of the differential amplifier and R_2 is the resistance $1/(Y_L + g_{ds1})$ at the output node. The zero can be ignored, since it lies far outside the frequency range of interest. The dominant pole p_1 contains two time constants in its denominator. The first is due to C_{gs} and C_{gd} of M_1, where the latter is Miller-amplified. The second time constant is due to R_2 and the load capacitance C_L, which can be large.

By changing the size of C_L, we can vary the second time constant and make it larger or smaller than the first time constant. The effect on the frequency response is illustrated in Figure 5.15, assuming the circuit designed in Example 5.2. Plot (a) shows the two poles as C_L is swept from 1 pF up till 1 μF. The dominant pole doesn't vary much when C_L is smaller than 0.1 nF, since the main time constant is set by the amplifier's internal node. However, the non-dominant pole grows steadily with the load capacitance. When C_L gets larger than 10 nF, the trends interchange. The output node now sets the dominant pole, whereas the internal node is responsible for the non-dominant pole. Consequently, when C_L is very small or very large, the LDO behaves like a first-order circuit, but when the two poles get close, the amplifier behaves like a second-order system that has poor phase margin. In plot (b), we see the consequences on the closed-loop frequency response (transfer function from v_{ref} to v_{out}). When the poles get close, peaking appears.

Building on this insight, we now consider the frequency response of the transfer function from the supply (v_{dd}) to the regulator output (v_{out}). This metric resembles the so-called line regulation ($\Delta V_{OUT}/\Delta V_{DD}$, often specified as a ratio of percentages). Note that this is the inverse of the PSR that we introduced earlier. It is more intuitive to work with transfer, rather than rejection, when frequency dependence is considered. Detailed analysis yields:

$$\frac{v_{out}}{v_{dd}} = \frac{N_2 s^2 + N_1 s + N_0}{D_2 s^2 + D_1 s + D_0},$$

where:

$$
\begin{aligned}
N_2 &= C_{gs1} C_{gd1} \\
N_1 &= C_{gs1} g_{ds1} + C_{gd1} \left(g_{m1} + g_{ds1} + g_{dsa} \right) \\
N_0 &= g_{ds1} g_{dsa} \\
D_2 &= C_L \left(C_{gs1} + C_{gd1} \right) + C_{gs1} C_{gd1} \\
D_1 &= C_L g_{dsa} + \left[\left(Y_L + g_{ds1} \right) \left(C_{gs1} + C_{gd1} \right) + C_{gd1} \left(g_{m1} + g_{dsa} - g_{ma} \right) \right] \\
D_0 &= \left(Y_L + g_{ds1} \right) g_{dsa} + g_{m1} g_{ma}.
\end{aligned}
\tag{5.14}
$$

Figure 5.16 plots the magnitude and phase of the above transfer function for various values of C_L, and assuming the sizing parameters from Example 5.2. At low frequencies (within the bandwidth of the feedback amplifier), the magnitude is equal to -48 dB, the reciprocal of PSR computed in Example 5.2. For $C_L = 10$ fF (negligibly small load capacitance), the magnitude begins to rise near the corner frequency of the amplifier, and it ultimately approaches a value near unity, due to the feedthrough via C_{gs1} and C_{gd1}. Increasing C_L helps counter this feedthrough and leads to a high-frequency roll-off and a peak at intermediate frequencies. In the extreme case of $C_L = 1$ µF, the magnitude decreases monotonically without peaking, while the peak still exists for $C_L = 10$ nF.

An attractive design point is the value for C_L that leads to a flat response (no peaking). To find this optimal value for C_L, we can inspect the poles and zeros of (5.14). The two poles depend on C_L. As the capacitance increases, they change from real to complex conjugate poles (which causes peaking) before they turn real again. On the other hand, the zeros do not depend on C_L. As already mentioned, we can neglect the zero associated with the gate-to-drain capacitance

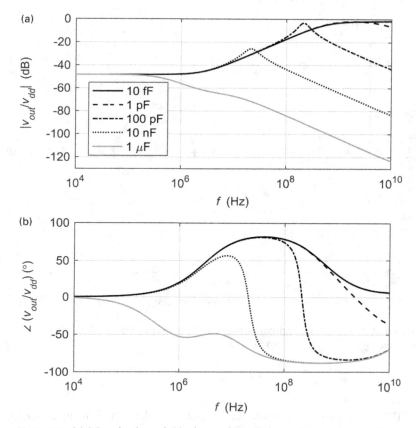

Figure 5.16 (a) Magnitude and (b) phase of (5.14) for various values of C_L.

feedthrough. The extraction of the other zero in the numerator of (5.14) is then straightforward:

$$|z| \cong \frac{g_{da}}{C_{gs1} + C_{gd1}\left(1 + \dfrac{g_{m1}}{g_{ds1}}\right)}. \tag{5.15}$$

A suitable design strategy is to design for real poles and place one of the poles on top of the zero. The remaining pole then sets the LDO's cutoff frequency. We will illustrate this approach in the following example.

Example 5.3 Sizing the LDO's Load Capacitance

Find the load capacitance that cancels the significant zero of the LDO's regulation factor circuit as described above. Compare the predicted transfer function with

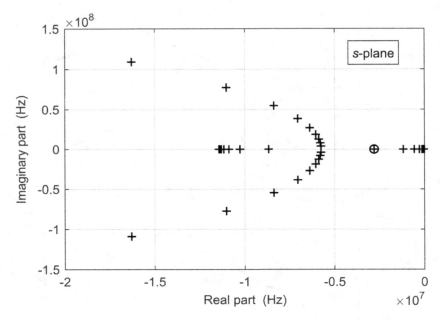

Figure 5.17 Locus of the regulation poles as C_L is swept. Also shown is the relevant zero (marked by "o"), given by (5.15).

a SPICE simulation. Examine the impact of slight variations in the differential amplifier's output conductance g_{dsa} on the frequency response.

SOLUTION

Figure 5.17 plots the locus of the two poles (marked by "+" symbols) in (5.14) as a function of C_L. Also shown is the position of the significant zero (marked by the "o" symbol). At the point where the poles enter the real axis, we have $C_L = 140$ nF. Increasing C_L further to 191 nF places one of the poles on top of the zero. The remaining pole defines the LDO's cutoff frequency, which we compute to be 9.6 MHz.

Figure 5.18 shows the computed and SPICE-simulated frequency response for $C_L = 191$ nF. The 3-dB corner frequency of the SPICE-simulated magnitude is 8.67 MHz, which is close to the analytical prediction above. Two additional curves show what happens when g_{dsa} departs from its nominal value by ±20%. An interesting situation occurs when g_{dsa} is less than the nominal value. The low-frequency regulation factor improves slightly, owing to the increase of the regulator's low-frequency loop gain. Since the load capacitance is kept constant, a small bump appears near the cutoff frequency.

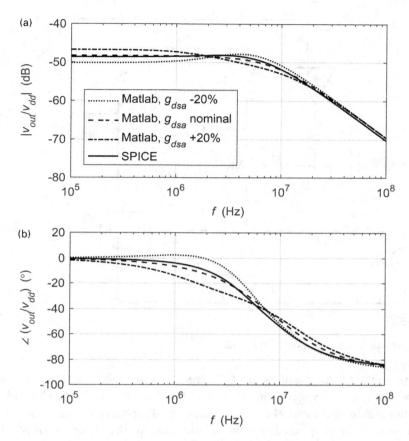

Figure 5.18 (a) Magnitude and (b) phase of (5.14) for $C_L = 191$ nF.

5.4   RF Low-Noise Amplifier

Figure 5.19 shows a simplified version of the radio frequency (RF) low-noise amplifier (LNA) described in [5]. The circuit not only amplifies, but also performs a single-ended input to differential output conversion (active balun). To achieve this, it combines two signal paths with opposite polarities via a common-gate (CG) and a common-source (CS) stage. Interestingly, this arrangement also cancels the thermal noise and distortion from M_1. The design flow discussed below therefore looks at multiple objectives:

1. achieving a symmetrical (balanced) output;
2. matching the input impedance;
3. minimizing the thermal noise contribution from M_2;
4. minimizing the distortion due to M_2.

How does the noise and distortion cancellation mechanism work? The thermal noise current of M_1 causes voltage noise at v_{in} and at the drain node of M_1. These noises are fully correlated with opposite polarity, and with a magnitude ratio that is

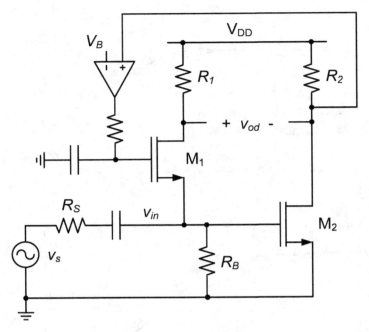

Figure 5.19 LNA circuit described in [5].

equal to the voltage gain of the CG stage. This noise at v_{in} is also amplified by the (negative) CS stage gain, and appears at the drain of M_2. If the CG and CS stage voltage gains are equal, the thermal noise of M_1 appears as common-mode output noise that can be ignored. The same holds true for the distortion of M_1, since distortion can be assimilated with an unwanted drain current component. The only remaining noise sources are those due to the CS stage and the resistors. Minimizing the thermal noise from M_2 is thus crucial for achieving a low-noise figure.

5.4.1   Sizing for Low-Noise Figure

We begin by focusing on the circuit's noise figure and make the following assumptions for our initial design. First, we choose the same gate lengths for both transistors, i.e. $L_1 = L_2 = 100$ nm. Next, we assume equal voltage drops across R_1 and R_2 (V_{R1} and V_{R2}), which leads to zero differential output at the quiescent point. Furthermore, we set these voltage drops to 0.4 V, which leads to a reasonable drain-source voltage for M_1. Note that $V_{DS1} = V_{DD} - V_{R1} - V_{GS2}$ must be kept above V_{Dsat1}. We will return to this point later.

The voltage gain of the CS stage is given by:

$$A_{v2} = -\frac{g_{m2}}{g_{ds2} + \dfrac{1}{R_2}} = -\frac{\left(\dfrac{g_m}{I_D}\right)_2}{\left(\dfrac{g_{ds}}{I_D}\right)_2 + \dfrac{1}{V_{R2}}}. \tag{5.16}$$

In this expression, $(g_m/I_D)_2$ is our main sizing parameter, and $(g_{ds}/I_D)_2$ follows from the choice of this inversion level:

```
gds_ID2 = lookup(nch,'GDS_ID','GM_ID',gm_ID2,'VDS',VDS2,'L',L);
```

V_{DS2} in the above command is set to $V_{DD} - V_{R2}$. We can also calculate the corresponding gate-to-source voltage V_{GS2}, the drain current density J_{D2} and normalized gate capacitance C_{gg}/W_2, as we will need these for later calculations:

```
VGS2   = lookupVGS(nch,'GM_ID',gm_ID2,'VDS',VDS2,'L',L);
JD2    = lookup(nch,'ID_W','GM_ID',gm_ID2,'VDS',VDS2,'L',L);
Cgg_W2 = lookup(nch,'CGG_W','GM_ID',gm_ID2,'VDS',VDS2,'L',L);
```

We consider now the CG stage, whose voltage gain is:

$$A_{v1} = \frac{\left(\dfrac{g_{ms}}{I_D}\right)_1 + \left(\dfrac{g_{ds}}{I_D}\right)_1}{\left(\dfrac{g_{ds}}{I_D}\right)_1 + \dfrac{1}{V_{R1}}}. \tag{5.17}$$

In this expression, g_{ms} is the source transconductance, equal to $g_m + g_{mb}$. Translated into Matlab code, this gives:

```
VDS1 = VDD - VR - VGS2;
for k = 1:length(gm_ID2),
   gm_W1(:,k)  = lookup(nch,'GM_W','VDS',VDS1(k),...
   'VSB',VGS2(k),'L',L);
   gmb_W1(:,k) = lookup(nch,'GMB_W','VDS', VDS1(k),...
   'VSB',VGS2(k),'L',L);
   gds_W1(:,k) = lookup(nch,'GDS_W','VDS', VDS1(k),...
   'VSB',VGS2(k),'L',L);
   JD1(:,k) = lookup(nch,'ID_W','VDS',VDS1(k),...
   'VSB',VGS2(k),'L',L);
end
A1 = (gm_W1 + gmb_W1+ gds_W1)./(gds_W1 + JD1/VR);
```

Because the input resistance of the LNA must match R_S (assumed to be 50 Ω) we also have:

$$\frac{1}{R_S} = \frac{1}{R_B} + \frac{A_{v1}}{R_1} = I_{D1}\left(\frac{1}{V_{GS2}} + \frac{|A_{v2}|}{V_{R1}}\right), \tag{5.18}$$

where the substitution of $|A_{v2}|$ for A_{v1} is justified since we are designing for a balanced output. We now know I_{D1}, since all other parameters in (5.18) are fixed. Having I_{D1}, we can compute R_1 and R_B and all pertinent parameters of M_1:

```
R1   = VR./ID1;
RB   = R1./(R1/RS - A2);
W1   = ID1./JD1
Css1 = W1.*Css1_W;
gm1  = W1.*gm1_W;
```

Table 5.7 LNA parameters versus the transconductance efficiency of M_2.

$(g_m/I_D)_2$ (S/A)	10	12	14	16	18	20
V_{GS2} (V)	0.5810	0.5370	0.5028	0.4752	0.4518	0.4311
$|A_{v2}|$	5.30	6.27	7.21	8.13	9.04	9.93
I_{D1} (mA)	1.895	1.625	1.428	1.277	1.158	1.060
R_1 (Ω)	158.3	184.6	210.1	234.9	259.2	283.0
R_B (Ω)	306.5	330.4	352.1	372.1	390.3	406.7
V_{GS1} (V)	0.6899	0.6387	0.5984	0.5658	0.5383	0.5145
W_1 (μm)	63.6	83.2	109.8	141.5	183.2	237.2
C_{ss1} (pF)	0.0659	0.0841	0.1082	0.1357	0.1704	0.2136

Using $A_{v1} = |A_{v2}|$, we find V_{GS1} by running an interpolation:

```
for k = 1:length(gm_ID2)
  VGS1(k,1) = interp1(A1(:,k), nch.VGS, A2(k));
end
```

Similar interpolations yield J_{D1} and C_{ss1}/W and so on. We list all variables determined so far in Table 5.7, considering a range of $(g_m/I_D)_2$.

The LNA's noise figure is set by I_{D2} and W_2, which we have not yet determined. To compute the noise figure, we add the noise factors from all contributors. In the Matlab code below, F1 is for the CG stage (ideally zero), F2 for the CS stage, and F3 and F4 account for the resistances R_1, R_2 and R_B. The variable A_v represents the voltage gain from v_{in} to v_{od}, i.e. $(A_{v1} + |A_{v2}|)$.

```
Denom = RS/4 * Av^2;
F1 = gam*gm1*(R1 - Av*RS/2)^2/Denom;
F2 = gam*gm2*(R2/(1 + gds2*R2))^2/Denom;
F3 = (R1 + R2)/Denom;
F4 = RS/RB;
F  = 1 + F1 + F2 + F3;
NF = 10*log10(F)
```

The variable gam captures the thermal noise factor of the MOSFET (see Section 4.1.1):

```
kB = 1.3806488e-23;
gam = lookup(nch, 'STH_GM', 'VGS', VGS2, 'L',L)/4/kB/nch.TEMP;
```

We illustrate the resulting sizing procedure for M_2 in the following example.

Example 5.4 Sizing the LNA for a Given Noise Figure

Size the circuit of Figure 5.19 so that its noise figure does not exceed 2.5 dB. Design for an input resistance of 50 Ω, $V_{DD} = 1.2$ V, $V_{R1} = V_{R2} = V_R = 0.4$ V and

$L = 100$ nm for both transistors. Use the data from Table 5.7 and validate the design using a SPICE simulation.

SOLUTION

We consider a set of I_{D2} that are multiples of I_{D1}. The resulting noise figures are plotted in Figure 5.20 versus the ratio I_{D2}/I_{D1}, and we consider the same $(g_m/I_D)_2$ values as in Table 5.7. We note that increasing I_{D2} improves the noise figure, regardless of the transconductance efficiency.

There are many ways to achieve NF ≤ 2.5 dB. To investigate further, we consider noise figures from 2.5 to 1.8 dB and find the corresponding $(g_m/I_D)_2$ and I_{D2}/I_{D1} pairs. Since we already know I_{D1}, we can compute and plot the absolute drain currents I_{D2} and widths W_2 for these points (see Figure 5.21). Notice that for every NF, one can identify a $(g_m/I_D)_2$ value that minimizes the width.

An important aspect in selecting the final design point is the bandwidth of the LNA. To get a feel for the RC product at the input node, we consider $R_{in} = R_S/2$ and C_{in}, the sum of C_{ss1} and C_{gg2} (neglecting the Miller effect, for simplicity). Figure 5.22 plots the resulting 3-dB frequency (f_{cin}), which shows that making $(g_m/I_D)_2$ large adversely affects the bandwidth of the LNA.

In Table 5.8, we consider an example taking NF equal to 2.3 dB and $(g_m/I_D)_2 = 14$ S/A. The table considers two alternative resistor voltage drops (V_R). Using $V_R = 0.5$ V enhances the gain, but is not recommended, since the drain voltage of the CG stage is too close to the drain saturation voltage. Also, the reduced transit frequency (f_{T1}) of the CG stage device would begin to have an impact on the

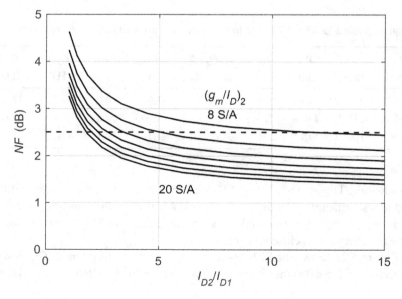

Figure 5.20 LNA noise figure versus I_{D2}/I_{D1} and for various $(g_m/I_D)_2$.

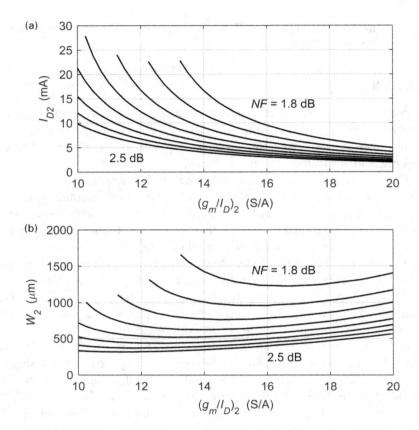

Figure 5.21 (a) I_{D2} and (b) W_2 versus $(g_m/I_D)_2$ for noise figures from 2.5 dB to 1.8 dB (0.1 dB steps).

Table 5.8 LNA design parameters for $NF = 2.3$ dB and $(g_m/I_D)_2 = 14$ S/A, considering various options for V_R.

V_R (V)	I_{D1} (mA)	I_{D2} (mA)	W_1 (μm)	W_2 (μm)	R_1 (Ω)	R_2 (Ω)	R_B (Ω)	V_{GS2} (V)	A_v	f_{cin} (GHz)	f_{T1} (GHz)	f_{T2} (GHz)
0.5	1.58	4.73	234.4	413.7	316	106	321	1.076	10.7	4.42	14.6	21.2
0.4	1.50	5.24	142.7	448.8	267	76.2	337	1.095	9.08	4.72	22.0	21.9
0.3	1.43	6.90	109.7	578.5	210	43.5	352	1.101	7.21	4.02	25.9	22.5

bandwidth. The case of $V_R = 0.3$ V does not suffer from these issues, but the gain reduces significantly. We decide on $V_R = 0.4$ V for our final design.

Table 5.9 compares the final design parameters with SPICE simulation data. The numbers are in close agreement.

Figure 5.23 shows the SPICE-simulated NF versus frequency. The design objective of 2.5 dB is met between 10 MHz and 2 GHz. Below 10 MHz, $1/f$ noise

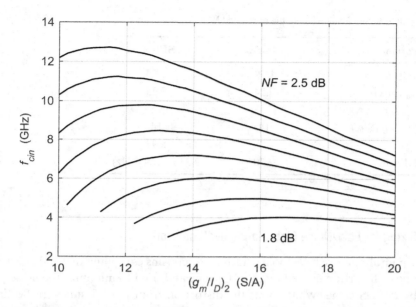

Figure 5.22 Approximate cutoff frequency caused by the input capacitance of the LNA.

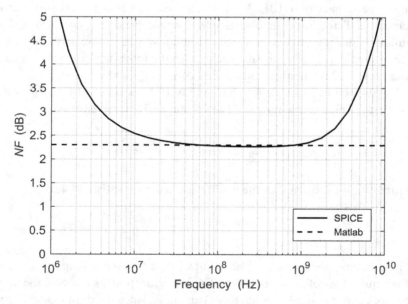

Figure 5.23 SPICE simulation of the noise figure.

takes over, while capacitive effects significantly deteriorate the noise figure above 2 GHz. The observed upper corner frequency can be increased by minimizing the device capacitances. We will show this in Example 5.5.

Table 5.9 SPICE validation of the designed LNA.

Parameter	Matlab	SPICE
I_{D1} (mA)	1.50	1.50
I_{D2} (mA)	5.24	5.25
$(g_m/I_D)_1$ (S/A)	13.49	13.55
$(g_m/I_D)_2$ (S/A)	14	14
A_v	9.08	8.94
NF (dB)	2.30	2.31 (at 0.9 GHz)

5.4.2 Sizing for Low-Noise Figure and Low Distortion

Up until now, we have looked at the mechanisms cancelling thermal noise and non-linear distortion caused by the CG stage and tried to minimize the noise produced by the CS stage. What about the distortion from the CS stage? In Section 4.2.2, we already dealt with this topic and pointed out that with short-channel devices, both the gate and drain nonlinearity contributions must be considered, owing to the large impact of DIBL. We also showed that it is possible to null fractional harmonic distortion components under well-defined bias conditions. Though the nulling requires stringent component tolerances (see Figure 4.19), we can still make use of partial cancellation near the null. In the following discussion, we will aim to minimize the HD_2 component of the LNA.

We showed in Example 4.5 that nulling HD_2 places the following constraint on the first-order coefficient (a_1) of the CS stage's Taylor expansion:[3]

$$a_1 = \frac{-x_{11} + \sqrt{x_{11}^2 - g_{m2} \cdot g_{ds2}}}{g_{ds2}}.$$ (5.19)

Moreover, we showed that the load resistance must satisfy (4.42):

$$Y = 1/R = -\frac{g_{m1}}{a_1} - g_{ds1}.$$ (5.20)

These same expressions apply here, but we note that the nulling in Examples 4.5 and 4.6 required V_R of 0.5 V or more. The larger drop helps increase the voltage gain, but we cannot make V_{R1} equal to V_{R2} anymore, since the drain-to-source voltage of M_1 ($V_{DD} - V_{GS2} - V_{R1}$) would be getting too small to keep it saturated. To accommodate this imbalance in the voltage drops, we can modify the design as shown in Figure 5.24 [5]. This circuit uses an AC coupling capacitance (C_{AC}) between the

---

[3] The index numbers in (5.19) and (5.20) refer to differentiation orders, not transistor labels. The latter are designated by indices outside parenthesis when needed.

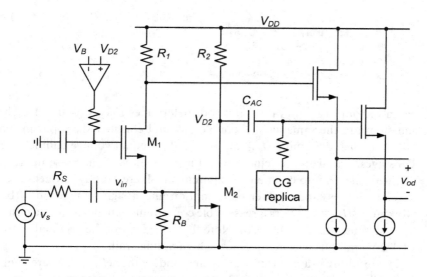

Figure 5.24 LNA circuit with AC coupling and output buffers [5].

LNA core and a set of output buffers. The box labeled "CG replica" is a sized copy of the CG stage, and it replicates the quiescent point voltage at the drain of M_1. This enables zero differential output at DC, despite the difference in V_{RI} and V_{R2}. In the example that follows, we will size this LNA for minimum second-order fractional harmonic distortion.

Example 5.5 Sizing the LNA for Minimum HD_2

Redesign the LNA from Example 5.4 to minimize HD_2, while maintaining a noise figure of 2.4 dB up to 2.4 GHz.

SOLUTION

To extend the low-noise figure to higher frequencies, we decide to shorten the channel lengths to 80 nm (from 100 nm). For all other design aspects, we follow the same steps as in Example 5.4, keeping in mind that V_{RI} and V_{R2} are now distinct quantities. We choose $V_{RI} = 0.3$ V so that a reasonable signal swing can be accommodated at the drain of M_1, leading to a large 1 dB compression point. The choice of V_{R2} sets V_{DS2}, and is fixed by the HD_2 nulling conditions. We use Matlab code similar to that in Example 4.5 to find the proper value:[4]

```
Vds = .2: .02: .64;
for k = 1:length(gm_ID2),
```

---

[4] The full sizing procedure is available in the Matlab file Sizing_LNA1.m.

```
UGS = lookupVGS(nch,'GM_ID',gm_ID2(k),'VDS',Vds,'L',L);
y   = blkm(nch,L,Vds,UGS);
A1  = (y(:,:,6)-sqrt(y(:,:,6).^2-y(:,:,4).*y(:,:,5)))./y(:,:,5);
U   = diag(y(:,:,3)./(y(:,:,1)./A1 - y(:,:,2)));
z(k,:) = interp1(VDD-Vds'-U, [UGS (VDD-U) diag(A1)], for 0);
end
```

For a vector of $(g_m/I_D)_2$ values, this code determines V_{DS2}, V_{GS2} and A_{v2}. The remaining steps are the same as in Example 5.4, and consider the input matching constraint (which determines I_{D1}) and the need for equal voltage gain for the CG and CS stages. The sizing concludes with the evaluation of the noise figure versus I_{D2}/I_{D1}, which leads into conflicting requirements. Increasing I_{D2} reduces the noise figure, but causes larger power consumption and a significant drop in bandwidth. Making $I_{D2}/I_{D1} = 11$ offers a reasonable compromise and leads to the design parameters summarized in Table 5.10. Note that V_{G1} is above the nominal supply voltage of 1.2 V and thus requires a separate high-voltage rail.

The resulting drain currents and transconductance efficiencies are compared to a SPICE simulation in Table 5.11. We observe close agreement.

The SPICE-simulated noise figure is plotted in Figure 5.25 as a function of frequency. At 2.4 GHz, the NF is better than the desired 2.4 dB.

As the last step of this investigation, we plot the second-order distortion as a function of input amplitude in Figure 5.26. At an input of –30 dBm (10 mV peak) we see $HD_2 = -90.8$ dB, which confirms that we are operating close to the HD_2 null. However, more generally, this plot reveals a common practical problem with distortion cancellation: it is effective only for small signals. For inputs larger than –15 dBm, uncancelled high-order terms lead to an increase in the second harmonic, and

Table 5.10 LNA design.

W_1	W_2	R_1	R_2	R_B	V_{G1}	V_{GS2}	V_{DS2}	f_{cin}	f_{T1}	f_{T2}
(µm)	(µm)	(Ω)	(Ω)	(Ω)	(V)	(V)	(V)	(GHz)	(GHz)	(GHz)
163.1	256.4	196.5	40.65	443.7	1.278	0.678	0.515	7.02	24.57	73.42

Table 5.11 SPICE validation of the designed LNA.

Parameter	Matlab	SPICE
I_{D1} [mA]	1.53	1.53
I_{D2} [mA]	16.8	16.8
$(g_m/I_D)_1$ (S/A)	14.26	14.27
$(g_m/I_D)_2$ (S/A)	7.38	7.34
A_v	6.98	6.96
NF (dB)	2.29	2.14 (at 1 GHz)
		2.21 (at 2.4 GHz)
IIP2 (dBm)	∞	37.7 dBm

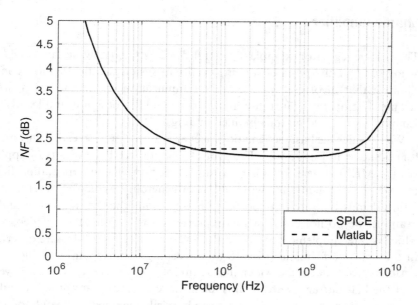

Figure 5.25 Simulated noise figure (SPICE).

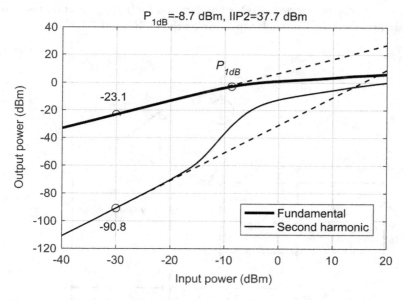

Figure 5.26 SPICE simulation of the distortion for a single-tone input at 100 MHz (the frequency dependence is weak in the band of interest).

thus the distortion cancellation does in fact not help much in accommodating large blocker signals (despite the very large second-order intercept point (IIP2) seen in the plot).

5.5   Charge Amplifier

The voltage regulator example of Section 5.3 illustrated how we can use g_m/I_D-based sizing to place the poles of a complex circuit systematically. We now consider an example that involves supply current minimization in presence of a noise constraint. The operational transconductance amplifier design examples discussed in Chapter 6 will build on the obtained findings.

We consider the basic schematic of a charge amplifier (also called current integrator) as shown in Figure 5.27. This type of circuit is used in a variety of applications, such as MEMS interfaces [6], X-ray detectors, photon counters and particle detectors [7]. For simplicity, and without loss of generality, we neglect several implementation details that are unimportant for the results that we want to establish. For example, we do not consider the implementation details of the wideband buffer, which is often implemented as a source follower. The buffer plays a non-critical role in the noise performance (due to the preceding voltage gain) and the bandwidth performance (due to the wideband nature of a follower). In addition, we assume that the bias current source (I_D) is ideal. In reality, this current source adds noise and parasitic capacitance, but these can be easily included and would only obscure the results in this introductory discussion. Finally, we do not consider the voltage biasing of the gate node (v_G). One basic option for biasing this node is to add a large resistor across the feedback capacitor, as shown in Figure 5.27.

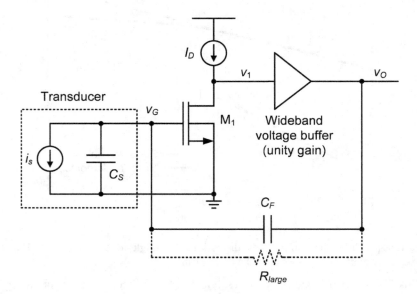

Figure 5.27 Charge amplifier circuit.

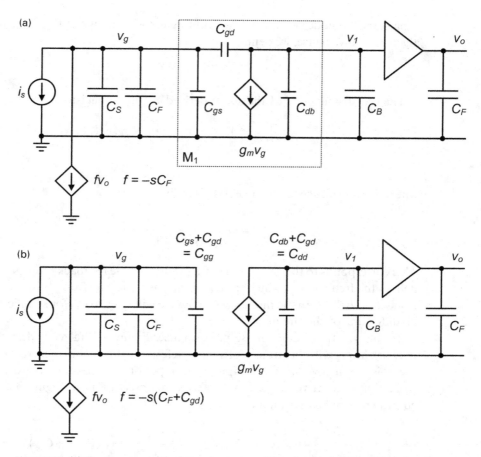

Figure 5.28 (a) Two-port model of the charge amplifier circuit. (b) Simplified circuit considering the additional shunt-shunt feedback by C_{gd}.

5.5.1   Circuit Analysis

We now analyze the frequency response and input-referred noise of the circuit. The obtained expressions will be used subsequently to drive the design and optimization process. To find the frequency response of the circuit, we consider the amplifier's small-signal two-port model as shown in Figure 5.28(a), which is obtained using the "shunt-shunt" feedback circuit approximation (see [3]). Here, we have included the small-signal model of M_1, as well as the input capacitance of the buffer (C_B). Note that we are neglecting the output resistance (r_{ds}) of M_1 for simplicity, and since we are primarily interested in the circuit's high-frequency behavior.

To simplify, we note that C_{gd} provides shunt-shunt feedback, just like C_F in Figure 5.27. Since $v_o = v_1$, this modifies the feedback gain (f) to the sum of C_F and C_{gd}. Furthermore, to include proper loading, C_{gd} must be added as a shunt capacitance to nodes v_g and v_l. The resulting model is shown in Figure 5.28(b). For algebraic convenience, we define:

$$C_{Ftot} = C_F + C_{gd} \qquad C_1 = C_{dd} + C_B.$$

$$(5.21)$$

The transfer function of the above-derived circuit model is:

$$\frac{v_o}{i_s} = \frac{1}{sC_{Ftot}} \frac{1}{1 - \dfrac{s}{p}},$$

$$(5.22)$$

where the magnitude of the pole p is given by:

$$|p| = \omega_c = \frac{C_{Ftot}}{C_F + C_S + C_{gg}} \frac{g_m}{C_1}.$$

$$(5.23)$$

This corresponds to the gain-bandwidth product of the feedback loop. Note that ω_c defines the frequency at which the circuit begins to deviate from its ideal integrator behavior ($1/sC_{Ftot}$). In the discussion below, we will refer to ω_c (for simplicity) as the "bandwidth" of the circuit.

To analyze the circuit's noise performance, Figure 5.29(a) considers the input network along with the transistor's input-referred voltage noise generator. To refer the noise to the overall circuit input, we can perform a Thévenin to Norton transformation and arrive at the input-referred representation of Figure 5.29(b). The equivalent input current noise is:

$$\frac{\overline{i_n^2}}{\Delta f} = \frac{\overline{v_n^2}}{\Delta f} \omega^2 \left(C_S + C_F + C_{gg} \right)^2 = \frac{4kT\gamma_n}{g_m} \omega^2 \left(C_S + C_F + C_{gg} \right)^2.$$

$$(5.24)$$

In this result, we are neglecting the flicker noise of the transistor, which is typically justifiable in broadband circuits (see Section 4.1.4).

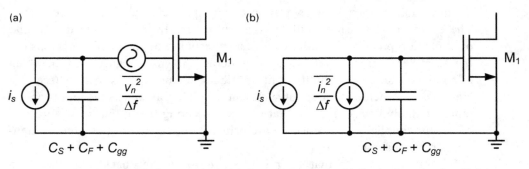

Figure 5.29 (a) Relevant sub-circuit for noise analysis. (b) Equivalent circuit with input current noise generator.

In a detector application, the noise performance of a charge amplifier is typically specified in terms of its effective noise charge (ENC) at the input. The ENC is found by measuring the RMS noise at the output, and referring this noise to the input by dividing it by the measured "charge impulse gain" of the amplifier. The charge impulse gain is found by applying a current pulse of a certain shape and duration (a charge packet) and measuring the resulting output voltage (after filtering). Regardless of the pulse and filtering details, however, the noise performance of the circuit is fundamentally linked to the input referred noise power spectral density (PSD), and one can convert from the PSD to ENC using tables available in the literature (see [8]). For simplicity, we will therefore consider only the PSD in our discussion.

To gain some basic insight about the achievable noise performance, it is useful to solve the gain-bandwidth expression (5.23) for g_m and substitute into (5.24). This yields:

$$\frac{\overline{i_n^2}}{\Delta f} = 4kT\gamma_n \frac{\omega^2}{\omega_c}(C_S + C_F + C_{gg})\frac{C_{Ftot}}{C_1}. \tag{5.25}$$

From this result, we see that given a fixed specification for ω_c, the input noise PSD is fully defined by the total capacitance at the input node and the ratio C_{Ftot}/C_1. Another way to interpret (5.25) is to realize that the right-hand side contains a noise charge variance term $kT(C_S + C_F + C_{gg})$ in units of Coulomb squared. The frequency terms in front of it convert from charge to current and to a power spectral density.

Furthermore, note that since in practice $C_{Ftot}/C_1 \cong C_F/C_B$, the multiplier on the right-hand side of (5.25) can be viewed buffer's fan-out (FO), i.e. the ratio between its load and input capacitance (see Figure 5.28). For wideband operation, the buffer FO is typically limited to values in the range of 2...5, implying that there is not much design freedom in this parameter.

Figure 5.30 plots the input noise PSD at $\omega = \omega_c$ (a convenient reference point) as a function of $C_S + C_F + C_{gg}$ assuming $C_{Ftot}/C_1 = 3$, $\gamma_n = 0.8$ and for various values of $f_c = \omega_c/2\pi$. We see that for practical capacitance values and bandwidths that are common in high-speed circuits, the input noise PSD takes on values between 1 and 100 pA/rt-Hz.

In practice, the design of a charge amplifier is often severely constrained by the fact that the capacitance C_S is fixed by the transducer (for example, a good fraction of a picofarad for a high-speed diode and its leads). From the above result, we see that this places a direct bound on the achievable noise performance. With C_S fixed, the design task boils down to the proper sizing of C_F and C_{gg}. Sizing C_{gg} means that we need to find the width of the transistor, and its bias current to generate a proper amount of g_m (to satisfy the bandwidth requirement).

At first glance, one might think that making C_{gg} as small as possible is the best design choice. However, since small C_{gg} also implies small W and consequently small g_m, it turns out that this design choice leads to sub-optimum performance. To analyze this tradeoff quantitatively, we now consider several optimization scenarios.

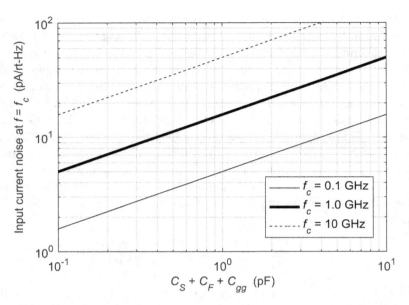

Figure 5.30 Input noise PSD as a function of the total capacitance at the input node.

5.5.2   Optimization Assuming Constant Transit Frequency

We first consider the case where the transit frequency (ω_T) of the transistor is assumed to be constant. In practice, this condition may be imposed by the need to operate the device at the maximum ω_T that it can offer (near the peak in strong inversion). On the other hand, this constraint may also be imposed by a need for very low V_{Dsat}, thereby mandating weak-inversion operation with large g_m/I_D and a correspondingly low and bounded value of ω_T. From a device sizing perspective, constant ω_T requires that the current is scaled proportional to the width, maintaining a constant current density throughout the optimization.

To quantify the noise performance under this assumption, we make use of (2.47) and substitute $g_m = \omega_T C_{gg}$ into (5.24), which gives:

$$\frac{\overline{i_n^2}}{\Delta f} = \frac{4kT\gamma_n}{\omega_T C_{gg}}\, \omega^2 \left(C_S + C_F + C_{gg}\right)^2 . \tag{5.26}$$

Therefore, under the assumption of constant γ_n and ω_T, the noise is minimized when the ratio $(C_F + C_S + C_{gg})^2/C_{gg}$ is minimized. Setting the first derivative of this term equal to zero gives $C_{gg} = C_F + C_S$. The existence of this optimum is noted in [6], [9].

An issue with this design point is that it does not guarantee the optimum tradeoff between noise, bandwidth and current consumption. It merely defines the point for which we obtain the lowest possible noise for a fixed value of ω_T.

5.5.3 Optimization Assuming Constant Drain Current

Consider now the constraint of constant bias current I_D. From a sizing perspective, constant I_D means that the current density (and inversion level) varies as we change the device width (and thus C_{gg}) during the optimization. To express g_m as a function of C_{gg} in (5.24) we can make use of (4.8), which states that

$$\omega_T \frac{g_m}{I_D} = \frac{g_m}{C_{gg}} \cdot \frac{g_m}{I_D} \cong 3 \frac{\mu}{nL^2}(1-\rho) \tag{5.27}$$

and thus:

$$g_m \cong \sqrt{3 \frac{\mu}{nL^2}(1-\rho)I_D C_{gg}} . \tag{5.28}$$

In this expression, ρ is the normalized transconductance efficiency, defined in (2.31). This parameter approaches one in weak inversion, and zero in strong inversion. Substituting (5.28) into (5.24) now yields:

$$\frac{\overline{i_n^2}}{\Delta f} \cong \frac{4kT\gamma_n}{\sqrt{3 \frac{\mu}{nL^2}(1-\rho)I_D}} \omega^2 \frac{\left(C_S + C_F + C_{gg}\right)^2}{\sqrt{C_{gg}}} . \tag{5.29}$$

If we assume that the device operates in strong inversion, ρ drops out of the equation. If we now further assume that γ_n and the mobility μ are constant, setting the first derivative of (5.29) equal to zero gives $C_{gg} = (C_F + C_S)/3$. This optimum was noted in [7]. However, as shown in [8], the optimum shifts considerably in moderate inversion (where ρ is non-zero and not constant) and also in strong inversion, due to the mobility degradation found in modern devices. Consequently, it is best to locate the optimum using numerical data, which is conveniently done using the lookup tables used throughout this book. We illustrate this in the following example.

Example 5.6 Charge Amplifier Optimization (Constant I_D)

Consider the given charge amplifier with $I_D = 1$ mA and $C_F + C_S = 1$ pF. Plot the noise level as a function of $C_{gg}/(C_F + C_S)$, $L = 60, 100, 200$ and 400 nm. Tabulate the values of $C_{gg}/(C_F + C_S)$ as well as the transistor's g_m/I_D, f_T and W at the points of minimum noise.

SOLUTION

The code shown below sweeps the transistor width and computes g_m and C_{gg} for the given drain current. The obtained values are then used to compute

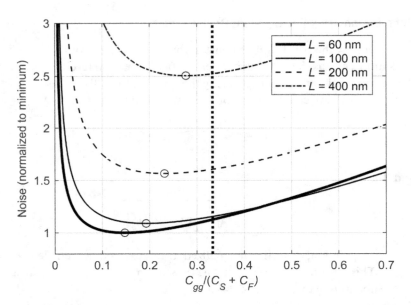

Figure 5.31 Input-referred noise as a function of $C_{gg}/(C_S + C_F)$ and for different channel lengths. The vertical line marks the expected location of the optima in strong inversion.

$(C_F + C_S + C_{gg})^2/g_m$, which is the factor that scales the noise (see (5.24)). The resulting plots for each gate length are shown in Figure 5.31.

```
% Parameters
Cf_plus_Cs = 1e-12;
ID = 1e-3;
W = 5:1000;
L = [0.06 0.1 0.2 0.4];
% Compute relative noise level
Cgg = [W; W; W; W].*lookup(nch,'CGG_W', 'ID_W', ID./W, 'L', L);
gm = [W; W; W; W].*lookup(nch,'GM_W', 'ID_W', ID./W, 'L', L);
Noise = (Cf_plus_Cs + Cgg).^2./gm;
```

The first observation from Figure 5.31 is that the optima are relatively shallow, i.e. the noise performance is not very sensitive to the exact value of $C_{gg}/(C_F + C_S)$. Furthermore, we see that the actual optimum for $L = 60$ nm is located at $C_{gg}/(C_F + C_S) = 0.14$, which is quite far away from the (approximate) analytical prediction of 1/3. Sizing the device per the square-law result would lead to a 15% noise penalty and a device size approximately 2.4 times larger than needed (recall that C_{gg} is proportional to the device width).

Table 5.12 summarizes the transistor parameters at the optima of the plotted curves. As the channel length is increased, the location of the optima gradually approaches the square-law prediction, but never reaches it exactly, since even the device with $L = 400$ nm does not follow the square-law equations accurately

Table 5.12 Summary of optimum parameters.

L (nm)	Optimum parameters			
	$C_{gg}/(C_S + C_F)$	g_m/I_D (S/A)	f_T (GHz)	W (µm)
60	0.14	17.1	18.5	169.1
100	0.19	17.0	14.0	168.7
200	0.23	12.6	8.6	117.4
400	0.28	8.4	4.8	73.0

(although it operates in strong inversion with $g_m/I_D = 8.4$ S/A). It is also worth noting that the noise penalty for a longer channel is not as large as predicted by the square-law analysis. Per (5.28), g_m scales with $1/L$ and hence the noise at the optimum should scale with L, or a factor of four for the above example. However, from the plot, we see that the noise scales only by a factor of 2.5, making it clear that mobility degradation, moderate inversion, etc., play a significant role in the observed trends.

As the channel length scales from 60 nm to 400 nm, the optimum device width decreases, the current density increases, and the inversion level changes from moderate to strong inversion. Note, however, that the absolute values of these inversion levels depend on the given parameters (I_D and $C_F + C_S$).

The above example illustrates the value of using a lookup table based approach. However, the constraint of fixed I_D rarely corresponds to a scenario seen in today's practical applications. Instead, the typical optimization problem is to minimize the current given fixed noise and bandwidth specifications. We will therefore investigate this scenario in the next subsection.

5.5.4 Optimization Assuming Constant Noise and Bandwidth

To investigate the optimum design point with fixed noise and bandwidth specifications, we begin by decomposing the drain current as:

$$I_D = g_m \cdot \frac{I_D}{g_m}. \tag{5.30}$$

To include the noise constraint, we solve (5.24) for g_m and substitute, which gives:

$$I_D = \frac{4kT\gamma_n}{\dfrac{i_n^2}{\Delta f}} \omega^2 \left(C_F + C_S + C_{gg}\right)^2 \cdot \frac{I_D}{g_m}. \tag{5.31}$$

Next, to consider a bandwidth constraint, we substitute $g_m = \omega_T C_{gg}$ into the expression for ω_c per (5.23) and solve for C_{gg}:

$$C_{gg} = \frac{C_S + C_F}{\dfrac{C_{Ftot}}{C_1} \dfrac{\omega_T}{\omega_c} - 1}.$$

(5.32)

This result contains important insight. As the ratio of ω_c/ω_T approaches the fan-out of the buffer (which is close to C_{Ftot}/C_1), the required device size, represented by C_{gg}, increases rapidly. This has an adverse effect on the supply current, as seen from (5.31). The tradeoff becomes evident after substituting (5.32) into (5.31), which results in:

$$I_D = \frac{4kT\gamma_n}{\dfrac{i_n^2}{\Delta f}} \omega^2 (C_S + C_F)^2 \left(\frac{1}{1 - \dfrac{C_1}{C_{Ftot}} \dfrac{\omega_c}{\omega_T}} \right)^2 \cdot \frac{I_D}{g_m}.$$

(5.33)

For a given noise specification, constant γ_n, C_S, and C_F,[5] the drain current is minimized when the following term is minimized:

$$K = \left(\frac{1}{1 - \dfrac{C_1}{C_{Ftot}} \dfrac{\omega_c}{\omega_T}} \right)^2 \cdot \frac{I_D}{g_m}.$$

(5.34)

If the bandwidth (ω_c) and the buffer fan-out is fixed, minimizing the current boils down to finding the best tradeoff between ω_T and g_m/I_D. Large ω_T reduces the bracketed term in (5.34), but large ω_T also dictates small g_m/I_D, which increases the multiplier outside the bracket. An optimum exists where the two trends are balanced.

The location of the optimum depends on how exactly the transistor trades g_m/I_D for ω_T. The best strategy for finding the optimum is to evaluate (5.34) using numerical lookup data, as done in Example 5.6. However, as a simple analytical reference point, it is again useful to consider (5.27), and approximate:

$$\omega_T \frac{g_m}{I_D} \cong 3 \frac{\mu}{nL^2} (1 - \rho) \cong const.$$

(5.35)

After solving this expression for g_m/I_D, inserting into (5.34) and setting the derivative to zero, we obtain the following first-order optimum:

$$\omega_T = 3 \frac{C_1}{C_{Ftot}} \omega_c.$$

(5.36)

---

[5] We will see in Example 5.7 and Example 5.8 that the required C_F is somewhat dependent on the chosen inversion level, but it is fair to approximate it as constant.

This result confirms the general intuition that the device ω_T must scale with the desired bandwidth and the circuit loading. We also see that the larger we can make the buffer's fan-out, the smaller the ω_T requirement for the input device becomes. In other words, more work done by the buffer reduces the requirement on the input device. Optimizing the buffer fan-out together with the input device is an interesting optimization problem that is outside the scope of this discussion. In any case, as already discussed, the range of practical fan-outs for a high-speed buffer tends to be limited.

As a final step, we can substitute (5.23) into (5.36) to eliminate ω_c and solve for C_{gg}:

$$C_{gg} = \frac{C_S + C_F}{2}.$$ (5.37)

Therefore, we note that under the assumption of fixed noise and bandwidth, the optimum device sizing lies between the values seen in the previous subsections ($C_S + C_F$ and ($C_S + C_F$)/3).

In the following example, we investigate the location of the optima using actual device data. As before, we expect deviations from the above results due to mobility degradation and moderate inversion behavior.

Example 5.7 Charge Amplifier Optimization (Constant Noise and Bandwidth)

Find the optimum inversion level for the input device of the charge amplifier. Assume $C_{Ftot}/C_1 = 3, f_c = 3$ GHz and $L = 60, 100, 200$ and 400 nm. At the points of minimum current, tabulate the transistor's $g_m/I_D, f_T$, and the ratio $C_{gg}/(C_S + C_F)$.

SOLUTION

To solve this problem, we sweep g_m/I_D and compute the corresponding ω_T using the following code:

```
gm_ID = 5:0.1:25;
wT = lookup(nch,'GM_CGG', 'GM_ID', gm_ID, 'L', L);
```

We then evaluate (5.34) using this data and for the given parameters, which yields Figure 5.32 and Table 5.13. The right column in the table is computed using (5.32). Similar to our findings from Chapter 3, we see from the plots that going to the shortest possible channel length is not beneficial; the current minima for $L = 60$ nm and $L = 100$ nm are essentially the same. As we increase the channel length further, however, the device must be pushed into strong inversion to satisfy the requirements. This results in smaller g_m/I_D and increased current.

As far as the observed f_T and $C_{gg}/(C_S + C_F)$ values are concerned, we observe relatively large discrepancies with respect to the first-order predictions of (5.36)

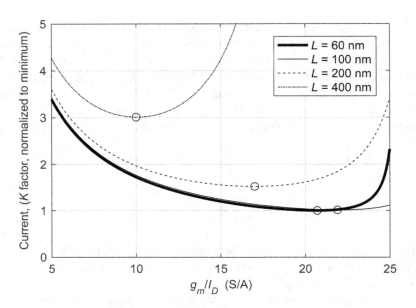

Figure 5.32 Relative drain current as a function of g_m/I_D.

Table 5.13 Summary of optimum parameters.

L (nm)	Optimum parameters		
	g_m/I_D (S/A)	f_T (GHz)	$C_{gg}/(C_S + C_F)$
60	20.7	9.6	0.116
100	21.9	7.5	0.155
200	17.0	5.1	0.247
400	10.0	3.9	0.345

and (5.37) (3 GHz and 0.5, respectively). These are again attributed to moderate inversion behavior and mobility degradation. As expected, the results come closest for $L = 400$ nm (in strong inversion).

The optimization methodology discussed above allows us to identify the optimum inversion level of the transistor for a given set of capacitor ratios. With this information, it is straightforward to size the input transistor for absolute noise and bandwidth specifications. We will illustrate this using another example.

Example 5.8 Charge Amplifier Sizing

Design the charge amplifier circuit to achieve $f_c = 3$ GHz and an input noise of 50 pA/rt-Hz at f_c, while maintaining the minimum possible current consumption.

Assume $C_S = 1$pF, and $C_{Ftot}/C_1 = 3$ and $L = 100$ nm. Validate the design using SPICE simulations.

SOLUTION

To solve this problem, we can re-use the data from Table 5.12, for $L = 100$ nm. As a first step, we then use (5.25) to solve for the capacitances:

$$C_S + C_F = \frac{\dfrac{\overline{i_n^2}}{\Delta f} 4kT\gamma_n \omega_c}{\left(1 + \dfrac{C_{gg}}{C_S + C_F}\right)\dfrac{C_{Ftot}}{C_1}}.$$

Assuming $\gamma_n = 0.7$ this yields $C_S + C_F = 3.30$ pF, $C_F = 2.30$ pF, and $C_{gg} = 511$ fF. Note that the computed value of C_F is only very weakly dependent on the exact value of the ratio $C_{gg}/(C_S + C_F)$. This is because at the optimum point, C_{gg} is relatively small compared to $C_S + C_F$, and the noise is then mainly scaled by the latter two capacitances (see (5.25)).

We can now also compute $g_m = \omega_T C_{gg} = 24$ mS. Using the g_m/I_D from Table 5.12, we then find $I_D = 1.09$ mA. To determine the device width, we look up the current density:

```
JD = lookup(nch,'ID_W', 'GM_ID', gm_ID, 'L', L)
```

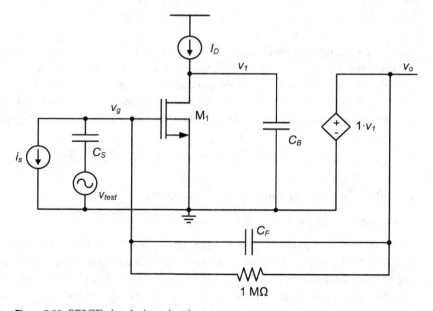

Figure 5.33 SPICE simulation circuit.

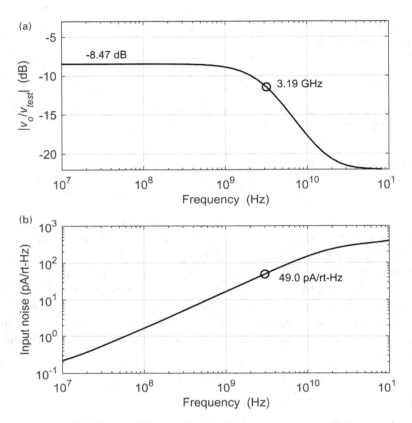

Figure 5.34 SPICE simulation results. (a) Frequency response. (b) Input noise versus frequency.

This yields 2.2 μA/μm and we therefore arrive at $W = I_D/J_D = 495$ μm. Note that we assumed the default V_{DS} (0.6 V) in the above call to the lookup function. The actual V_{DS} will be different, but that should not affect the result significantly.

To validate the design, we set up a simulation schematic as shown in Figure 5.33. One more parameter that is needed for the simulation is the value of C_B, given by:

$$C_B = C_1 - C_{gd} = C_{Ftot}\frac{C_1}{C_{Ftot}} - C_{gd} = \left(C_F + C_{gd}\right)\frac{C_1}{C_{Ftot}} - C_{gd}.$$

To find C_{gd}, we use the following lookup command:

```
Cgd = W*lookup(nch,'CGD_W', 'GM_ID', gm_ID, 'L', L)
```

which gives $C_{gd} = 164$ fF and consequently $C_B = 658$ fF. The actual fanout of the buffer is $C_F/C_B = 3.5$. This data can now be used to implement the buffer and (if needed) the assumed value of C_{Ftot}/C_1 could be fine-tuned to help optimize it.

After running an operating point analysis in SPICE, we find that $g_m = 23.6$ mS. This matches the expected value of 24 mS closely. The small discrepancy is due to the fact that $V_{DS} = 427$ mV in the simulation, instead of the 600 mV assumed during sizing.

To measure f_c, we inject an input signal via the v_{test} voltage source shown in Figure 5.33 and run an AC simulation. Setting up the input this way is preferred over stimulating i_s, since it is easier to see and measure the corner frequency by looking at the voltage transfer. The ideal mid-band voltage gain is $C_S/(C_F + C_{gd}) = 0.405 = -7.84$ dB.

The simulated results plotted in Figure 5.34 show a mid-band voltage gain of -8.47 dB and $f_c = 3.19$ GHz. Both discrepancies are due to the fact that we neglected the output resistance of the M_1 in all analytical expressions. The simulated input-referred current noise at $f = 3$ GHz is 49 pA/rt-Hz, which is very close to the design goal.

The device sizing in the previous example was done with the constraint of achieving the minimum possible supply current. This led to a relatively large device size, since the transistor ended up being biased close to weak inversion. In this context, it is worth re-visiting Figure 5.32 (plot for $L = 100$ nm). From the curve, we see that lowering g_m/I_D to 18 S/A will result only in a small current penalty. On the other hand, the current density, and hence device width, should change considerably. This may be an interesting tradeoff to explore if the layout area is critical. We will look at the numbers in the following example.

As a final comment, also note that exploring the current versus area tradeoff is attractive only for the short channels. For example, the plot for $L = 400$ nm in Figure 5.32 is not shallow and there is little freedom in moving away from the optimum. This case corresponds to a scenario where we are pushing the circuit close to the maximum achievable bandwidth, and this narrows the range of feasible and attractive design points.

Example 5.9 Charge Amplifier Re-sizing for Smaller Area

Re-design the charge amplifier circuit of Example 5.8 given the same specifications, but choosing $g_m/I_D = 18$ S/A as the transistor's inversion level. Quantify the change in drain current and device width.

SOLUTION

We begin by finding the device f_T using the lookup function, which gives 12.5 GHz. As a next step, we can find $C_{gg}/(C_F + C_S)$ using (5.32), which yields 0.087. From here, the flow of calculations is the same as in Example 5.8. We obtain $I_D = 1.33$ mA, $W = 272$ μm and $C_F = 2.50$ pF. Relative to Example 5.8, the current increased by 22% and the width decreased by 45%. The current increase is somewhat larger than expected from Figure 5.32 due to the increase in C_F (about 9%) but the obtained current-area tradeoff may still be quite attractive for a practical application.

The main take-home from this section is that noise- and bandwidth-constraint optimization boils down to balancing the tradeoff between g_m/I_D and f_T. The examined circuit is sensitive to this tradeoff primarily due to the transistor's gate capacitance (C_{gg}), which affects both the noise and bandwidth. Making this capacitance negligible is not a design option, because one would simultaneously minimize g_m, which is the main feature that we are trying to extract from the transistor. Instead, one must balance C_{gg} and g_m, which is equivalent to finding the optimum f_T and g_m/I_D.

We have shown that for a noise- and bandwidth-constrained charge amplifier, the optimum point is easily found by sweeping lookup data through a simple analytical expression (5.34). In Chapter 6, we will see that essentially the same approach can be used to size operational transconductance amplifiers.

5.6   Designing for Process Corners

All the examples studied so far in this book were based on nominal and fixed device parameters. For a practical design, however, the designer must consider variations in the fabrication process, supply voltage and temperature (collectively referred to as "PVT").

While the impact of supply voltage and temperature changes are relatively easy to test and quantify in simulation, accounting for variations in the MOSFET behavior is more difficult, simply because it is defined by a large number of parameters. The classical (and simplified) way to deal with process variations is through the introduction of extreme process corners, representing "slow, nominal, and fast" parameter sets [2]. This naming is chosen based on the speed of a digital inverter in each case. For example, a parameter set with large threshold voltage and low mobility is considered as the "slow" corner.

The corner parameter sets are typically generated by the semiconductor foundry and reflect the expected worst-case variations seen in mass production. The most basic and common way to quantify process variations is to look at a transistors' threshold voltage (V_T) and current factor ($\beta = \mu C_{ox} W/L$). The variations in these parameters for the technology used in this book are quantified in Section A.1.6.

Dealing with process variations in terms of shifts in V_T and β is mostly useful for design based on the square-law model. However, in this book we are using a sizing methodology based on g_m/I_D. Therefore, this section investigates the impact of process corners and temperature variations on the core metrics of our sizing methodology: g_m/I_D, f_T, g_m/g_{ds}. In addition, we will discuss options for considering process variations in g_m/I_D-based sizing and illustrate the proposed flow using an example.

5.6.1   Biasing Considerations

A critical aspect in a circuit's sensitivity to process and temperature variations is the biasing strategy. For our discussion, we will consider two extreme scenarios: constant current biasing and constant g_m biasing (see Section 5.1). To understand how

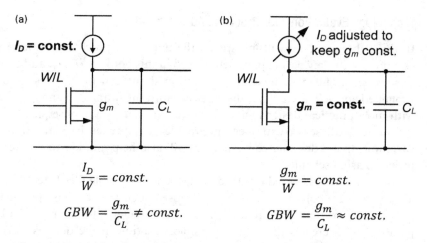

(a)

I_D = const.

W/L

g_m

C_L

$$\frac{I_D}{W} = const.$$

$$GBW = \frac{g_m}{C_L} \neq const.$$

(b)

I_D adjusted to keep g_m const.

W/L

g_m = const.

C_L

$$\frac{g_m}{W} = const.$$

$$GBW = \frac{g_m}{C_L} \approx const.$$

Figure 5.35 Intrinsic gain stage with (a) constant current biasing, and (b) constant g_m biasing. For our discussion, we assume C_L = constant (for simplicity).

the choice of biasing may affect the circuit behavior, consider the simple (but representative) example of an IGS (see Figure 5.35).

In an intrinsic gain stage with fixed drain current (Figure 5.35(a)), the current density (I_D/W) remains constant, regardless of process and temperature variations. However, the gain-bandwidth product (g_m/C_L) is strongly affected by process and temperature. In weak inversion, g_m is inversely proportional to n and U_T (kT/q), where n varies with process and U_T is proportional to absolute temperature. In strong inversion, g_m is (approximately) proportional to the square root of μC_{ox}. C_{ox} varies mainly with process, whereas the mobility μ varies with process and temperature (see also Section 4.3.3).

In an intrinsic gain stage with constant g_m bias (Figure 5.35(b)), the current density of the device adjusts over process and temperature to keep g_m and thus also g_m/W constant. Assuming constant C_L, this means that the gain-bandwidth product of the circuit will be nearly constant as well.

With the above observation in mind, why wouldn't one always use a constant g_m biasing scheme? There are two basic issues to consider, which will be discussed in more detail below. The first is that the current consumption of a constant g_m circuit will vary widely over process and temperature. The second issue is that due to the changes in current density, V_{Dsat} varies significantly and this may push certain devices out of saturation. This latter issue is particularly troublesome in low-voltage designs.

In summary, we see that neither constant current nor constant g_m may be an optimum choice. Many practical designs therefore operate between these two extremes. For example, the bias current generator can be designed to reduce variations in g_m over process and temperature, but it does not fully compensate for these variations. In some sense, this solution can be viewed as "spreading" the variability across all dimensions instead of trying to keep certain parameters strictly constant.

5.6.2   Technology Evaluation over Process and Temperature

In this subsection, we will investigate the impact of process variations and temperature on the key design parameters used in this book: g_m/I_D, f_T, and g_m/g_{ds}. To do this properly, one must make an assumption about how the devices are biased. As explained above, one can view the cases of constant I_D and constant g_m as extremes, with most practical designs lying somewhere in-between. Consequently, a characterization for these two extremes will provide a comprehensive view on the range of variations that a designer can expect. It will also help him/her to decide about the proper biasing scheme.

Our evaluation is based on the data in the lookup table files listed in Table 5.14. The worst-case corner conditions are chosen such that the fast parameter set is paired with low temperature and the slow parameter set is paired with high temperature. This creates the largest possible variation in the devices' current for a given inversion level (or, equivalently, some fixed value of $V_{GS} - V_T$). The threshold voltage also varies across corners but analog circuits are usually biased such that the threshold voltage plays no role in setting a circuit's bias current [2].

Figure 5.36 shows the corner variations of g_m/I_D for a 100-nm n-channel device. For the case of constant I_D biasing in plot (a), we observe a variation of approximately ±30%, which is only weakly dependent on the current density (and hence the inversion level). Note that the transconductance of the device will vary by the same percentage, since I_D is assumed to be constant.

For constant g_m biasing, the transconductance per width (g_m/W) is held constant and therefore defines the nominal inversion level. Hence, this variable is used on the x-axis (see Figure 5.36(b)). We now see that g_m/I_D varies by as much as ±50% over process and temperature. Since g_m is held constant, this means that $I_D = g_m/(g_m/I_D)$ will vary by about −50% ...+100%, a factor of 4! This disadvantage was already mentioned earlier. Also, we find that in the slow/hot corner, the sharp decline of g_m/I_D is due to the device entering the triode region; i.e. $V_{Dsat} = 2/(g_m/I_D)$ no longer exceeds the device's fixed $V_{DS} = 0.6$ V.

To investigate the V_{Dsat} issue further, Figure 5.37 plots this quantity for the same corner variations. For low inversion levels, i.e. small I_D/W and small g_m/W, the V_{Dsat} variations are of the order of 10...30 mV only. However, for the regions corresponding to strong inversion, we see an increase in V_{Dsat} of up to 200 mV for the constant g_m biasing in the slow/hot corner. Once again, this confirms that "strict" constant g_m biasing may be impractical for low-voltage design that operate in or near strong inversion. This can be viewed as another motivation for design in moderate inversion.

Table 5.14 Corner conditions and corresponding lookup tables.

Corner, temperature	Parameter file
Fast, −40°C	65nch_fast_cold.mat
Nominal, 27°C	65nch.mat
Slow, 125°C	65nch_slow_hot.mat

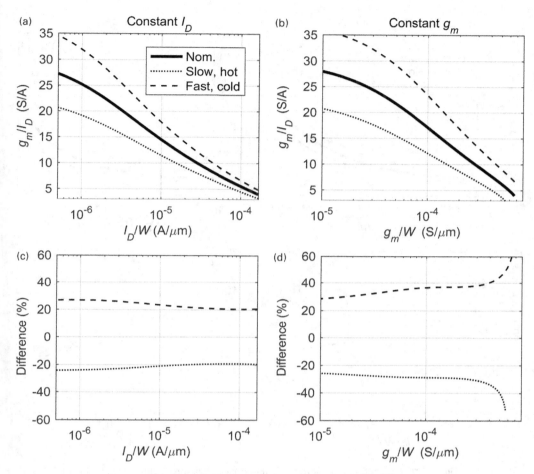

Figure 5.36 Transconductance efficiency (g_m/I_D) across corners for two biasing scenarios: (a) constant I_D and (b) constant g_m. Plots (c) and (d) show the percent differences relative to the nominal corner. The shown data is for an n-channel with $L = 100$ nm with $V_{DS} = 0.6$ V.

Next, we inspect the variations in transit frequency ($f_T = g_m/(2\pi C_{gg})$) in Figure 5.38. As expected, the variations are much larger (about ±25%) for the constant I_D scenario, since g_m varies significantly in this case. For constant g_m, the variations are small, since there is not much process and temperature variation in C_{gg}.

Finally, we create a similar plot for the device's intrinsic gain in Figure 5.39. We find that the largest variations are again seen at high inversion levels. For the case of constant g_m biasing, much of the variation at high g_m/W is due to the fact that V_{Dsat} is increasing and thus the device is pushed closer to the boundary between saturation and triode. Once again, operation in weak or moderate inversion looks advantageous from this perspective.

As an example of how the above-plotted corner variations play out in a realistic circuit, we consider the constant-g_m bias circuit studied in Section 5.1.

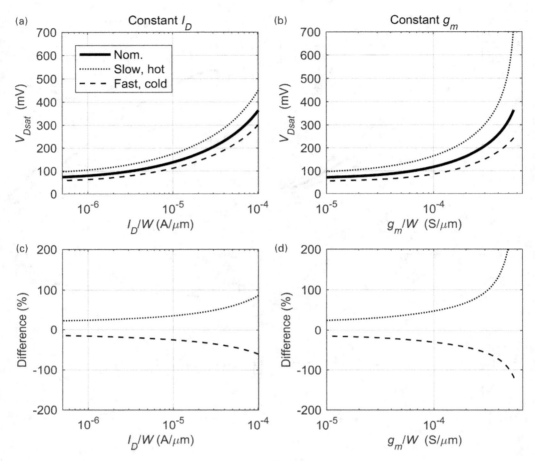

Figure 5.37 $V_{Dsat} = 2/(g_m/I_D)$ across corners for two biasing scenarios: (a) constant I_D and (b) constant g_m. Plots (c) and (d) show the percent differences relative to the nominal corner. The shown data is for an n-channel with $L = 100$ nm with $V_{DS} = 0.6$ V.

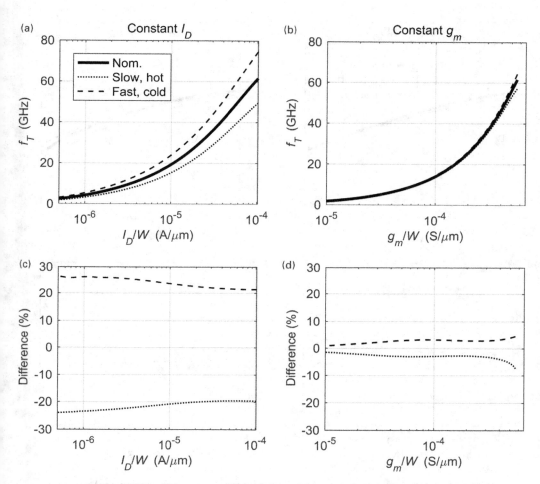

Figure 5.38 Transit frequency (f_T) across corners for two biasing scenarios: (a) constant I_D and (b) constant g_m. Plots (c) and (d) show the percent differences relative to the nominal corner. The shown data is for an n-channel with $L = 100$ nm with $V_{DS} = 0.6$ V.

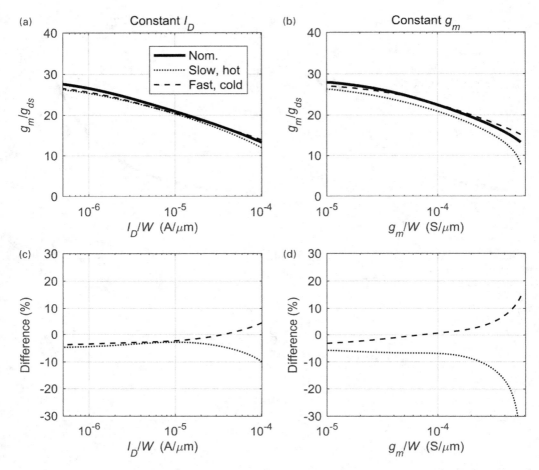

Figure 5.39 Intrinsic gain (g_m/g_{ds}) across corners for two biasing scenarios: (a) constant I_D and (b) constant g_m. Plots (c) and (d) show the percent differences relative to the nominal corner. The shown data is for an n-channel with $L = 100$ nm with $V_{DS} = 0.6$ V.

Example 5.10 Constant Transconductance Bias Circuit Performance across Process Corners

The constant-g_m bias circuit sized in Example 5.1 is subjected to fast/cold and slow/hot corner conditions. Evaluate the changes in the gate voltages, currents and the transconductance g_{m2}.

SOLUTION

The following parameters were used in Example 5.1: $I_D = 50$ μA, $R = 2$ kΩ, $V_{DD} = 1.2$ V, and all gate lengths equal to 0.5 μm. The sizing procedure led to $W_1 = 82.59$ μm, $W_2 = 15$ μm and $W_3 = 6.985$ μm with $V_{GS2} = 0.5137$ V and

Table 5.15 Comparison of calculated and SPICE-simulated corner variations.

	Fast/cold				Slow/hot			
	Matlab		SPICE		Matlab		SPICE	
V_{GS2} (mV)	483	−30 mV	484	−29 mV	530	+16 mV	530	+16 mV
I_{D1} (μA)	38.7	−22.5%	38.2	−23.5%	66.8	+33.7%	66.7	33.5%
I_{D2} (μA)	39.5	−21.0%	39.0	−22.0%	65.8	+31.5%	65.7	31.3%
g_{m2} (μS)	685	−4.0%	683	−4.3%	631	+4.2%	631	+4.2%

$g_{m2} = 658.8$ μS. Furthermore, recall that the circuit was designed such that the sum of the voltage drops across diode-connected transistors equals V_{DD}. This ensured that transistors of the same current mirror have identical drain voltages.

As the operating conditions change from nominal to fast/cold or slow/hot, V_{GS2}, I_{D2} and g_{m2} will change. To estimate the magnitudes of these changes in Matlab, we proceed as follows. We slightly shift V_{GS2} and find the corresponding change in I_{D2}. We then evaluate the gate voltage of M_4 and compute the corresponding I_{D3}. This yields the voltage drop across R, and since the gate and drain voltages of M_1 are known, we find I_{D1}. All that remains to do is to adjust V_{GS2} so that the currents I_{D3} and I_{D1} are equal. Once done, we can assess the changes in V_{GS2}, I_{D2} and g_{m2} with respect to nominal conditions. We compare the obtained results with SPICE simulations in Table 5.15.

From these results, we see that the sensitivity of g_{m2} to corner conditions is far less pronounced than that of I_{D2} and V_{GS2}. This confirms that the circuit studied in Section 5.1 is indeed a good choice when constant g_m is desired across corners.

5.6.3 Possible Design Flows

How can we account for the parameter variations across corners in our g_m/I_D-based design flow? There are two basic options to consider:

1. Identify the "worst case" parameter set (often slow/hot) and perform the sizing using this particular data set in Matlab. Since the sizing is done for the worst case, all other cases should automatically meet the design specifications (subject to verification). This approach was taken in [10].
2. Perform all sizing calculations using the nominal parameter set (used throughout this book) and "pre-distort" the design specification such that they will still be met in presence of corner variations (subject to verification).

The advantage of the second approach is that there is no need to extract and maintain multiple Matlab data files. Also, it tends to be more intuitive to work with a single set of nominal parameters, and these also correspond to what one will most likely measure in the lab. For these reasons, we recommend the second strategy.

The key question that remains is how one can estimate the amount of overdesign margin that should be applied to the worst-case performance targets. As we shall see in the example below, it is typically straightforward to estimate the required margins using the figures in Section 5.6.2. For example, we have seen that in a constant bias current design, g_m will vary about ±30%. Since bandwidths are typically directly proportional to g_m, it is straightforward to compute the required margin. The following example looks at this more closely, using the charge amplifier example from Section 5.5.

Example 5.11 Design of a Charge Amplifier with Corner Awareness

Repeat Example 5.7 and Example 5.8 with the goal of meeting the same specifications across process corners. Use a channel length of $L = 100$ nm. As before, assume $C_{Ftot}/C_1 = 3$ (approximate buffer fan-out), $f_c = 3$ GHz, $C_S = 1$pF and an input noise of 50 pA/rt-Hz at f_c. Assume that the circuit is biased with a constant current. Validate the design in SPICE.

SOLUTION

In the given circuit, the bandwidth scales with g_m (see (5.23)) and the noise is inversely proportional to g_m (see (5.24)). Under constant current bias, we expect g_m to decrease by about 30% in the slow/hot corner (see Figure 5.36). Consequently, we should overdesign the circuit (assuming nominal parameters) by a factor of about $1/0.7 = 1.4$. The nominal bandwidth target is therefore 4.2 GHz.

With this target defined, the rest of the problem boils down to repeating the steps outlined in Example 5.7 and Example 5.8 (using the usual nominal Matlab parameter set), and subsequently checking if the performance holds up across corners (using SPICE).

As a first step, we determine the inversion level that minimizes the current. We find that the optimum g_m/I_D equals 20.5 S/A. This value is slightly lower than the result of Example 5.7 (21.9 S/A), since we are now targeting a larger nominal bandwidth. The corresponding device f_T is 9.08 GHz (was 7.5 GHz in Example 5.7). With this data, we now re-run the calculations of Example 5.8, targeting a noise performance of 50 pA/rt-Hz at 4.2 GHz. The resulting component sizes are annotated in Figure 5.40. For comparison, the values from Example 5.8 are added in brackets.

It is interesting to note that the current is less than 10% larger than in Example 5.8, even though we increased the bandwidth by about 40%. This is once again explained by the argument that led to Example 5.9. Using the strict bias current minimum as the design point in Example 5.8 led to a very large device with significant self-loading. In the current example, we are designing for a larger bandwidth and this pushes the design toward a more reasonable tradeoff point, leading to a current increment that is not as large as one may expect.

Finally, we simulate the circuit in SPICE for all three corners: Slow/hot (125°C), nominal (27°C) and fast/cold (−40°C). The resulting plots are shown in Figure 5.41, and the data is summarized in Table 5.16.

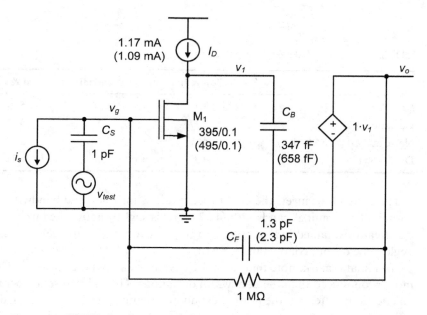

Figure 5.40 SPICE circuit schematic.

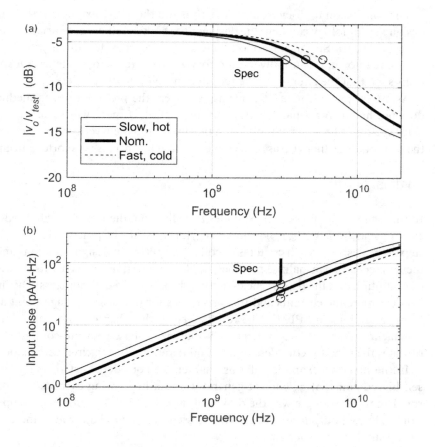

Figure 5.41 SPICE simulation results for (a) bandwidth and (b) noise across corners.

Table 5.16 Summary of SPICE simulation results.

	Slow & hot	Nominal	Fast & cold
f_c (GHz)	3.19	4.37	5.74
Deviation (%)	−27	0	+31
Noise at 3 GHz (pA/rt-Hz)	46.8	35.4	27.7
Deviation (%)	+32	0	−21

The design thus meets the target specifications for noise and bandwidth across corners. The nominal bandwidth (4.37 GHz) is slightly above the target (4.2 GHz). As already explained in Example 5.8, this is because the analytical expressions neglect the output conductance of M_1.

As a final remark, note that the noise deviations across the corners are slightly more pronounced than the bandwidth deviations. This is because temperature has an additional effect on the noise (beyond just changing the value of g_m), due to the kT term (see (5.24)).

The conclusion to draw from this section and the above example is that the g_m/I_D design methodology can help the designer to think about corner variations in a systematic manner. Specifically, once the worst-case corner deviations of g_m/I_D, f_T and g_m/g_{ds} are known, it is usually straightforward to "pre-distort" the design specifications such that the circuit has the required margins across corners.

As a final remark, note that one aspect where the proposed design methodology does not help is the usual SPICE verification of all corners, which is mandatory for chip tape-out. The designer must invest the time to run these verifications, regardless of the methodology that was used to create the design. There is no shortcut around this.

5.7   Summary

In this chapter, we presented examples that illustrate the use of normalized parameters such as I_D/W, g_m/I_D, g_m/g_{ds}, and g_m/C_{gg} for the sizing of practical circuits. The first step in the design of each circuit is to collect the pertinent design equations and frame them in terms of normalized parameters. Next, we optimize the circuits by leveraging our intuition and the dependencies among the normalized parameters. The final step is to compute the currents and transistor sizes for the desired design point and perform validations in SPICE. The results seen in this chapter confirm an important statement made in Chapter 1, namely that sizing based on pre-computed lookup obviates the need for countless tweaks and iterations during circuit simulation.

In the first two examples of this chapter, we considered DC-biasing circuits: a self-biased constant-g_m current generator and a high-swing cascoded current mirror. The third example was the design of a feedback LDO with large supply rejection and small output resistance. We assessed the LDO's bandwidth and computed the optimal load capacitance.

The fourth and fifth example concerned high-frequency circuits. We designed a two-transistor LNA that realizes an active balun and leverages partial noise cancellation. A further objective was to minimize distortion, using the insight gained from Chapter 4. In the fifth example, we examined charge amplifiers with high bandwidth and low noise. We reviewed optimization techniques under various constraints and sized a circuit for minimum current, given noise and bandwidth specs. An important observation from the optimization curves is that optima tend to be shallow, and that it is sometimes possible to gain large area savings for small penalties in the current consumption.

To conclude the chapter, we reviewed the implications of process corner variations in the light of the g_m/I_D methodology. We compared the benefits and drawbacks of constant-I_D and constant-g_m biasing and suggested appropriate design flow for corner-aware design. The proposed approach is based on "pre-distorting" the design specifications so that the sizing in Matlab can be carried out using nominal process parameters.

5.8   References

[1] T. H. Lee, *The Design of CMOS Radio-Frequency Integrated Circuits*, 2nd ed. Cambridge University Press, 2004.

[2] B. Murmann, *Analysis and Design of Elementary MOS Amplifier Stages*. NTS Press, 2013.

[3] P. R. Gray, P. Hurst, S. H. Lewis, and R. G. Meyer, *Analysis and Design of Analog Integrated Circuits*, 5th ed. Wiley, 2009.

[4] V. Gupta, G. A. Rincon-Mora, and P. Raha, "Analysis and Design of Monolithic, High PSR, Linear Regulators for SoC Applications," in *Proc. IEEE International SOC Conference*, 2004, pp. 311–315.

[5] S. C. Blaakmeer, E. A. M. Klumperink, D. M. W. Leenaerts, and B. Nauta, "Wideband Balun-LNA with Simultaneous Output Balancing, Noise-Canceling and Distortion-Canceling," *IEEE J. Solid-State Circuits*, vol. 43, no. 6, pp. 1341–1350, June 2008.

[6] B. E. Boser and R. T. Howe, "Surface Micromachined Accelerometers," *IEEE J. Solid-State Circuits*, vol. 31, no. 3, pp. 366–375, Mar. 1996.

[7] W. M. C. Sansen and Z. Y. Chang, "Limits of Low Noise Performance of Detector Readout Front Ends in CMOS Technology," *IEEE Trans. Circuits Syst.*, vol. 37, no. 11, pp. 1375–1382, Nov. 1990.

[8] G. De Geronimo and P. O'Connor, "MOSFET Optimization in Deep Submicron Technology for Charge Amplifiers," in *IEEE Symposium on Nuclear Science*, 2004, vol. 1, pp. 25–33.

[9] T. Chan Caruosone, D.A. Johns, and K. Martin, *Analog Integrated Circuit Design*, 2nd ed. Wiley, 2011.

[10] T. Konishi, K. Inazu, J. G. Lee, M. Natsui, S. Masui, and B. Murmann, "Design Optimization of High-Speed and Low-Power Operational Transconductance Amplifier Using gm/ID Lookup Table Methodology," *IEICE Trans. Electron.*, vol. E94–C, no. 3, pp. 334–345, Mar. 2011.

6   Practical Circuit Examples II

Switched-capacitor (SC) circuits are essential components in A/D converters, fil-
ters, sensor interfaces and many other mixed-signal blocks. In this chapter, we consider the generic switched-capacitor gain stage shown in Figure 6.1, and discuss
the design of its constituent operational transconductance amplifier (OTA) in the
charge redistribution phase (ϕ_2). We assume that the reader is familiar with the
general operation of SC circuits as covered in standard textbooks [1].

Our specific goal is to show how g_m/I_D-based design can be used to size the OTA
under given noise and settling speed constraints. The investigation begins with the
simplest possible OTA, a differential pair with idealized current source loads, to
establish a feel for the basic tradeoffs. Next, we consider the folded-cascode and
two-stage topologies, which are used more commonly due to their increased low-
frequency voltage gain. Finally, we show how to use lookup tables to size the circuit's switches.

6.1   Basic OTA for Switched-Capacitor Circuits

We begin by considering the simplest possible amplifier, shown in Figure 6.2, which
was already introduced in Example 3.11. In Section 6.1.1, we will analyze the SC
circuit of Figure 6.1 (during charge redistribution) using the small-signal model of
this amplifier. In Section 6.1.2, we then show that the optimum sizing under a noise
and bandwidth constraint is very similar to that of the charge amplifier considered
in Section 5.5. Finally, Section 6.1.3 considers the effect of slewing, and discusses
how it can be incorporated in the optimization process.

6.1.1   Small-Signal Circuit Analysis

To start our analysis, we construct a model for the SC circuit in the charge redistribution phase (ϕ_2). This model is shown in Figure 6.3 and incorporates the
differential half-circuit model of the amplifier of Figure 6.2, along with the transistors' thermal noise source. The subscript "d" denotes differential quantities,
i.e. $v_{od} = v_{op} - v_{om}$. The switching transient is modeled by a step at the input, and
the output is assumed to settle according to the amplifier's time constant. We
neglect the switch resistances, which should not limit or significantly affect the

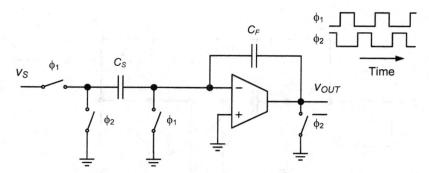

Figure 6.1 Conceptual model of a switched-capacitor gain stage (actual implementation is typically fully differential). During the clock phase $\phi 1$, the input is sampled. During $\phi 2$, the charge sampled on C_S is redistributed onto the feedback capacitor C_F.

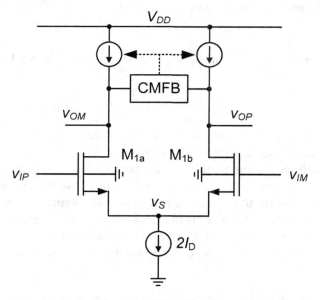

Figure 6.2 Basic differential OTA. The common mode feedback block (CMFB) forces the output common mode voltage to the desired value, typically near mid-supply.

settling dynamics in a properly designed circuit. For simplicity, we also neglect the noise contributed by the switches, which is justified if the on-resistances are sufficiently small [2]. An analysis that takes the switch noise into account is presented in [3].

For simplicity, we will initially neglect the drain-to-bulk capacitances (C_{db}) of the transistors. We will evaluate the discrepancies caused by this simplification in a later example. Similarly, we neglect the implementation details of the bias current sources and model them as ideal. Note, however, that the gate-to-drain capacitance

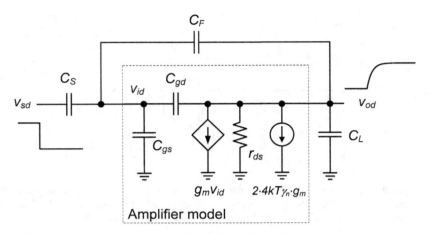

Figure 6.3 Simplified small-signal model of an SC circuit using the basic OTA of Figure 6.2. The factor of two in the noise term captures the contributions from both M_{1a} and M_{1b}.

(C_{gd}) appears in parallel to C_F and must be included. As in Section 5.5.1, we therefore define:

$$C_{Ftot} = C_F + C_{gd}. \tag{6.1}$$

We also include the output resistance (r_{ds}) of the transistors, since it sets the static gain error of the circuit (see detailed analysis below).

With this setup, the circuit can be analyzed in various ways. In principle, we could use a shunt-shunt two-port model as in Section 5.5, but it is more common, accurate and intuitive to analyze this circuit using a return ratio approach (see [4]). The loop gain of the circuit (denoted as L) is found by injecting a test current source in place of g_m and measuring the returning current in this controlled source. Following such an analysis, we obtain the following closed-loop transfer function:

$$A_{CL}(s) = \frac{v_{od}(s)}{v_{sd}(s)} = -G \frac{L_0}{1+L_0} \frac{1-\dfrac{s}{z}}{1-\dfrac{s}{p}}, \tag{6.2}$$

where:

$$G = \frac{C_S}{C_{Ftot}} \tag{6.3}$$

is the ideal closed-loop gain magnitude of the circuit and L_0 is the low-frequency loop gain (ideally infinite):

$$L_0 = \beta g_m r_{ds}. \tag{6.4}$$

The variable β is called the feedback factor, defined by the capacitive divider from the output back to the amplifier input:

$$\beta = \frac{v_{id}}{v_{od}} = \frac{C_{Ftot}}{C_{Ftot} + C_S + C_{gs}}. \tag{6.5}$$

For future use, we also define the maximum possible feedback factor for $C_{gs} = 0$ (ideal amplifier with no input capacitance):

$$\beta_{max} = \frac{C_{Ftot}}{C_{Ftot} + C_S} = \frac{1}{1+G}. \tag{6.6}$$

The zero (z) in (6.2) is located at $+g_m/C_{Ftot}$ (right half-plane) and is usually neglected, since it lies far beyond the circuit's closed-loop bandwidth. We can therefore approximate:

$$A_{CL}(s) = \frac{v_{od}}{v_{sd}} \cong -G\frac{L_0}{1+L_0}\frac{1}{1-\dfrac{s}{p}} = \frac{A_{CL0}}{1-\dfrac{s}{p}}. \tag{6.7}$$

The magnitude of this transfer function is sketched versus frequency in Figure 6.4, along with the circuit's loop gain. Since this is a first-order feedback system, the closed-loop corner frequency ω_c (equal to $|p|$) is well-approximated by the unity gain frequency of the loop gain. Assuming $g_m r_{ds} \gg 1$, detailed analysis yields:

$$|p| = \omega_c \cong \omega_u \cong \beta\frac{g_m}{C_{Ltot}}. \tag{6.8}$$

In this expression, the total load capacitance C_{Ltot} is given by:

$$C_{Ltot} = C_L + (1-\beta)C_{Ftot}. \tag{6.9}$$

The second term in this equation is due to loading from the feedback network, whereas C_L is an explicit load capacitance connected to the output (see Figure 6.3).

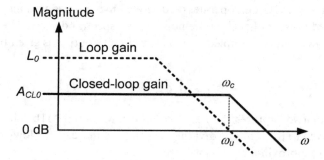

Figure 6.4 Illustration of the closed-loop gain and loop gain.

For instance, C_L could be the sampling capacitance of the next stage. With this setup in mind, we define the fan-out of the circuit as the ratio of this explicit load capacitance to its own sampling capacitance:

$$FO = \frac{C_L}{C_S} \qquad (6.10)$$

Since the circuit is well approximated by a single-pole transfer function, the response to a small differential input step ($v_{sd,step}$) applied at $t = 0$ is simply:

$$v_{od}(t) = v_{od,final}\left(1 - e^{-\frac{t}{\tau}}\right), \qquad (6.11)$$

where $\tau = 1/\omega_u$ and the decaying exponential in the brackets is called the dynamic settling error. SC circuits are typically specified in terms of the desired dynamic settling error (ε_d) at the desired settling time t_s. We can use (6.11) to translate these specs into a time constant requirement:

$$e^{-\frac{t_s}{\tau}} = \varepsilon_d \quad \Rightarrow \quad \tau = \frac{1}{\omega_u} = \frac{t_s}{\ln\left(\frac{1}{\varepsilon_d}\right)}. \qquad (6.12)$$

The final output voltage (for t $\rightarrow \infty$) is given by

$$v_{od,final} = -v_{sd,step} \cdot G\frac{L_0}{1 + L_0} = -v_{sd,step} \cdot G(1 + \varepsilon_s). \qquad (6.13)$$

The parameter ε_s is called the static gain error:

$$\varepsilon_s = \frac{L_0}{1 + L_0} - 1 \cong -\frac{1}{L_0}. \qquad (6.14)$$

Thus, the static gain error defines the low-frequency loop gain requirement for the amplifier.

An important difference to the charge amplifier from Chapter 5 is the way we quantify the noise. Since the SC circuit operates on discrete time samples, we must consider the total integrated noise, rather than the noise power spectral density. The analysis provided in [2] shows that the sampled noise at the circuit's output is given by:

$$\overline{v_{od}^2} = 2\frac{\gamma_n}{\beta}\frac{k_B T}{C_{Ltot}}. \qquad (6.15)$$

Here, the factor of two captures the noise from both M1a and M1b. The sampled noise can be referred to the input by dividing it with the square of the low-frequency gain term in (6.2) (approximately G^2 for large L_0).

6.1.2 Optimization Assuming Constant Noise and Bandwidth

Using the above equation set, we can now study the tradeoff between bandwidth, noise and current consumption. Following the same approach as in Section 5.5, we begin by writing:

$$I_D = g_m \cdot \frac{I_D}{g_m}. \tag{6.16}$$

As before, we carry out two substitutions to incorporate the bandwidth and noise constraints. First, we solve (6.8) for the transconductance and substitute. Second, we eliminate C_{Ltot} in the obtained expression using (6.15). This leads to:

$$I_D = \frac{2k_B T \gamma_n}{\overline{v_{od}^2}} \cdot \omega_u \cdot \frac{1}{\beta^2} \frac{I_D}{g_m}. \tag{6.17}$$

From this result, we see that minimizing I_D for a given bandwidth and noise specification (and assuming constant γ_n) boils down to minimizing

$$K = \frac{1}{\beta^2} \frac{I_D}{g_m}. \tag{6.18}$$

This result is very important for the remainder of this chapter and hence deserves an intuitive interpretation. Essentially, what (6.18) says is that we want to make both the feedback factor and g_m/I_D as large as possible. Unfortunately, the two quantities trade off with one another, and this gives rise to a specific value of g_m/I_D that minimizes (6.18). To see this, note first that making g_m/I_D larger leads to a more efficient translation of the invested current into g_m, and hence the factor K is inversely proportional to g_m/I_D. However, we also know from Chapter 3 that larger g_m/I_D implies lower ω_T, and hence larger gate capacitance for a given g_m. Larger gate capacitance, however, reduces the feedback factor β (see (6.5)). This requires higher g_m to maintain the bandwidth (see (6.8)) and thus counters and eventually negates the positive impact that we get from increasing g_m/I_D.

To capture the tradeoff between g_m/I_D and ω_T and its impact on β analytically, we now make additional algebraic substitutions. Starting from (6.5), we express β as:

$$\beta = \frac{1}{1 + G + \dfrac{C_{gs}}{C_{Ftot}}}. \tag{6.19}$$

The capacitance ratio in this expression is:

$$\frac{C_{gs}}{C_{Ftot}} = \frac{g_m / C_{Ftot}}{g_m / C_{gs}} = \frac{g_m / C_{Ftot}}{\omega_{Ti}}. \tag{6.20}$$

Here, we introduced $\omega_{Ti} = g_m/C_{gs}$ which is similar to the transistor's angular transit frequency $(\omega_T = g_m/C_{gg})$, but includes only the involved intrinsic capacitance C_{gs}. Note that ω_{Ti} is always slightly larger than ω_T, since $C_{gs} < C_{gg}$.

The numerator of (6.20) is:

$$\frac{g_m}{C_{Ftot}} = \frac{1}{\beta}\frac{\omega_u}{\omega_{Ti}}\left(FO \cdot G + (1-\beta)\right). \tag{6.21}$$

This equation follows from:

$$\omega_u = \frac{\beta g_m}{C_{Ltot}} = \frac{\beta g_m}{C_L + C_{Ftot}(1-\beta)} = \frac{\beta g_m / C_{Ftot}}{\dfrac{C_L}{C_{Ftot}} + (1-\beta)} \tag{6.22}$$

and

$$\frac{C_L}{C_{Ftot}} = \frac{C_L}{C_S}\frac{C_S}{C_{Ftot}} = FO \cdot G. \tag{6.23}$$

Using these expressions, substituting (6.21) into (6.20) and subsequently solving for β yields:

$$\beta = \frac{1-(1+FO \cdot G)\dfrac{\omega_u}{\omega_{Ti}}}{1+G-\dfrac{\omega_u}{\omega_{Ti}}}. \tag{6.24}$$

Equation (6.24) analytically captures the dependency of β on the transistor's transit frequency. Substituting it into (6.18) gives the result below, which must be minimized for minimum current:

$$K = \left(\frac{1+G-\dfrac{\omega_u}{\omega_{Ti}}}{1-(1+FO \cdot G)\dfrac{\omega_u}{\omega_{Ti}}}\right)^2 \frac{I_D}{g_m}. \tag{6.25}$$

While the closed-loop gain magnitude G in (6.25) is typically known a priori, the fan-out term (FO) is also subject to optimization. However, in certain cases, there exist known a-priori optima for the fan-out as well. For example, it has been shown that for a cascade of SC stages, the near-optimum value of FO is equal to $1/G$ [5].[1] Hence, in this case, the product of FO and G in (6.25) simply becomes equal to one. If the circuit is used in an application where this result does not apply, the fan-out

---

[1] The specific condition that leads to this result is that the noise from the sampling and amplification phase of the cascaded SC gain stages are the same. This is optimal since the cost for minimizing either noise component is the same.

becomes a design parameter that must be optimized and/or suitably chosen based on the given application constraints.

The main difference between (6.25) and (5.34) (for the charge amplifier) is due to the difference in the noise expressions and the feedback loading term in (6.9). However, we again see that once the capacitor ratios (closed-loop gain, fan-out) and the bandwidth are fixed, minimizing the drain current boils down to finding the best tradeoff between ω_{Ti} and g_m/I_D. We will illustrate typical tradeoff curves and dependencies using a numerical example.

Example 6.1 Optimization of the Basic OTA

Plot (6.24) and (6.25) for a basic n-channel OTA with $L = 100$ nm, $f_u = 1$ GHz, and (i) $G = 2$ and $FO = 0.5, 1, 2, 4$, as well as (ii) $G = 1, 2, 4, 8$ and $FO \cdot G = 2 =$ constant. Tabulate the following quantities at the minimum current locations: the transistor's g_m/I_D and f_{Ti}, the ratio of β and β_{max} (see (6.6)), as well as the ratio $C_{gs}/(C_S + C_{Ftot})$.

SOLUTION

We sweep g_m/I_D and use the lookup function to find the corresponding ω_{Ti}. Next, we evaluate (6.24) and (6.25) for the given parameters. For part (i), this yields the results given in Figure 6.5 and Table 6.1.

Figure 6.5(a) shows the progression of β as predicted by (6.24). As discussed earlier, larger g_m/I_D leads to larger devices, hence larger C_{gs}, and the feedback factor β must therefore monotonically decrease. Figure 6.5(b) plots (6.25). Recall that this is simply

$$K = \frac{1}{\beta^2} \frac{I_D}{g_m}.$$

K decreases initially due to the increasing g_m/I_D, but ultimately slopes up again due to the sharp drop in β. A shallow minimum exists in-between. It is interesting to note that the minima move toward strong inversion and become less shallow as we increase the fan-out, which means that we make the amplifier "work harder." Also, note that K, and thus the required current, increases rapidly for large g_m/I_D. This rapid increase is essentially a feasibility boundary. In the given technology, it is not possible to build an OTA with $f_u = 1$ GHz with devices biased deep in weak inversion.

Using the same approach as above, we obtain Figure 6.6 and Table 6.2 for part (ii). The most important observation from these results is that the normalized β curves show very little dependency on G and hence the optima essentially coincide. This is explained by examining the denominator of (6.24). The term ω_u/ω_{Ti} is small compared to $1 + G$ and the denominator is therefore approximately constant. Additionally, the numerator contains $FO \cdot G$, which is also kept constant

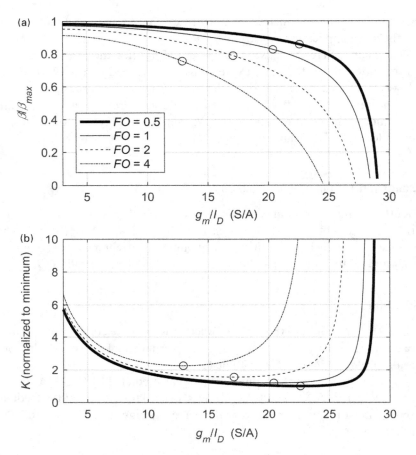

Figure 6.5 (a) Feedback factor β versus g_m/I_D. (b) Relative drain current (K factor, defined in (6.25)) versus g_m/I_D. FO is varied and $G = 2$ (fixed). The curves are normalized with respect to the minimum K for $FO = 0.5$.

Table 6.1 Optimum design parameters for each FO value and $G = 2$.

FO	Optimum Parameters			
	g_m/I_D (S/A)	f_{Ti} (GHz)	β/β_{max}	$C_{gs}/(C_S + C_{Ftot})$
0.5	22.6	12.0	0.857	0.167
1	20.4	15.6	0.825	0.212
2	17.1	22.3	0.788	0.270
4	12.9	35.6	0.755	0.325

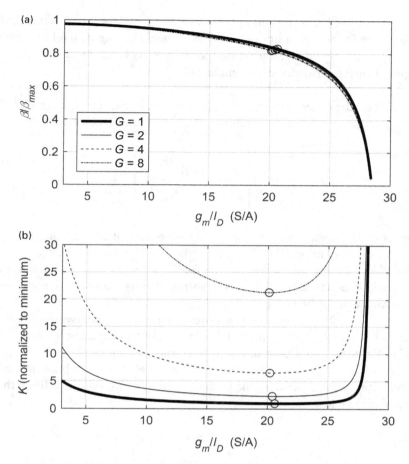

Figure 6.6 (a) Feedback factor β versus g_m/I_D. (b) Relative drain current (K factor defined in (6.25)) versus g_m/I_D. G is varied and $FO \cdot G = 2$ (fixed). The curves are normalized with respect to the minimum K for $G = 1$.

Table 6.2 Optimum design parameters as a function of G for $FO \cdot G = 2$ (fixed).

G	Optimum Parameters			
	g_m/I_D (S/A)	f_{Ti} (GHz)	β/β_{max}	$C_{gs}/(C_S + C_{Ftot})$
1	20.6	15.2	0.830	0.204
2	20.4	15.6	0.825	0.211
4	20.2	15.9	0.822	0.216
8	20.1	16.1	0.819	0.220

in this example. What remains is a similar roll-off in β for all cases due to the ω_u/ω_{Ti} term in the numerator of (6.24). Note also that Figure 6.6(b) captures the "cost" of generating larger closed-loop gain. As G increases, the minimum required current also increases significantly.

Based on the observation from the previous example, it makes sense to eliminate the ω_u/ω_{Ti} term from the numerator of (6.25) and approximate:

$$K \cong \left(\frac{1+G}{1-(1+FO\cdot G)\dfrac{\omega_u}{\omega_{Ti}}} \right)^2 \frac{I_D}{g_m}. \tag{6.26}$$

Algebraically, neglecting the ω_u/ω_{Ti} term is equivalent to dropping the β term in (6.9) in the derivation of (6.24). Physically, this means that we are dropping the dependence of the self-loading on the feedback factor and assume that the entire feedback capacitance (C_{Ftot}) loads the output. This is a conservative and acceptable approximation for first-order analysis.

Equation (6.26) is now almost identical to that of the charge amplifier, given by (5.34). The only differences are in the numerator (which is a constant) and the extra "1" in the ω_u/ω_{Ti} multiplier in the denominator. Using similar steps as in Section 5.5, and assuming strong inversion, we can therefore again obtain tractable first-order expressions describing the optimum sizing conditions. It follows that (6.26) is minimized when

$$\frac{\omega_{Ti}}{\omega_u} = 3(FO\cdot G+1) \tag{6.27}$$

and

$$C_{gs} = \frac{C_S + C_{Ftot}}{3}. \tag{6.28}$$

With this value of C_{gs}, it also follows that

$$\frac{\beta}{\beta_{max}} = \frac{3}{4}. \tag{6.29}$$

Looking back at the values from Example 6.1, we see that these predictions get reasonably close to the observed results. Figure 6.7 illustrates this by comparing (6.29) (vertical gray line) against the actual K factor curves and their optima. The match is particularly good for $FO = 4$, where the transistor is operated near strong inversion. Even for the other cases in moderate inversion with g_m/I_D as large as 22.6, using the β value of (6.29) for design would be quite acceptable due to the shallowness of the optima.

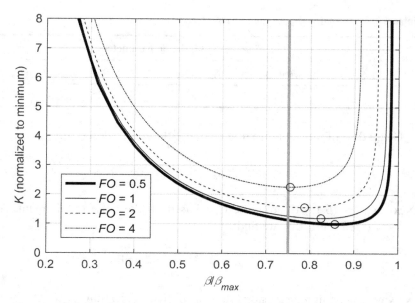

Figure 6.7 Relative drain current (K factor defined in (6.25)) versus normalized feedback factor. All parameters are as in Example 6.1, case (i).

The above expressions also provide valuable first-order guidance on the technology requirements for some given specifications. For example, (6.27) predicts that for $FO \cdot G = 2$, power efficient operation requires a transistor with an ω_{Ti} that is nine times larger than ω_u (the exact values seen in Example 6.1 are in the range of 7.6 to 8). Such estimates are useful for process technology selection and first-order feasibility studies.

Once we know the optimum inversion level and feedback factor of the transistor, it is straightforward to complete the design using the noise and bandwidth specifications. We will illustrate this in the following example.

Example 6.2 Sizing of the Basic OTA

Size the basic n-channel OTA circuit to achieve $f_u = 1$ GHz and a total integrated output noise of 100 μV_{rms} at minimum current consumption. Assume $G = 2$, $FO = 1$ and $L = 100$ nm. Determine the device width and all capacitor sizes. Calculate the static gain error and the time it takes for the dynamic settling error to decay below 0.1%. Validate the design using SPICE simulations.

SOLUTION

To solve this problem, we can re-use the data from Example 6.1, summarized for convenience in Table 6.3.

Table 6.3 Optimum design point from Example 6.1.

FO	Optimum parameters			
	g_m/I_D (S/A)	f_{Ti} (GHz)	β/β_{max}	$C_{gs}/(C_S + C_{Ftot})$
1	20.4	15.6	0.825	0.212

Using this information, we first compute the feedback factor:

$$\beta = 0.825\beta_{max} = 0.825 \cdot \frac{1}{3} = 0.275.$$

Next, we use (6.15) to compute C_{Ltot} based on the noise requirement. Assuming $\gamma_n = 0.7$ (see Figure 4.2), we find:

$$C_{Ltot} = 2\frac{\gamma_n}{\beta}\frac{k_BT}{v_{od}^2} = 2.1\text{pF}.$$

Now we can compute g_m using the bandwidth specification:

$$g_m \cong \frac{C_{Ltot}\omega_u}{\beta} = 48.2\,\text{mS}.$$

Using the g_m/I_D value from Table 6.3, we then find I_D = 2.36 mA. To compute the device width, we use the lookup function to determine the current density (J_D = 3.02 A/m), and arrive at W = 783 µm. Finally, to determine the capacitor sizes, we can use

$$C_{Ltot} = C_L + (1-\beta)C_{Ftot} = C_{Ftot}\frac{C_L}{C_{Ftot}} + (1-\beta)C_{Ftot} = C_{Ftot}\left(FO \cdot G + (1-\beta)\right).$$

This lets us solve for C_{Ftot} = 774 fF and using the given values for G and FO leads to $C_S = C_L$ = 1.55 pF. To find the static gain error, we first look up the transistor's intrinsic gain

```
gm_gds = lookup(nch,'GM_GDS', 'GM_ID', gm_ID, 'L', L)
```

which is 24.2. From here, we compute the static gain error using

$$\varepsilon_s = -\frac{1}{L_0} \cong -\frac{1}{\beta\frac{g_m}{g_{ds}}} = -15\%.$$

Note that this value is rather large and not necessarily practical for a real application. The time to settle to $\varepsilon_d < 0.1\%$ follows from (6.12):

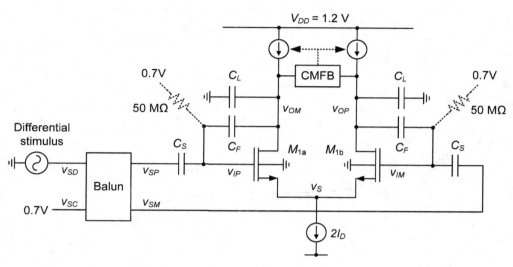

Figure 6.8 Simulation circuit.

$$t_s = \tau \cdot ln\left(\frac{1}{0.1\%}\right) = 6.9\tau = \frac{6.9}{\omega_u} = 1.10\,\mathrm{ns}.$$

To validate the design, we set up a simulation schematic as shown in Figure 6.8. The 50 MΩ resistors establish the gate bias voltages (not needed in the actual switched-capacitor circuit). The balun is used for convenience, to create the required differential input signals from a single-ended source. An ideal common-mode feedback (CMFB) circuit is used to set the output quiescent points to 0.8 V. One additional parameter that is needed for simulation is the value of $C_F = C_{Ftot} - C_{gd}$. To find C_{gd}, we use the following lookup command:

```
Cgd = W*lookup(nch,'CGD_W', 'GM_ID', gm_ID, 'L', L)
```

which gives $C_{gd} = 259$ fF and consequently $C_F = 515$ fF.

To measure the closed-loop bandwidth f_c, we inject an AC input as the differential stimulus, which results in the response shown in Figure 6.9(a). The expected low-frequency voltage gain is $C_S/C_F(1 + \varepsilon_s) = 1.7 = 4.61$ dB. The simulation result matches this number closely. The value of f_c (1.05 GHz) is also very close to the expected value. At first glance, this is somewhat surprising since we have neglected the transistors' drain-to-bulk capacitance, which can be estimated using

```
Cdb = W*lookup(nch,'CDD_W', 'GM_ID', gm_ID, 'L', L) - Cgd
```

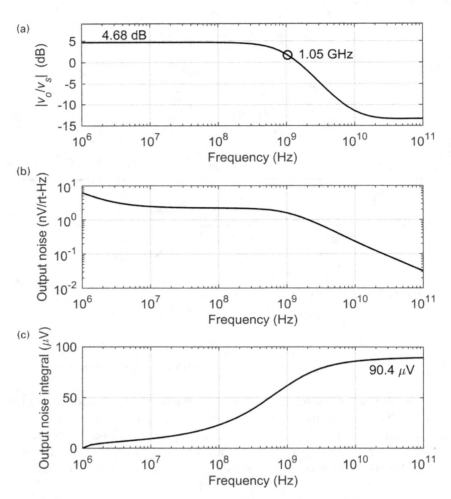

Figure 6.9 AC and noise simulation results. (a) Magnitude of closed-loop voltage gain. (b) Output noise spectral density. (c) Running integral of the noise spectral density in (b).

which gives 230 fF (about 10% of C_{Ltot}). This is significant, and should lead to a bandwidth degradation of about 10%. However, this degradation is compensated by r_{ds}, which we also neglected in the calculation of f_u. Overall, the initial decision to neglect C_{db} is well-justified, since it simplified the calculations without causing a large error.

The simulated total integrated RMS noise voltage (see Figure 6.9(c)) is about 10% smaller than expected. This is again explained by the neglected drain-bulk capacitance.

The result of a transient simulation with an input step of 10 mV (applied at t = 1 ns) is shown in Figure 6.10. The circuit settles to the value expected from the static gain error and the dynamic settling error decays below 0.1% in 1.06 ns. This matches the computed estimate of 1.1 ns very closely.

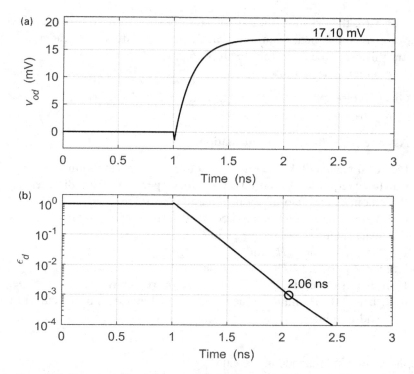

Figure 6.10 Transient simulation results. (a) Output voltage step. (b) corresponding dynamic setting error.

As we saw in the previous example, neglecting the junction capacitance C_{db} had a somewhat significant effect on the observed total integrated noise. In this context, it is important to note that this particular discrepancy can be resolved easily, without affecting any other specification. To achieve this, one can use the following noise scaling approach:

1. Calculate the ratio (actual noise power)/(desired noise power). Call this number S.
2. Multiply all capacitances, device widths and drain currents by S.

This method maintains constant current densities (hence constant g_m/I_D, f_T) and constant ratios of g_m/C, and therefore the frequency response remains unaffected. However, the noise, which is inversely proportional to the capacitances, is properly adjusted. For Example 6.2, the scale factor would be $S = 0.9^2 = 0.81$. The current therefore reduces by 19% if we allow the RMS noise voltage level to rise to the specified 100 μV.

An additional aspect relates to the observation that we have already made in Section 5.5. If we insist on operating at the exact optimum, and the target speed is relatively low compared to the technology limits (consider sensor circuits as an example), one will end up with a low inversion level and correspondingly large device sizes. To some extent, this is also the case in Example 6.2, where we arrived at $W = 783$ μm. Since the optimum in such a case is known to be shallow (see

Figure 6.6), the designer should consider moving away from the optimum, use a lower g_m/I_D and benefit from significantly smaller widths. The reader is encouraged to try this using the data from Example 6.2.

In summary, the findings from this section lead to the following typical design flow:

- Given: Noise spec, settling time spec, ideal closed-loop gain, low-frequency loop gain, fan-out.
- Assume $\beta = \dfrac{3}{4}\beta_{max}$ for all calculations.
- Pick channel length based on low-frequency loop gain requirement.
- Calculate required C_{Ltot} based on the noise spec. This fixes all other capacitances (based on β, ideal closed-loop gain and fan-out).
- Calculate g_m based on the required bandwidth (settling time).
- Calculate required f_{Ti} (using g_m and C_{gs}).
- Use lookup to find g_m/I_D and calculate I_D.
- Use lookup to find I_D/W and calculate W.
- If the inversion level is low and the widths are excessively large, consider reducing g_m/I_D below the optimum value to trade small increases in the current for significant width reductions.

6.1.3   Optimization in Presence of Slewing

The previous two subsections considered the settling of the amplifier for small inputs and perfectly linear circuit behavior. However, since practical SC circuits rely on relatively large swings (to maximize the signal-to-noise ratio), large-signal effects must usually be considered for a better prediction of the setting time.

The problem can be understood by considering the large-signal I-V characteristic of the amplifier's input pair ($M_{1a,b}$ in Figure 6.2), shown in Figure 6.11. We can linearly approximate this transfer function only when the differential input lies well within $2I_D/g_m$. If the signal goes outside this range, the differential drain current saturates, leading to a constant rate of change in the output voltage. This phenomenon is known as "slewing" (see standard text books, such as [1], [4]).

Compared to continuous-time circuits, the analysis of slewing in SC circuits is complicated by the fact that one must consider the transient behavior, instead of

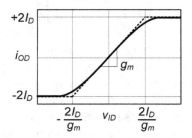

Figure 6.11 I-V characteristic of a differential pair. v_{ID} is the differential input voltage and i_{OD} is the differential output current ($i_{D1a} - i_{D1b}$ in Figure 6.2).

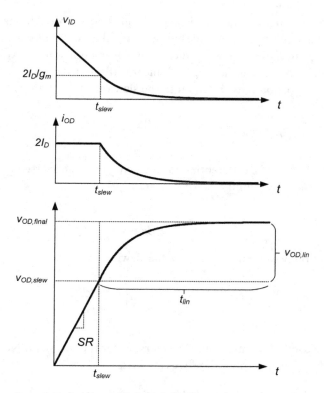

Figure 6.12 Settling with initial slewing and subsequent linear settling.

steady-state waveforms. As illustrated in Figure 6.12, SC circuits slew only during the initial transient and later return into the linear regime. At $t = 0$, when the input step is applied, the input pair may leave its linear region (as shown) and the differential pair current saturates at $2I_D$. However, as time progresses, the input signal reduces and the amplifier eventually returns back into the linear region for the remainder of the setting process. The transition between slewing and linear settling is difficult to describe with closed-form expressions. Therefore, it is common to assume an approximate piecewise model in which the transition occurs abruptly, corresponding to the dashed line in Figure 6.11.

To compute the time required for initial slewing (t_{slew}) and subsequent linear settling (t_{lin}), we follow a derivation similar to the one presented in [6] for a two-stage OTA. The main difference here is that we instead consider the basic single-stage OTA of Figure 6.2.

During the initial slewing, all the differential pair's tail current ($2I_D$) is supplied to the load capacitance and therefore the rate of change in the output, called the slew rate, is given by:

$$SR = \frac{2I_D}{C_{Ltot}} = \frac{2I_D}{\tau \beta g_m}. \qquad (6.30)$$

The result on the right-hand side of this expression follows from the bandwidth expression in (6.8), using $\tau = 1/\omega_u$. As we shall see later, this substitution leads to an elegant result that combines the slewing and linear settling portions. Next, we realize that at the transition between slewing and linear settling ($t = t_{slew}$), the derivative of the linear settling portion must be equal to the slew rate (see Figure 6.12):

$$\frac{d}{dt}\left[v_{OD,lin}\left(1 - e^{-\frac{t}{\tau}}\right)\right]_{t=0} = \frac{v_{OD,lin}}{\tau} = SR.$$

(6.31)

In this expression, $v_{OD,lin}$ is the differential output swing that is traversed during linear settling. Using (6.31), we find:

$$v_{OD,lin} = \tau \cdot SR = \frac{2I_D}{\beta g_m}.$$

(6.32)

With this information, we can now compute the slewing time:

$$t_{slew} = \frac{v_{OD,slew}}{SR} = \frac{v_{OD,final} - v_{OD,lin}}{\dfrac{v_{OD,lin}}{\tau}} = \tau\left(\frac{v_{OD,final}}{v_{OD,lin}} - 1\right),$$

(6.33)

where $v_{OD,final}$ is the total differential output excursion for the transient. Now, to calculate the remaining linear settling time, we can write

$$t_{lin} = \tau \cdot \ln\left(\frac{1}{\varepsilon_d} \cdot \frac{v_{OD,final}}{v_{OD,lin}}\right).$$

(6.34)

Here, the term multiplying $1/\varepsilon_d$ inside the bracket term accounts for the fact that the linear settling portion spans only a fraction of the overall transient. In the previous subsections, where we ignored slewing, this multiplier was unity. The total settling time (t_s) now follows from adding the above-computed terms:

$$t_s = t_{slew} + t_{lin} = \tau\left(X - 1 + \ln\left(\frac{1}{\varepsilon_d} \cdot X\right)\right),$$

(6.35)

where we defined for convenience:

$$X = \frac{v_{OD,final}}{v_{OD,lin}} = v_{OD,final} \cdot \frac{\beta}{2}\frac{g_m}{I_D}.$$

(6.36)

Based on this result, the required unity gain frequency is:

$$\omega_u = \frac{1}{\tau} = \frac{1}{t_s}\left(X - 1 + \ln\left(\frac{1}{\varepsilon_d} \cdot X\right)\right).$$

(6.37)

For fixed specifications of settling time (t_s) and dynamic settling error (ε_d), the required amplifier bandwidth depends on the slewing parameter X, which involves

g_m/I_D and the feedback factor β. These dependencies not only make it impossible to derive closed-form optima, but also require special attention in devising a workable numerical optimization flow, since both g_m/I_D and β are subject to optimization themselves. We will approach this problem using a two-dimensional numerical search, which is described below. The main idea of this algorithm is to assume some β value smaller than β_{max}, and calculate the resulting circuit parameters (g_m and C_{gs}). Using these parameters, the actual β value is found. If the assumed β value and the actual β value match, the computed design point is feasible.

1. Sweep β (from some fraction of β_{max} to β_{max}) using a for-loop.
2. For each β_k in the for-loop, and for a vector of g_m/I_D within a reasonable range (weak to strong inversion) compute the following:
   a. C_{Ltot} using (6.15), the noise specification and β_k
   b. C_{Ftot} using (6.9), C_{Ltot} and β_k
   c. X using (6.36), $v_{od,final}$, β_k and the g_m/I_D vector
   d. ω_u using (6.37), t_s, ε_d and X
   e. g_m using (6.8), C_{Ltot} and β_k
   f. I_D and f_{Ti} using the g_m/I_D vector
   g. C_{gs} using g_m and f_{Ti}
   h. The actual β values along the g_m/I_D vector used for the above computations.
3. In the vector of actual β values, find the closest match with β_k. If a close match exists, this corresponds to a physical design point that can be considered/plotted.

We will now illustrate this algorithm using an example.

Example 6.3 Sizing of the Basic OTA Circuit in Presence of Slewing

Size the basic OTA circuit to achieve minimum current consumption for $t_s = 1.1$ ns (0.1% settling accuracy) and an output noise of 100 μV_{rms}. Assume $G = 2$, $FO = C_L/C_S = 1$, $L = 100$ nm and $v_{OD,final} = 10$ mV (small-signal operation), 800 mV, and 1600 mV. Determine the device width and all capacitor sizes for $v_{OD,final} = 800$ mV and validate the design using SPICE simulations. Note: The given differential output swing of 1600 mV is not practical for a 1.2 V supply, but is included here for illustrative purposes.

SOLUTION

The described approach is implemented using the Matlab code below, with an outer for-loop added for $v_{OD,final}$.

```
% Search parameters
vodfinal = [0.01 0.8 1.6];
gm_ID = (5:0.01:28)';
beta = (0.25*beta_max:0.001:beta_max)';
% pre-compute wti
```

```
wti = lookup(nch, 'GM_CGS', 'GM_ID', gm_ID, 'L', L);
for i = 1:length(vodfinal);
   for j = 1:length(beta)
      % compute CLtot based on noise
      CLtot = 2*kB*T*gamma./beta(j)/vod_noise^2;
      CFtot = CLtot./(CL_CFtot + 1-beta(j));
      % compute X and drain current
      X = vodfinal(i)*beta(j)./2*gm_ID;
      X(X<1) = 1;
      ID = CLtot/beta(j)./gm_ID/ts.*(X-1 - log(ed*X));

      % compute gm and Cgs
      gm = gm_ID.*ID;
      Cgs = gm./wti;

      % compute actual beta and find self-consistent point
      beta_actual = CFtot./(CFtot*(1+G) + Cgs);
      m = interp1(beta_actual,1:length(beta_actual),beta(j), ...
         'nearest', 0);
      if(m)
        gm_ID_valid(j,i) = gm_ID(m);
        ID_valid(j,i) = ID(m);
        X_valid(j,i) = X(m);
      end
   end
end
```

The results are shown in Figure 6.13. As expected, the small-signal case (labelled SS) gives the minimum possible current and corresponds to the same result seen in Example 6.2. The cases with slewing require larger currents and the optima lie at slightly lower values of g_m/I_D. This outcome is intuitive, since the extra time spent on slewing shortens the linear settling time, requiring a faster circuit with higher bandwidth and higher device f_T (smaller g_m/I_D). For an output swing of 800 mV, the slewing time is about 16% of the overall transient and it increases to about 32% for 1600 mV swing. The latter value is impractical for a circuit operating within a 1.2 V supply, but it helps show the trend toward larger slewing times for larger signals.

At the optimum point for $v_{OD,final} = 800$ mV, we find the following parameter values:

```
gm_ID =   19.5500
beta =  0.2773
ID =    0.0026
CLtot = 2.0909e-12
```

From here, we can compute all remaining parameters as done in Example 6.2, which gives:

```
W =    713.5175
CFtot = 7.6796e-13
CS =   1.5359e-12
```

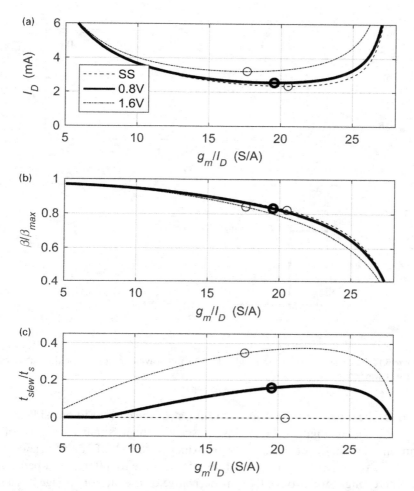

Figure 6.13 Sweep for drain current optimization. Three different differential output signal swings are considered: 10 mV (SS = small-signal), 0.8 V and 1.6 V. (a) Drain current. (b) Normalized feedback factor. (c) Normalized slewing time.

```
CL =    1.5359e-12
CF =    5.3203e-13
```

The expected slew rate and slewing time are:

```
SR =     2.4488e+09
tslew =  1.7606e-10
```

Figure 6.14 shows a SPICE simulation of the circuit's step response, with the input sized to yield an 800-mV output step. We see that the settling time is very close to the expected value (1.1 ns).

The observed slew rate is smaller than expected for two reasons: (1) extra junction capacitance (about 10% of C_{Ltot}), which was not included in the analysis,

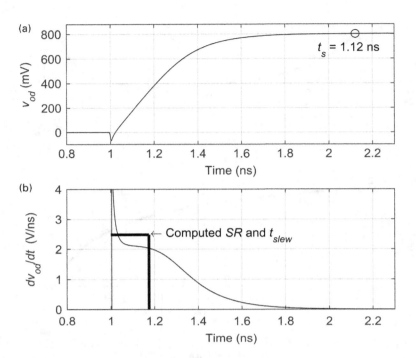

Figure 6.14 Transient simulation results. (a) Output waveform. (b) Time derivative of the output waveform.

and (2) the differential pair does not fully steer the current to one side as assumed in the model of Figure 6.11. A closer inspection reveals that about 10% of the current flows in the device that is approximated to be "off" during slewing. This explains the overall discrepancy of about 20% (due to differential operation).

Interestingly, we also see that the derivative of the output voltage has a spike initially, and this partly makes up for the reduced slew rate plateau. The spike is present because the tail node of the differential pair is forced upward as the input step is applied, and this contributes extra dynamic tail current via the parasitic capacitance at that node. Some additional speedup is due to r_{ds}, which shortens the linear settling time as already observed in Example 6.2.

In summary, this section illustrated suitable design procedures for an SC gain stage based on a basic differential pair OTA. We saw that closed-form (first-order) optima can be found only if small-signal operation is assumed. Once we apply large signals that cause slewing, such optima are no longer analytically tractable. However, the g_m/I_D-based design approach allows us to find the optima using a numerical sweep, making it possible to minimize the current considering the slewing and linear settling portions simultaneously.

While we saw that it is possible to find the exact optimum design point using a 2-D sweep, one can also devise a simpler design flow that is based on assuming

$\beta = 0.75 \, \beta_{max}$. This is once again justified, since the optima tend to be quite shallow. The recommended flow is then similar to the case without slewing (considered earlier), but has two extra steps (in bold):

- Given: Noise spec, settling time spec, ideal closed-loop gain, low-frequency loop gain, fan-out.
- Assume $\beta = 0.75 \, \beta_{max}$ for all calculations.
- Pick channel length based on low-frequency loop gain requirement.
- **Budget a reasonable amount of slewing time, for example $t_{slew}/t_s = 0.3$ (subject to iteration).**
- Calculate required C_{Ltot} based on noise spec. This fixes all other capacitances as well (based on β, ideal closed-loop gain and fan-out).
- Calculate g_m based on required bandwidth (settling time).
- Calculate required f_T (using g_m and C_{gs}).
- Use lookup to find g_m/I_D and calculate I_D.
- **Calculate the actual slewing time and adjust assumed t_{slew}/t_s (if needed).**
- Use lookup to find I_D/W and calculate W.

6.2 Folded-Cascode OTA for Switched-Capacitor Circuits

The previous section considered the simplest possible OTA implementation to establish the fundamentals. We now discuss a more common circuit implementation based on the folded-cascode topology shown in Figure 6.15. As before, we assume that this circuit is operated within a typical switched-capacitor circuit, as shown in Figure 6.1. The reader may refer to any standard text book (such as [4]) to review the basic operation and biasing of the circuit in Figure 6.15.

Before we can begin to design and optimize this circuit, we must refine the equations derived in the previous section to include additional artifacts. For brevity, we consider only linear settling for the time being, and discuss modifications required for slewing at the end of this section.

6.2.1 Design Equations

The most significant deviation from the basic OTA in Section 6.1 is due to the non-dominant pole caused by the folded cascodes:

$$\omega_{p2} \cong \frac{g_{m3} + g_{mb3}}{C_{dd1} + C_{ss3} + C_{dd2}} \cong \frac{g_{m3} + g_{mb3}}{C_{ss3} + 2C_{dd2}}. \tag{6.38}$$

In the final approximation on the right-hand side, we assumed that the total drain capacitances of M_1 and M_2 are comparable. As we will see below, this allows us to estimate ω_{p2} before the size of M_1 is known.

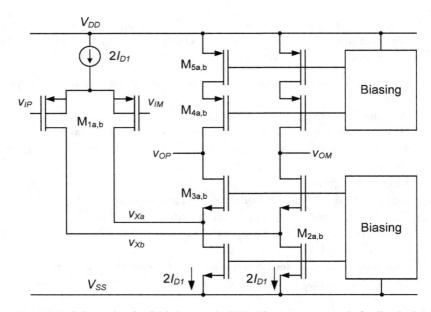

Figure 6.15 Schematic of a folded-cascode OTA (the common-mode feedback circuit is not shown). The bulks of M_4 and M_5 are connected to V_{DD}, while the bulks of M_2 and M_3 are tied to V_{SS}.

For large phase margin, the non-dominant pole is usually placed beyond the unity gain frequency of the loop, as shown in Figure 6.16. Also, shown in this figure is the non-dominant pole's effect on the exact unity gain frequency. To interpret SPICE simulation results properly, it is worth distinguishing the actual unity gain frequency (ω_u) and the value predicted from a straight line analytical approximation (ω_{ul}). The actual unity gain frequency will always be somewhat smaller that ω_{ul}, due the "bending" of the magnitude response caused by the second pole. We will enumerate the ratio ω_{ul}/ω_u as part of our analysis below.

With a second pole in the loop transfer function, it can be shown that the circuit's closed-loop transfer function is given by

$$A_{CL}(s) = \frac{v_{od}}{v_{sd}} = \frac{A_{CL0}}{1 + \dfrac{s}{\omega_0 Q} + \dfrac{s^2}{\omega_0^2}}, \tag{6.39}$$

where A_{CL0} is the closed-loop low-frequency gain as in (6.7), and

$$\omega_0 \cong \sqrt{\omega_{ul}\omega_{p2}} \quad Q \cong \sqrt{\frac{\omega_{ul}}{\omega_{p2}}}. \tag{6.40}$$

Table 6.4 enumerates the relationship between the relative non-dominant pole position ω_{p2}/ω_{ul} and other relevant metrics of the second-order system. The case of

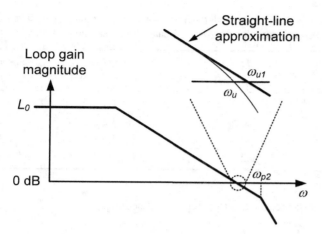

Figure 6.16 Loop gain magnitude for the folded-cascode OTA. The non-dominant pole leads to a slightly smaller unity gain frequency (ω_u) than the predicted value from a straight-line approximation (ω_{u1}).

Table 6.4. Parameters of the second-order transfer function as a function of non-dominant pole location. The limit case of $\omega_{p2}/\omega_{u1} \to \infty$ corresponds to a first-order system.

ω_{p2}/ω_{u1}	Q	ω_u/ω_{u1}	Phase margin (°)
1	1	0.786	51.8
2	0.707	0.910	65.5
3	0.577	0.953	72.4
4	**0.500**	**0.972**	**76.3**
5	0.477	0.981	78.9
6	0.408	0.987	80.7
7	0.378	0.990	81.9
8	0.354	0.992	82.9
9	0.333	0.994	83.7
10	0.316	0.995	84.3
∞	—	1	90

$\omega_{p2}/\omega_{u1} = 4$ ($Q = 0.5$) results in a critically damped step response. This corresponds to the fastest possible settling without overshoot [7] and is the preferred choice for SC circuits that are designed for maximum speed. Designing for $\omega_{p2}/\omega_{u1} < 4$ is not recommended to avoid overshoot, which is difficult to manage if the goal is to create a robust design. On the other hand, designing for $\omega_{p2}/\omega_{u1} > 4$ is acceptable, but comes with a minor loss of settling speed. In this context, it is useful to compare the settling time for the critically damped case with the limit case of a first order system ($\omega_{p2}/\omega_{u1} \to \infty$).

Table 6.5 Required number of settling time constants for a first-order system and a second-order system with critical damping ($\omega_{p2}/\omega_{u1} = 4$). The rightmost column quantifies the speed advantage of the second-order system.

Dynamic settling error (ε_d)	t_s/τ ($\omega_{p2}/\omega_{u1} \to \infty$)	t_s/τ ($\omega_{p2}/\omega_{u1} = 4$)	Speedup (%)
10%	2.3	1.9	15.5
1%	4.6	3.3	27.9
0.1%	6.9	4.6	33.1
0.01%	9.2	5.9	36.2

For the critically damped case, the step response is:

$$v_{od}(t) = v_{od,final}\left(1 - \left(1 + \frac{2t}{\tau}\right)e^{-\frac{2t}{\tau}}\right) = v_{od,final}\left(1 - \varepsilon_d(t)\right), \tag{6.41}$$

where $\tau = 1/\omega_{u1}$. We can solve this expression numerically for the settling time and compare to the first-order expression of (6.11). Table 6.5 compares the required number of time constants as a function of the dynamic settling error for both cases.

We now turn our attention to the unity gain frequency of the loop, which is similar to (6.8), but more precisely given by:

$$\omega_{u1} = \beta\frac{\kappa g_{m1}}{C_{Ltot}}. \tag{6.42}$$

We will inspect the terms in this equation one by one and highlight the differences with respect to the basic OTA. The κ term captures the current division at the folded-cascode nodes ($v_{Xa,b}$ in Figure 6.15):

$$\kappa \cong \frac{g_{m3} + g_{mb3}}{g_{m3} + g_{mb3} + g_{ds1} + g_{ds2}}$$

$$\cong \frac{1}{1 + \dfrac{g_{ds1}}{g_{m1}}\dfrac{g_{m1}}{g_{m3} + g_{mb3}} + \dfrac{g_{ds2}}{g_{m2}}\dfrac{g_{m2}}{g_{m3} + g_{mb3}}}$$

$$\cong \frac{1}{1 + \dfrac{g_{ds1}}{g_{m1}}\dfrac{g_{m1}}{g_{m3}} + 2\dfrac{g_{ds2}}{g_{m2}}}. \tag{6.43}$$

The factor of two in the final expression comes from assuming $g_{m2} = 2g_{m3}$, which will hold when M2 and M3 have the same channel length and $W_2 = 2W_3$.

To get a feel for the value of κ, we recall from Chapter 2 that the intrinsic gain g_m/g_{ds} for short-channel devices in our 65-nm process is of the order of 10 (for reasonably high V_{DS}, in moderate inversion). If g_{m1} and g_{m3} in the final approximation of (6.43) are comparable, it then follows that κ can be as low as 0.7. We will use this number as a conservative estimate in some of our initial calculations.

The total load capacitance in (6.42) is defined as:

$$C_{Ltot} = \left[C_L + (1-\beta) C_F \right] (1 + r_{self}),$$

(6.44)

where r_{self} is given by:

$$r_{self} = \frac{C_{dd3} + C_{dd4}}{C_L + (1-\beta) C_F}.$$

(6.45)

Compared to (6.9), C_{Ftot} has now become C_F, since the gate-drain capacitance of the input pair is no longer connected in parallel. The parameter r_{self} represents the self-loading of the amplifier, which can be significant in high-speed circuits with small on-chip load capacitances. We will see this in the numerical example below.

The feedback factor β in (6.42) also differs somewhat from that of the basic OTA (see (6.5)):

$$\beta = \frac{C_F}{C_F + C_S + C_{in}}.$$

(6.46)

Here, C_{in} is the input capacitance of the folded-cascode OTA, which we will approximate as:

$$C_{in} \cong C_{gs1} + C_{gb1} + C_{gd1} \left(1 + \frac{g_{m1}}{g_{m3}} \right) = C_{gg1} + C_{gd1} \frac{g_{m1}}{g_{m3}}.$$

(6.47)

The bracketed term in this expression is due to the Miller multiplication of the gate-drain capacitance, and the "1" inside the bracket is absorbed into C_{gg1} in the final result on the right-hand side. The term due to C_{gd1} may not be significant, but including it in the optimization can help improve the results.

Based on (6.46), the expressions involving the maximum possible β become:

$$\beta_{max} = \frac{C_F}{C_F + C_S} = \frac{1}{1+G}$$

(6.48)

and thus:

$$\beta = \beta_{max} \frac{1}{1 + \dfrac{C_{in}}{C_F + C_S}}.$$

(6.49)

The low-frequency loop gain of the circuit is

$$L_0 = \beta \kappa g_{m1} R_o,$$

(6.50)

where R_o is the OTA's output resistance. Neglecting the g_{mb} terms for simplicity, R_o is approximately given by:

$$\frac{1}{R_o} \cong \frac{g_{ds4}}{1 + \dfrac{g_{m4}}{g_{ds5}}} + \frac{g_{ds3}}{1 + \dfrac{g_{m3}}{g_{ds1} + g_{ds2}}}. \tag{6.51}$$

This expression mixes parameter ratios from different devices and is therefore hard to use for design. We make the following assumptions to arrive at a useful first-order approximation: (1) All transistors have the same g_m/I_D, which implies $g_{m1} = g_{m2}/2 = g_{m3} = g_{m4} = g_{m5}$. (2) The output conductance of M_1 is close to $g_{ds2}/2$ (note that M_1 carries half of M_2's current). Applying these simplifications to (6.51) and inserting into the reciprocal of (6.50) then yields:

$$\frac{1}{L_0} \cong \frac{1}{\beta \kappa} \left(\frac{1}{\left(1 + \dfrac{g_{m5}}{g_{ds5}}\right) \dfrac{g_{m4}}{g_{ds4}}} + \frac{1}{\left(1 + \dfrac{1}{3} \dfrac{g_{m2}}{g_{ds2}}\right) \dfrac{g_{m3}}{g_{ds3}}} \right). \tag{6.52}$$

The advantage of this simplified expression is that it links the low-frequency loop gain more directly to the individual g_m/g_{ds} ratios. This coarse approximation is reasonable since L_0 is not a parameter that must be precisely controlled. It only needs to be larger than a certain minimum and is often overdesigned by some margin.

The total integrated noise of the folded-cascode OTA is similar to that of the basic topology in (6.15):

$$\overline{v_{od}^2} = \frac{\alpha}{\beta} \frac{k_B T}{C_{Ltot}}. \tag{6.53}$$

The main difference is that there is excess noise due to the current sources in the folded-cascode branch. The factor α has a similar format to the bracketed term in (4.9), but contains two excess noise terms, due to M_2 and M_5:

$$\alpha = 2\gamma_1 \left(1 + \frac{\gamma_5}{\gamma_1} \frac{(g_m/I_D)_5}{(g_m/I_D)_1} + 2 \frac{\gamma_2}{\gamma_1} \frac{(g_m/I_D)_2}{(g_m/I_D)_1} \right). \tag{6.54}$$

For simplicity, we neglected the noise contributed by the cascode devices in the above expressions.[2]

With these modifications to the basic equation set, the problem of finding the optimum inversion level and feedback factor is again not analytically tractable. However, we will see in the next section that the optimal feedback factor is still

---

[2] The noise from the cascode devices is irrelevant at low frequencies, but becomes significant at high frequencies. Since high-frequency noise folds back into the signal band of SC circuits, cascode devices can contribute about 10–20% excess noise in typical scenarios.

close to the first-order analytical results derived in Section 6.1.2. Therefore, these guidelines remain useful for sanity checks and initial parameter guesses.

6.2.2   Optimization Procedure

Owing to the larger circuit size, the optimization of the folded-cascode OTA is more complex than that of the basis OTA studied previously. However, using a divide and conquer approach lets us manage this complexity by breaking the task into multiple steps. The first step is to design the cascode stack ($M_2 - M_5$) based on the circuit's output swing and low-frequency loop gain requirements. Next, we investigate the optimum inversion level for the input pair, much like it was done in the previous section for the basic OTA. Finally, we can combine the two parts to complete the sizing of the entire circuit and run SPICE simulations to validate the design.

We begin with the design of the cascode stack, noting that its design must consider the output swing (which bounds V_{Dsat} of $M_2 - M_5$), the low-frequency loop gain (per (6.52)) and the non-dominant pole (per (6.38)). These specifications are strongly affected by the sizing of $M_2 - M_5$, while the input pair parameters have only a small impact that can be neglected in a first-pass design. Let us consider an example.

Example 6.4 Sizing of the Folded-Cascode Output Branch

Pick the inversion levels and channel lengths of the output branch devices (M_2, M_3, M_4 and M_5 in Figure 6.15) so that the circuit can accommodate a differential peak-peak output swing of 0.8 V, while achieving a low-frequency loop gain $L_0 > 50$ with $G = C_S/C_F = 2$. Estimate the non-dominant pole frequency for the chosen design values.

SOLUTION

We begin by inspecting the implications of the output swing requirement. If the output common mode is at $V_{DD}/2 = 0.6$ V, each half circuit output will swing from 0.4 V to 0.8 V, leaving about 400 mV of saturation voltage for the cascode stacks. If we split this voltage equally, then each transistor has a minimum V_{DS} of 200 mV. From Chapter 2, we know that this restricts the g_m/I_D of the transistors to values larger than 10 S/A. On the other hand, we may want to make g_m/I_D as small as possible to extract the highest possible ω_T, for large ω_{p2} and commensurately large unity gain frequency. As a compromise that leaves some margin, we decide on $g_m/I_D = 15$ S/A for all transistors in the output branches. This design choice can be revisited later, but there is not much flexibility; one can essentially only sacrifice margin.

We can now decide about the channel lengths in the cascode stack based on (6.52). The code below computes the low-frequency loop gain versus channel

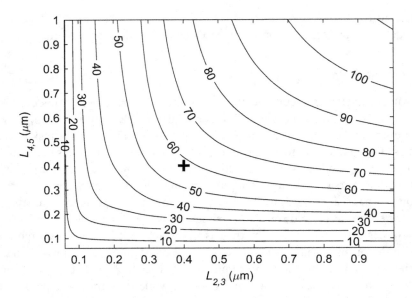

Figure 6.17 Estimate for the low-frequency loop gain (L_0) as a function of channel lengths. The "+" sign marks the chosen design point.

length for the cascode devices. For simplicity, it is assumed that all n-channels and all p-channels have the same length ($L_2 = L_3 = L_{2,3}$ and $L_4 = L_5 = L_{4,5}$). This is yet another design choice that can be re-visited later, if needed.

```
% Design specifications and assumptions
G = 2;
beta_max = 1/(1+G);
beta = 0.75*beta_max;   % first-order optimum
kappa = 0.7;            % conservative estimate
gm_ID = 15;
% Channel length sweep
L = linspace(0.06, 1, 100); L23=L; L45=L;
gm_gds2 = lookup(nch, 'GM_GDS', 'GM_ID', gm_ID,...
  'VDS', 0.2, 'L', L23);
gm_gds3 = lookup(nch, 'GM_GDS', 'GM_ID', gm_ID,...
  'VDS', 0.4, 'L', L23);
gm_gds4 = lookup(pch, 'GM_GDS', 'GM_ID', gm_ID,...
  'VDS', 0.4, 'L', L45);
gm_gds5 = lookup(pch, 'GM_GDS', 'GM_ID', gm_ID,...
  'VDS', 0.2, 'L', L45);
```

The result of this sweep is shown in Figure 6.17. To meet the requirement of $L_0 > 50$, we decide to use $L_{2,3} = L_{4,5} = 0.4$ μm. This choice leads to the point marked by a "+".

With the channel lengths in the output branch fixed, we can now also estimate the non-dominant pole frequency using (6.38).

```
% Chosen length
L23 = 0.4;
% Resulting device parameters
gmb_gm3 = lookup(nch,'GMB_GM','GM_ID',gm_ID,'VDS',0.4,...
 'VSB',0.2,'L', L23);
gm_css3 = lookup(nch,'GM_CSS','GM_ID',gm_ID,'VDS',0.4,...
 'VSB',0.2, 'L',L23);
cdd_css3 = lookup(nch,'CDD_CSS','GM_ID',gm_ID,'VDS',0.4,...
 'VSB',0.2,'L',L23);
cdd_w3 = lookup(nch,'CDD_CSS','GM_ID',gm_ID,'VDS',0.4,...
 'VSB',0.2,'L',L23);
cdd_w2 = lookup(nch,'CDD_CSS','GM_ID',gm_ID,'VDS',0.2,...
 'L',L23);
% Nondominant pole frequency
fp2=1/2/pi*gm_css3*(1+gmb_gm3)/(1+2*cdd_css3*2*(cdd_w2/cdd_w3));
```

The factor of two in front of the $(C_{dd}/W)_2$ and $(C_{dd}/W)_3$ ratio in the last line follows from $W_2 = 2W_3$. The result of the above computation is $f_{p2} = 1.45$ GHz. For comparison, the transit frequency of M_3, which is the main device that defines f_{p2}, is 1.96 GHz.

As we have seen from the above example, the low-frequency loop gain and swing requirements essentially set the non-dominant pole frequency. Since we require $\omega_{p2}/\omega_{ul} \geq 4$, this also restricts ω_{ul} and the achievable settling time (which is inversely proportional to ω_{ul}).

To proceed with our design, we continue with the choices made in Example 6.4 and assume that we will design the circuit for a given settling time. The remaining goal is then to minimize the power dissipation given a certain noise budget. The general flow that we will follow to size the circuit is as follows:

1. Calculate the required total load capacitance using (6.53) to meet the noise specification. This also sets the values of the feedback capacitances for a given closed-loop gain (G) and fan-out ($FO = C_L/C_S$).
2. Calculate the required g_{m1} using (6.42) and the desired unity gain frequency, ω_{ul}.
3. Given a value for $(g_m/I_D)_1$, we can now compute I_{D1}. This fixes all currents and device widths in the circuit, since we have already chosen g_m/I_D values for the cascode stack.

Although the above sizing scheme is relatively simple, it comes with a similar problem that we already encountered in Section 6.1.3. In the very first step, the computation of the total load capacitance requires knowledge of both β and $(g_m/I_D)_1$, whose optimum values we don't know a priori. There are several ways in which we can resolve this problem. One option is simply to assume $\beta/\beta_{max} = 0.75$, since we already know that this will lead to near-minimum current. Another option could be to ignore the dependency of the noise on the exact value of $(g_m/I_D)_1$. This is what we did in the derivation of the gain expression in (6.52), where we assumed that all g_m/I_D in the circuit are similar. However, the big difference between the gain and noise specification is that we do not want to overdesign the noise by a significant amount, since this would be very costly in terms of power. A third option is to perform a two-dimensional computation as already done in Section 6.1.3. Since this is quite straightforward and makes no approximations, we will follow this same approach in our example below.

Another issue that must be addressed is self-loading. We have already seen in Example 6.2 that the extrinsic capacitances of the devices connected to the output can be a significant fraction of the total load. We expect the issue to be more significant in the present design, since we employ relatively long channels in the output branch, which means that the width must be commensurately large. To address this issue, we invoke the approach already presented in Section 3.1.7. We will perform initial calculations assuming $r_{self} = 0$, and then iteratively re-compute all values using estimates of r_{self} from the sized circuit.

In summary, this leads to the following algorithm:

1. Begin by assuming that self-loading is negligible, i.e. $r_{self} = 0$.
2. Sweep β (from some fraction of β_{max} to β_{max}) using a for-loop.
3. For each β_k in the for-loop, and for a vector of g_m/I_D within a reasonable range (weak to strong inversion) compute the following:

    a. excess noise factor α using (6.54);
    b. C_{Ltot} using (6.53) and the noise specification; this also fixes C_S, C_F and C_L based on the fan-out, closed-loop gain specifications and the r_{self} estimate;
    c. current division factor κ using (6.43);
    d. g_{m1} using (6.42) and the unity gain frequency ω_{u1};
    e. I_{D1} and f_{Ti} using the g_m/I_D vector;
    f. C_{gg1} using g_m and f_{Ti}. This also lets us compute C_{gd1} and C_{in} per (6.47);
    g. the actual β values using (6.46), evaluated along the g_m/I_D vector that was used for the above computations.

4. In the vector of actual β values, find the closest match with β_k. If a close match exists, it corresponds to a physical design point that can be plotted.
5. Pick the design point that minimizes the current and compute r_{self}. If r_{self} is significant, return to step 2 and repeat all calculations. Re-iterate as needed to converge to a design point that properly accounts for r_{self}.

We will illustrate this approach through the following example.

Example 6.5 Optimization of the Folded-Cascode OTA

Find the optimum inversion level for the input pair of the folded-cascode OTA. Assume the parameters for the cascode stack that were already established in Example 6.4. Design for a 0.1-% settling time of 5 ns, a total integrated differential output noise of 400 μV_{rms}, $G = 2$ and $FO = 0.5$. Begin by computing the required unity gain frequency of the loop and the expected phase margin.

Consider channel lengths of 100, 200, 300 and 400 nm for the input pair, assume $\gamma = 0.7$ for all transistors and account for self-loading as needed.

SOLUTION

The unity gain frequency estimate is computed using the first-order expression from (6.11). The result is saved into the structure "s," which collects all design specifications.

```
% Compute required unity gain frequency
s.ts = 5e-9;
s.ed = 0.1e-2;
s.ful = 1/2/pi * log(1/s.ed)/s.ts
```

This calculation leads to $f_{ul} = 220$ MHz. With the non-dominant pole from Example 6.4, we find $f_{p2}/f_{ul} = 6.6$, and thus the circuit is expected to have about 81 degrees of phase margin (see Table 6.4). This large phase margin justifies the use of (6.11) in the above calculation, which assumes that the second pole is infinitely far away.

Next, we set up two-dimensional sweeps as described earlier. Since there will be a need to iterate around these sweeps, it is most convenient to group all the calculations into a function. In the code below, the function "folded_cascode" takes as its arguments the transistor types of M_1 and M_2 as well as structures that contain the specifications (s) and other design parameters (d). Within this function, we then perform the sweeps as described, and find the points that correspond to physical designs. Around this function, we set up a for-loop that lets us compare the results for different channel lengths. Note that we initially ignore self-loading and set $r_{self} = 0$.

```
% Parameter setup
L1 = [0.1 0.2 0.3 0.4];
d.rself = 0;
d.gm_ID1 = (3:0.01:27)';
d.beta = beta_max*(0.2:0.001:1)';
%---------------------------------------------------------
% Channel length sweep
for i = 1: length(L1)
    d.L1 = L1(i);
    [m1(i) p(i)] = folded_cascode(pch, nch, s, d);
end
```

The folded_cascode function outputs two structures that summarizes the parameters of M_1 (e.g. I_D) and other computed parameters (e.g. C_{Ltot}). Running the for-loop as shown lets us plot the graphs shown in Figure 6.18. We make the following observations:

- As in all previous examples, the current optima in Figure 6.18(a) are relatively shallow.

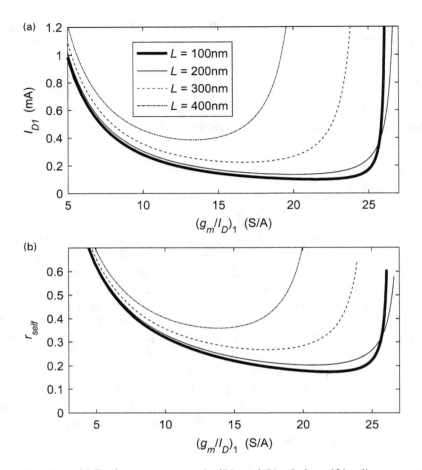

Figure 6.18 (a) Drain current versus $(g_m/I_D)_1$ and (b) relative self-loading across the sweep.

- Using shorter channels allows us to reduce the current, but there is diminishing return beyond $L = 200$ nm.
- For long channels, the self-loading capacitance becomes very large, nearly 40% of C_{Ltot} for $L = 400$ nm. Note that this value is the result of an initial iteration based on the guess $r_{self} = 0$. Once we begin to factor the observed value of r_{self} into the next iteration, the current would have to grow substantially to maintain the proper bandwidth.

Based on these observations, we decide to use $L = 200$ nm and now consider self-loading to arrive at our final design. In the code below, we handle self-loading in the same way as proposed in Section 3.1.7. We set up a for-loop that is initialized with $r_{self} = 0$ for the first iteration. In the second and all following iterations, the actual r_{self} that was computed in the previous step is used as the current estimate. In each iteration, we record all parameters of interest at the

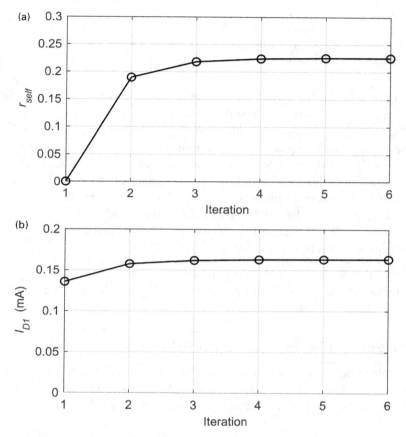

Figure 6.19 (a) Self-loading capacitance along the sizing iterations. (b) Corresponding drain current of the input pair transistors.

minimum current along the g_m/I_D sweep (similar to the minima we have seen in Figure 6.17).

```
% Search parameter setup
d.L1=0.2;
rself = zeros(1,6);
d.gm_ID1 = (5:0.01:27)';
d.beta = beta_max*(0.2:0.001:1)';
% Self-loading sweep
for i = 1:length(rself)
    d.rself = rself(i);
    [m1 p] = folded_cascode(pch, nch, s, d);
    % Find minimum current point and record parameters
    [ID1(i) m] = min(m1.ID);
    gm_ID1(i) = m1.gm_ID(m);
```

```
    cltot(i) = p.cltot(m);
    beta(i) = d.beta(m);
    % Use actual self-loading at optimum as guess for next iteration
    rself(i+1) = p.rself(m);
end
```

Figure 6.19 illustrates the results of the iterative sizing procedure, which converges at the fourth step. The final design values are: $(g_m/I_D)_1 = 18.7$ S/A, $I_{D1} = 163$ µA, $\beta/\beta_{max} = 0.728$, $C_{Ltot} = 508$ fF and $r_{self} = 0.23$. It is interesting to note that despite the complex equation set pertaining to this circuit, the relative feedback factor β/β_{max} for the final design is still very close to the first-order prediction of 0.75, given in (6.29). However, the transit frequency of the input pair is about 2.5 GHz, which is about 11 times larger than f_{ul}, while (6.27) predicts a factor of 6. The discrepancy can be attributed to self-loading (which makes the input pair "work harder") and the fact that the devices operate in moderate inversion (strong inversion was assumed in the derivation of (6.27)).

Now that we have found the optimum inversion level of the input pair and the required current, it is straightforward to complete the sizing of the circuit. The following example illustrates this.

Example 6.6 Sizing the Folded-Cascode OTA

Complete the sizing of the folded-cascode OTA using the final design point identified in Example 6.5: $(g_m/I_D)_1 = 18.7$ S/A, and $I_{D1} = 163$ µA. Validate the circuit against the design specifications using SPICE simulations.

SOLUTION

Since all g_m/I_D and I_D are fixed, it is straightforward to determine the widths of all transistors via their current densities:

```
ID_W1 = lookup(pch, 'ID_W', 'GM_ID', gm_ID1_opt, 'L', d.L1);
ID_W2 = lookup(nch, 'ID_W', 'GM_ID', d.gm_IDcas,...
  'L', d.Lcas, 'VDS', 0.2);
ID_W5 = lookup(pch, 'ID_W', 'GM_ID', d.gm_IDcas,...
  'L', d.Lcas, 'VDS', 0.2);
W1 = ID1_opt/ID_W1;
W2 = 2*ID1_opt/ID_W2;
W3 = W2/2;
W5 = ID1_opt/ID_W5;
W4 = W5;
```

The final step is to compute the feedback and load capacitances with the help of (6.44):

```
CF = CLtot./(s.FO*s.G + 1-beta_opt)/(1+rself);
CS = s.G*CF;
CL = s.FO*CS;
```

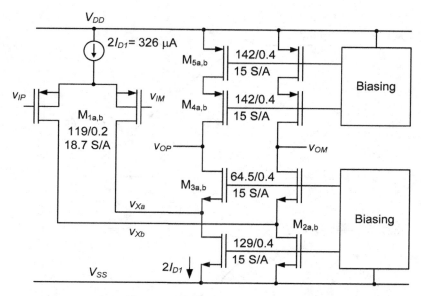

Figure 6.20 Final sizing of the folded-cascode OTA. All dimensions are in microns.

This gives $C_F = 224$ fF, $C_S = 448$ fF and $C_L = 224$ fF. The transistor geometries are summarized in Figure 6.20. Since the device sizes are relatively large, it is worth recalling the discussion on the area-power tradeoff in Section 5.5.4. Since the current optima are shallow (see Figure 6.18(a)), it is possible to move to smaller g_m/I_D and consequently smaller device sizes with only a small penalty in current. The interested reader may explore this option.

We now turn our attention to the SPICE verification of the circuit. Figure 6.21 shows the results of a loop gain simulation for which the circuit was placed into its capacitive feedback network (see Figure 6.1). The observed unity gain frequency (207.93 MHz) is close to the f_{ul} value of 220 MHz that was computed in Example 6.5. The phase margin is very close to the expected value of 81°. Both of these small deviations are readily explained by the many approximations that we made to arrive at low-complexity design equations. The low-frequency loop gain is 38.9 dB (88), which meets our objective of exceeding 50. The simulated gain is larger mostly because of the conservative calculation followed in Example 6.4.

Figure 6.22 shows the results of a SPICE transient simulation. The circuit settles in 4.39 ns to within 0.1% of the final value. This number is about 12% smaller than the desired value (5 ns) even though the loop unity gain frequency is somewhat smaller than the initial target. This is explained by the speed-up from the second pole, which was ignored in the settling time calculation of Example 6.5.

Figure 6.23 shows the results of a SPICE noise simulation. The total integrated noise is very close to the design target (400 μV). This is mainly because the noise equations that we applied do not involve any significant approximations. The

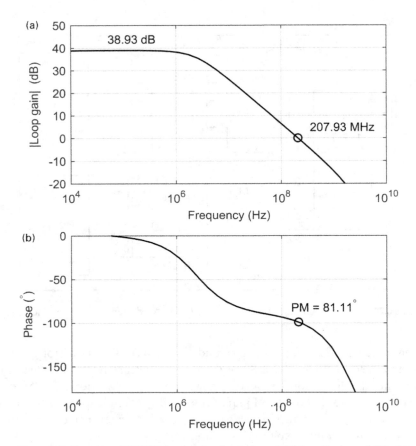

Figure 6.21 Results of SPICE loop gain simulation. (a) Magnitude and (b) phase versus frequency.

main uncertainty comes from the thermal noise factors. We assumed $\gamma = 0.7$ for all devices, which is a reasonable number for moderate inversion (see Chapter 4). With extra work, it is possible to look up exact γ values along the design sweeps and for each device's bias conditions.

In summary, we observe that the design meets the key requirements (settling time, low-frequency loop gain and noise) with some margin, and there is no need for adjustments. One interesting modification to try is to use an n-channel input device. This would make it easier to achieve large β, but the non-dominant pole will now be set by p-channel devices. For circuits with only moderate speed requirements, this may be an attractive option.

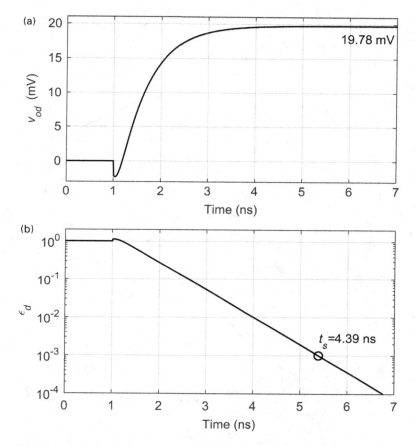

Figure 6.22 Results of SPICE transient simulation. (a) Overall transient versus time. (b) Dynamic settling error (relative to final settling value) versus time.

More generally speaking, a practical application may require a much larger loop gain, for example 1000 to 10,000. We could make some improvements in this spec by using even longer channels or by employing the well-known technique of gain boosting [8], [9]. This would add auxiliary amplifiers within the cascode stack and thereby boost the output resistance. The design of these auxiliary amplifiers can be done in almost exactly the same way as the examples provided in this chapter. The main difference is that noise performance won't play a major role.

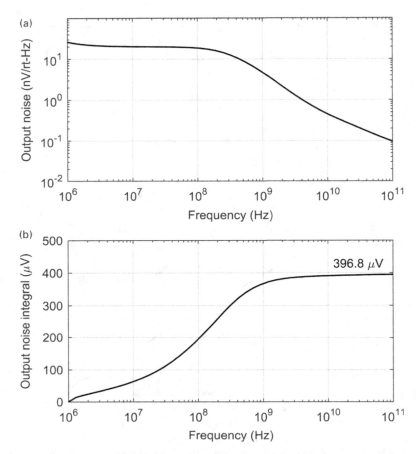

Figure 6.23 Results of SPICE noise simulation. (a) Output noise power spectra density. (b) running integral of the power spectral density.

The above example reaffirms the fact that the g_m/I_D methodology can be efficiently applied to circuits that have no closed-form design solution. This is typically accomplished by sweeping the inversion levels of key devices in the signal path, and enumerating the critical performance parameters along the way. To resolve transcendental dependencies, one can perform two-dimensional sweeps (using β and g_m/I_D in the above example), and look for self-consistent points to extract valid design points.

6.2.3 Optimization in Presence of Slewing

A logical next step in the design of the folded-cascode amplifier design of this section is the consideration of slewing. However, as mentioned initially, slewing can be handled in the same way as done in Section 6.1.3, and we will therefore only outline the (minor) differences.

For the folded-cascode OTA, the slew rate is:

$$SR = \frac{2\kappa I_{D1}}{C_{Ltot}}.$$

(6.55)

As before, we can eliminate C_{Ltot} based on the bandwidth equation, given by (6.42) for the folded-cascode OTA:

$$C_{Ltot} = \frac{\beta g_{m1}\kappa}{\omega_{u1}}.$$

(6.56)

With $\tau = 1/\omega_{u1}$, this leads us to the same result as for the basic OTA in (6.29):

$$SR = \frac{2I_{D1}}{\tau\beta g_{m1}}.$$

(6.57)

From here, all remaining steps of the analysis of Section 6.1.3 apply, and we arrive at the same expression for the unity gain frequency requirement in presence of slewing:

$$\omega_{u1} = \frac{1}{\tau} = \frac{1}{t_s}\left(X - 1 + \ln\left(\frac{1}{\varepsilon_d}\cdot X\right)\right),$$

(6.58)

where:

$$X = v_{OD,final}\cdot\frac{\beta}{2}\left(\frac{g_m}{I_D}\right)_1.$$

(6.59)

To account for slewing in Example 6.5, the only required change is to compute X before step 3d, and compute g_{m1} using the unity gain frequency computed via (6.58). Since the performed sweep in Example 6.5 is already two-dimensional in β and g_m/I_D, the computation of X can be readily inserted without changing the structure of the algorithm.

6.3 Two-Stage OTA for Switched-Capacitor Circuits

The two-stage OTA is another circuit that is frequently employed in switched-capacitor circuits. Since this topology stacks fewer devices, it can typically achieve a larger output swing than a folded-cascode OTA, which can be a significant advantage in low-voltage CMOS. On the other hand, the two-stage amplifier comes with additional poles and zeros that must be properly dealt with. This not only complicates the design, but usually also leads to constraints and inefficiencies that counter the advantage of larger output swing.

For our treatment on this section, we will assume the circuit given in Figure 6.24. It is stabilized using Miller compensation and the feedforward zero caused by the compensation capacitance C_C is moved to infinity by letting $R_Z = 1/g_{m2}$. The

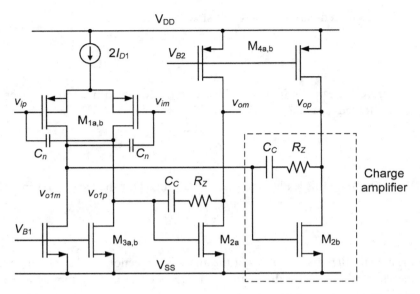

Figure 6.24 Schematic of a two-stage OTA. The bias voltage generators (for V_{B1} and V_{B2}) and the required common-mode feedback circuit are not shown (for simplicity).

resulting circuit has three poles, but the third pole can typically be neglected with this design choice (see equations below). The neutralization capacitors C_n serve to eliminate the Miller effect at the input of the first stage. Detailed treatments on the pole locations, neutralization and biasing considerations for this circuit are available in standard IC design textbooks such as [4].

As indicated via the dashed box in Figure 6.24, the second stage of the circuit can be viewed as a charge amplifier, which was analyzed in detail in Chapter 5. We will leverage some results from this analysis to optimize the circuit.

6.3.1   Design Equations

We assume that the amplifier is inserted into the same feedback configuration as in Figure 6.1. Thanks to neutralization via C_n, the feedback factor is well-approximated by:

$$\beta \cong \frac{C_F}{C_F + C_S + C_{gg1}}. \tag{6.60}$$

The low-frequency loop gain of the circuit is

$$L_0 = \beta \cdot g_{m1} R_1 \cdot g_{m2} R_2, \tag{6.61}$$

where:

$$R_1 = \frac{1}{g_{ds1} + g_{ds3}} \qquad R_2 = \frac{1}{g_{ds2} + g_{ds4}}. \tag{6.62}$$

The total load capacitance of the circuit has a similar form as (6.44):

$$C_{Ltot} = C_L + (1-\beta)C_F + C_{self2} = \left[C_L + (1-\beta)C_F\right](1 + r_{self2}), \tag{6.63}$$

where C_{self2} and r_{self2} represent the absolute and normalized self-loading for the second stage:

$$C_{self2} = C_{db2} + C_{dd4} \quad \text{and} \quad r_{self2} = \frac{C_{self2}}{C_L + (1-\beta)C_F}. \tag{6.64}$$

The capacitance to ground at nodes $v_{olm,p}$ is

$$C_1 = C_{gs2} + C_{self1} = C_{gs2}\left(1 + r_{self1}\right), \tag{6.65}$$

where the terms added to C_{gs2} capture self-loading:

$$C_{self1} = C_{dd1} + C_{dd3} \quad \text{and} \quad r_{self1} = \frac{C_{self1}}{C_{gs2}}. \tag{6.66}$$

The unity gain frequency of the feedback loop is given by the product of L_0 and the dominant pole of the amplifier, which leads to [4]:

$$\omega_{u1} = \frac{\beta \cdot g_{m1} R_1 \cdot g_{m2} R_2}{R_1\left(C_1 + C_c\left(1 + g_{m2}R_2\right)\right) + R_2\left(C_{Ltot} + C_c\right)}$$

$$= \frac{\beta g_{m1}}{C_c}\left(\frac{1}{1 + \dfrac{1 + \dfrac{C_1}{C_c}}{g_{m2}R_2} + \dfrac{1 + \dfrac{C_{Ltot}}{C_c}}{g_{m2}R_1}}\right) \cong \frac{\beta g_{m1}}{C_c}. \tag{6.67}$$

Note that the final approximation, which is typically advertised in standard textbooks, can be quite inaccurate when the $g_m R$ products are small. If C_1, C_c and C_{Ltot} are comparable (we will see later that this is the case) and $g_m R \sim 10$, the approximation error will be of the order of 40%! Consequently, our optimization below will work with the more accurate expression that takes the finite $g_m R$ terms into account.

The angular non-dominant pole frequency of the amplifier is [4]:

$$\omega_{p2} \cong \frac{g_{m2}}{C_1 + \dfrac{C_1 C_{Ltot}}{C_c} + C_{Ltot}} = \frac{g_{m2}}{C_1}\frac{1}{1 + \dfrac{C_{Ltot}}{C_c} + \dfrac{C_{Ltot}}{C_1}}. \tag{6.68}$$

Since C_1 contains C_{gs2}, it is clear that the non-dominant pole cannot lie beyond the transit frequency of M_2. When C_{Ltot} is comparable to C_c and C_1, which is a typical design outcome (see the example below), the non-dominant pole lies between one-third to one-fifth of f_{T2}. The circuit's third pole is located at [4]:

$$\omega_{p3} = \frac{1}{R_Z C_1} = \frac{g_{m2}}{C_1}. \tag{6.69}$$

By comparing with (6.68), we see that this frequency will always be significantly larger than ω_{p2}. The third pole will therefore reduce the phase margin by a few degrees, but this usually does not need to be considered in the design flow.

As far as noise is concerned, one can show that [10]:

$$\overline{v_{od}^2} = 2\frac{1}{\beta}\frac{k_B T}{C_C}\gamma_1\left(1+\frac{\gamma_3}{\gamma_1}\frac{g_{m3}}{g_{m1}}\right) + 2\frac{k_B T}{C_{Ltot}}\left(1+\gamma_2\left(1+\frac{\gamma_4}{\gamma_2}\frac{g_{m4}}{g_{m2}}\right)\right). \tag{6.70}$$

The first term is the noise from stage 1, the second term is from stage 2. The prefactors of two are due to the fully differential architecture.

Clearly, given this complex set of intertwined equations, there is once again no closed form solution for the optimal design point. However, we will again see that we can identify near-optimum choices through sweeps in Matlab.

6.3.2   Optimization Procedure

Most of the available literature on two-stage amplifiers focuses on analysis, and there exists very little information about suitable sizing strategies. An exception is Sansen's book on Analog Design Essentials [11], which identifies the degrees of freedom in two-stage design and pays particular attention to the capacitor ratios in the circuit (C_{gs2}/C_C and C_{Ltot}/C_C). For example, it is stated that C_C should be of the order of three times C_{gs2} and somewhat smaller than C_L. These guidelines are all derived from stability considerations, but unfortunately do not take the circuit's noise performance into account.

In the approach outlined below, we will also work with capacitor ratios to untangle the complex design equations, but additionally include a noise constraint. In this context, one specific result that we can re-use from Chapter 5 is the optimal design point for a noise- and bandwidth-constrained charge amplifier stage. According to (5.37), the optimum ratio between the gate capacitance and the total capacitance in the charge amplifier's feedback network is 1/2. In the context of our circuit, where we view the second stage as a charge amplifier (see Figure 6.24), this guideline translates into[3] $C_{gs2}/C_C = 0.5$ (neglecting self-loading, for simplicity). This result helps us eliminate one degree of freedom (for example, in (6.68)) and thus simplifies the design process significantly. In the end, we can always re-visit this assumption and perturb the value to see if our design improves for slightly smaller or larger values.

In this same spirit, we will decide on other parameter ratios a priori, knowing that our choices may not be exactly optimal, but reasonable. In summary, we declare the

---

[3] The stated ratio neglects self-loading at the output of stage 1. Strictly speaking, the ratio between C_{gs} and $C_C + C_{self1}$ equals ½ in the optimum. However, since C_{self1} is usually much smaller that C_C, we neglect this extra term. We will see later that the optimum is indeed somewhat smaller than 0.5; which makes sense in light of the neglected term.

following secondary design variables, which are subject to optimization, but do not appear as the "primary knobs" in the optimization:

- The channel lengths of all transistors (L_1, L_2, L_3 and L_4). The choice of these is directly constrained by the desired low-frequency loop gain, and we usually must pick them with some margin in mind (like in Example 6.4). One opportunity for optimization is to go for an unequal gain partitioning between stage 1 and stage 2. However, such a decision typically makes sense only after the designer has gained insight about which one of the two stages limits the design. In our design below, an approximately equal gain partitioning was assumed.
- The transconductance ratios between active loads and signal path devices (g_{m3}/g_{m1} and g_{m4}/g_{m2}). These ratios appear in the noise expression of (6.70) and it is desirable to minimize them. However, as analyzed in Chapter 4, reducing these ratios comes with a loss in swing, giving rise to a shallow optimum for the net dynamic range. Based on this result, it is clear that not much can be gained by optimizing the g_m ratios. A reasonable pick will suffice in practice.
- Finally, as already explained above, we will assume $C_{gs2}/C_C = 0.5$ unless otherwise stated.

Based on these choices, we can now follow a general design flow that is similar to the one developed for the folded-cascode amplifier:

1. Calculate the required compensation capacitance using (6.70) to meet the noise specification.
2. Calculate the required g_{m1} using (6.67) and the desired unity gain frequency, ω_{ul}.
3. Calculate the required g_{m2} using (6.68) and the desired non-dominant pole frequency, ω_{p2}.

However, upon detailed inspection of these steps, we see that we require the feedback factor β as well as the ratio C_{Ltot}/C_C to complete the noise calculation. We therefore declare these parameters as sweep variables and perform a two-dimensional search, as done before. The main difference to the folded-cascode example is that all points of this sweep are potentially feasible design points. In other words, there is no need to search for self-consistent points. Lastly, we once again need to address self-loading via outer iterations (if significant).

In summary, this leads to the following algorithm:

1. Begin by neglecting self-loading, i.e. $r_{self1} = r_{self2} = 0$.
2. Sweep β (from some fraction of β_{max} to β_{max}) and C_{Ltot}/C_C. We know that C_{Ltot} and C_C must be approximately of the same order [11], so sweeping the ratio around unity should provide useful results.
3. Compute the following for each value of β and C_{Ltot}/C_C:
   a. C_C using (6.70) and the noise specification; this fixes all other capacitances: C_{Ltot}, C_F, C_S, C_L, C_{gs2}, C_I and C_{gg1};
   b. g_{m1} using (6.67) and the unity gain frequency f_{ul}. Use estimates for the $g_m R$ products to achieve improved accuracy;
   c. $(g_m/I_D)_1$ given g_{m1} and C_{gg1}; this also fixes I_{D1};

Table 6.6 Summary of target specifications.

Description	Variable(s)	Value
Low-frequency loop gain	L_0	>50
Ideal closed-loop gain magnitude	$G = C_S/C_F$	2
Fan-out	$FO = C_1/C_S$	0.5
Total integrated output noise	$\overline{v_{od}^2}$	400 μV_{rms}
Settling time (0.1%)	t_s	5 ns

  d. g_{m2} using (6.68) and based on the desired f_{p2};
  e. $(g_m/I_D)_2$ given g_{m2} and $C_1 = C_{gs2}(1 + r_{self1})$; this also fixes I_{D2};
  f. the total current $I_{Dtot} = I_{D1} + I_{D2}$.

4. Pick the design point in the two-dimensional space that minimizes the total current and compute r_{self1} and r_{self2}. If self-loading is significant, return to step 2 and repeat all calculations. Re-iterate as needed to converge to a design point that properly accounts for self-loading.

We will illustrate this approach through the following example, in which we target the same specifications as in Section 6.2. These parameters are repeated in Table 6.6 for convenience.

Example 6.7 Optimization of the Two-Stage OTA

Find the design parameters for the two-stage OTA that achieve the specifications listed in Table 6.6 with minimum current consumption. Assume the following channel lengths, which were chosen to meet the low-frequency loop gain requirement with some margin: $L_1 = L_4 = 150$ nm (p-channel devices) and $L_2 = L_3 = 200$ nm (n-channel devices). Furthermore, design for $g_{m3}/g_{m1} = 1$, $g_{m4}/g_{m2} = 0.5$ and assume $\gamma = 0.8$ for all transistors.

SOLUTION

The first decision to make is about the location of the non-dominant pole. As discussed in Section 6.2.1, using $f_{p2}/f_{ul} = 4$ will lead to the fastest settling and a phase margin of about 76°. However, since this is a first design iteration, we use a more conservative choice of $f_{p2}/f_{ul} = 6$, which will lead to a phase margin of about 80°. Furthermore, to create some margin for the expected settling time, we once again use the (pessimistic) first-order expression from (6.11), and hence target $f_{ul} = 220$ MHz (same as in Example 6.5).

  Next, we set up a two-dimensional sweep as described above. Since there will be a need to iterate around these sweeps, it is again most convenient to group all the calculations into a function. Similar to Example 6.5, the code below employs the function "two_stage," which takes as its arguments the transistor types of $M_1 - M_4$ as well as structures that contain the specifications (s) and other known design parameters (d). Within this function, we perform a vectorized computation along the β dimension and leave the sweep of C_{Ltot}/C_C to an outer for-loop. For simplicity, we initially ignore self-loading and thus set $r_{self1} = r_{self2} = 0$. In addition, we start with the above-discussed heuristic of $C_{gs2}/C_C = 0.5$ (subject to further optimization).

```
% Design decisions and estimates
d.L1 = 0.15; d.L2 = 0.20; d.L3 = 0.20; d.L4 = 0.15;
d.gam1 = 0.8; d.gam2 = 0.8; d.gam3 = 0.8; d.gam4 = 0.8;
d.gm3_gm1 = 1; d.gm4_gm2 = 0.5;
d.cgs2_cc = 0.5;
d.rself1 = 0;
d.rself2 = 0;
% Search range for main knobs
cltot_cc = linspace(0.2, 1.5, 100);
d.beta = beta_max*linspace(0.4, 0.88, 100)';
for j=1:length(cltot_cc)
    d.cltot_cc = cltot_cc(j);
    [m1, m2, m3, m4, p] = two_stage(pch, nch, nch, pch, s, d);
    ID1(j,:) = m1.id;
    ID2(j,:) = m2.id;
    gm_ID1(j,:) = m1.gm_id;
    gm_ID2(j,:) = m2.gm_id;
end
```

Running this code yields the contour plot shown in Figure 6.25. The minimum current of 318 μA is achieved for $\beta/\beta_{max} = 0.84$ and $C_{Ltot}/C_C = 0.56$. To gain further

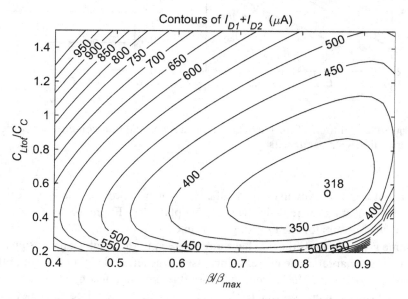

Figure 6.25 Contours of total drain current versus the sweep parameters.

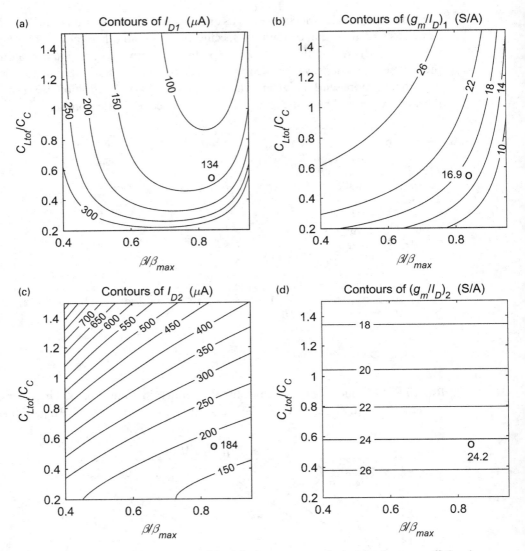

Figure 6.26 Contours of the individual drain currents and transconductance efficiencies versus the sweep parameters.

insight about the involved tradeoffs, it is worth looking at the inversion levels and individual stage currents in the same 2D plane (see Figure 6.26).

The contours for the first stage's drain current in Figure 6.26(a) resemble the same general trends that we have already seen in the single-stage designs of the previous chapters. The currents increase to the left and right of some optima for β/β_{max}. To the left, the current increases due the small feedback factor value, and to the right it increases due to the small g_m/I_D that is required to achieve a large feedback factor (smaller capacitance requires higher level of inversion). We see this latter trend for $(g_m/I_D)_1$ confirmed in Figure 6.26(b); increasing β/β_{max} pushes us onto smaller contour values for $(g_m/I_D)_1$.

Figure 6.26(a) also indicates a significant dependence of I_{D1} on the C_{Ltot}/C_C ratio. When C_{Ltot} is small, much of the noise budget is consumed by the second stage (see (6.70)), placing a significantly higher noise burden on the first stage (C_C must increase, requiring larger g_{m1} to maintain the bandwidth).

The noise tradeoff also plays a major role in the contours of the stage 2 current, shown in Figure 6.26(c). As β/β_{max} increases, I_{D2} decreases, since more noise can be allocated to stage 2 (and thus C_{Ltot} can be small). On the other hand, I_{D2} increases with C_{Ltot}/C_C since a larger load makes it more difficult to maintain the non-dominant pole frequency (see (6.68)). From Figure 6.26(d), we see indeed that $(g_m/I_D)_2$ decreases along this direction. It is clear from (6.68) that larger C_{Ltot} requires larger g_{m2}/C_1 (hence larger (g_m/C_{gs2})) and therefore smaller $(g_m/I_D)_2$. The contours are flat with respect to β since the feedback factor plays no role in (6.68). The inversion level of the second stage device is set by C_{Ltot}/C_C and C_{gs2}/C_C (which is set to 0.5 in this iteration).

At the optimum point, we find the following self-loading ratios: $r_{self1} = 0.29$ and $r_{self2} = 0.40$. This indicates that self-loading is particularly pronounced in the second stage, and must be addressed. We thus re-run the program iteratively, as in Example 6.5 and obtain the self-loading trajectories shown in Figure 6.27. Interestingly, the self-loading of the first stage decreases. This is because M_2 becomes larger to address the self-loading at the output. This increases C_{gs2} relative to the parasitics at the stage 1 output, yielding smaller

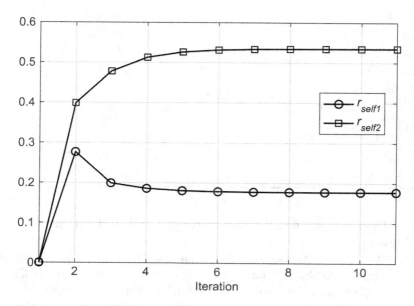

Figure 6.27 Self-loading factors along the sizing iterations.

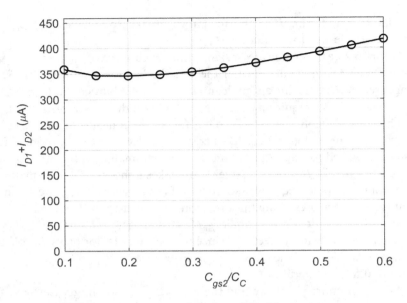

Figure 6.28 Minimum current as a function of C_{gs2}/C_C.

Table 6.7 Summary of the final design parameters for the two-stage OTA.

Parameter	Value
C_{gs2}/C_C	0.3
β/β_{max}	0.81
C_{Ltot}/C_C	0.53
r_{self1}	0.40
r_{self2}	0.28
$(g_m/I_D)_1$ (S/A)	15.2
$(g_m/I_D)_2$ (S/A)	20.6
C_S (fF)	198
C_C (fF)	416
I_{D1} (µA)	157
I_{D2} (µA)	196
$I_{D1} + I_{D2}$ (µA)	353

r_{self1}. Upon convergence, the total current increases from 318 µA to 387 µA (not plotted).

At this stage, we have a design that should meet the specifications. However, before turning to SPICE validation, it is worth re-visiting the choice of $C_{gs2}/C_C = 0.5$. This number was derived in Chapter 5 assuming strong inversion, but M$_2$ actually operates close to weak inversion (see Figure 6.26(d)). Setting C_{gs2}/C_C to smaller values may drive the sizing algorithm to a more favorable point. To investigate, we add yet another for-loop around the entire optimization (including the self-loading iterations shown in Figure 6.27) and sweep C_{gs2}/C_C from 0.1 to 0.6. The result is shown in Figure 6.28, which indicates that the tradeoff is remarkably flat. For our final design, we decide on $C_{gs2}/C_C = 0.3$, which leads to the parameters summarized in Table 6.7.

As the final step, we will now determine all device sizes and validate the design against SPICE simulations. This is done in the following example.

Example 6.8 Sizing the Two-Stage OTA

Complete the sizing of the two-stage OTA using the final design point established in Example 6.7. Validate the circuit against the design specifications of Table 6.6 using SPICE simulations.

SOLUTION

Since all meta parameters are fixed, it is straightforward to calculate all device widths. The function "two_stage" already contains the required code, an excerpt of which is given below:

```
m3.gm_id = m1.gm_id.*d.gm3_gm1;
m4.gm_id = m2.gm_id.*d.gm4_gm2;
m1.W = m1.id./lookup(dev1, 'ID_W', 'GM_ID',m1.gm_id, 'L',m1.L);
m2.W = m2.id./lookup(dev2, 'ID_W', 'GM_ID',m2.gm_id, 'L',m2.L);
m3.W = m1.id./lookup(dev3, 'ID_W', 'GM_ID',m3.gm_id, 'L',m3.L);
m4.W = m2.id./lookup(dev4, 'ID_W', 'GM_ID',m4.gm_id, 'L',m4.L);
```

Another component value that we must compute is the neutralization capacitance C_n, which is equal to C_{gd1}:

```
m1.cgd = m1.W.*lookup(dev1, 'CGD_W', 'GM_ID',m1.gm_id, 'L',m1.L);
p.cn = m1.cgd;
```

Finally, we compute the explicit compensation capacitance that must be added in parallel to C_{gd2}. This is further illustrated in Figure 6.29. The gate-drain capacitance of M_2 by itself already acts as a Miller capacitance, and its value should therefore be subtracted to arrive at the capacitance that is added explicitly (called C_{Cadd}):

```
m2.cgd = m2.W.*lookup(dev2, 'CGD_W', 'GM_ID',m2.gm_id, 'L',m2.L);
p.cc_add = p.cc - m2.cgd;
```

With these calculations, we arrive at the final circuit depicted in Figure 6.29.

We now look at the SPICE verification results. Figure 6.30 shows the results of a loop gain simulation for which the circuit was placed into its capacitive feedback network (see Figure 6.1). The observed unity gain frequency (203.37 MHz) is close to the f_{ul} value of 220 MHz that was assumed in Example 6.7. The phase

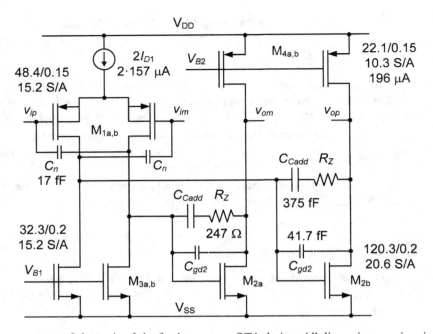

Figure 6.29 Schematic of the final two-stage OTA design. All dimensions are in microns. The bias voltage generators (for V_{B1} and V_{B2}) and the required common-mode feedback circuit are not shown (for simplicity).

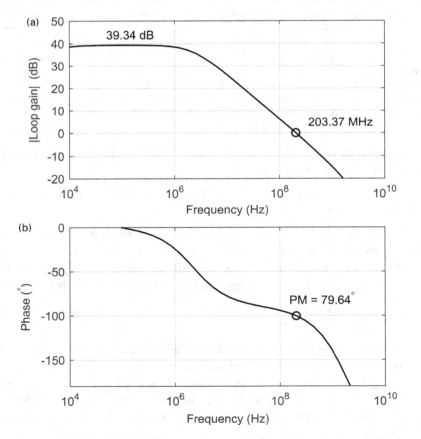

Figure 6.30 Results of SPICE loop gain simulation. (a) Magnitude and (b) phase versus frequency.

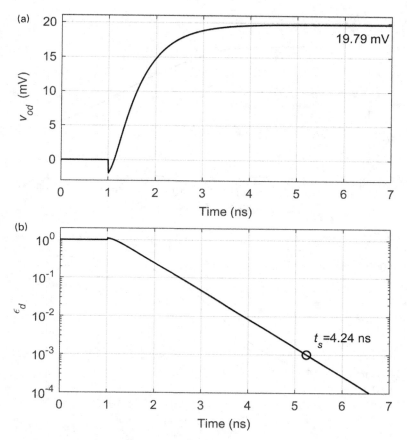

Figure 6.31 Results of SPICE transient simulation. (a) Overall transient versus time. (b) Dynamic settling error (relative to final settling value) versus time.

margin is very close to the expected value of 80°. Both of these small deviations are readily explained by analytical approximations that we made, as well as various voltage dependencies that were not accounted for in the calculations (for example, we assumed $V_{DS} = 0.6$ V for all junction capacitance estimates). The low-frequency loop gain is 39.4 dB (93), which meets our objective of exceeding 50.

Figure 6.31 shows the results of a SPICE transient simulation. The circuit settles in 4.24 ns to within 0.1% of the final value. This number is about 15% smaller than the specification (5 ns) even though the loop unity gain frequency is below the initial target. This is explained by the fact that we ignored the speed-up from the second pole in our conservative settling time calculation.

Figure 6.32 shows the results of a SPICE noise simulation. The total integrated noise is 9% above the design target of 400 μV. This is due to the approximate nature of the noise expression, the approximate γ estimates, as well as the ignored contribution from flicker noise. There are several ways in which this discrepancy can be resolved.

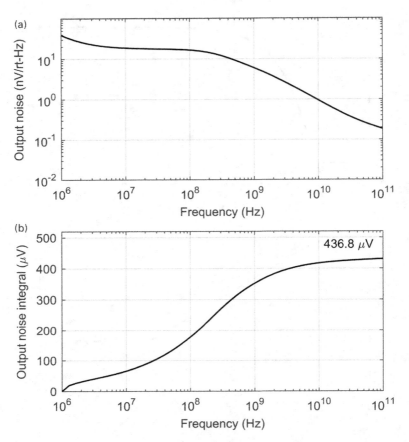

Figure 6.32 Results of SPICE noise simulation: (a) Output noise power spectra density. (b) running integral of the power spectral density.

The first option is to increase the size of all components, all currents and all widths by 9%. The design will now meet the noise spec exactly (with all other specs unchanged), but with somewhat larger power and area. The second option is to slightly increase the compensation capacitor. This will trade in some of the excess settling speed for reduced noise. Since these adjustments are rather trivial, we do not validate them here.

In summary, we observe that the presented design performs very closely to its key specs. However, further adjustments and optimizations are always possible. For instance, one could now consider optimizing the device lengths and transconductance ratios (g_{m3}/g_{m1}, etc.). In addition, it would again be interesting to swap device polarities (use n-channels in stage 1 and p-channels in stage 2).

It is worth noting that the total current consumption of the above design is 706 μA, while the folded-cascode design required 652 μA for essentially the same specifications. These numbers are remarkably close and reflect the fact that the tradeoffs in the two circuits are quite similar. However, the two-stage circuit will be preferred where large output swing is needed, whereas the folded-cascode design offers a wider input common-mode range [4].

6.3.3   Optimization in Presence of Slewing

A detailed discussion of the slewing behavior in a two-stage OTA is presented [6]. The main difference relative to a single-stage topology is that the two-stage OTA's positive and negative half circuits may not necessarily slew at the same rate. To see this, consider the schematic of the positive stage 2 half circuit in Figure 6.33. Depending on the polarity of the input during slewing, the first stage's differential pair will either sink (Case 1) or source (Case 2) its bias current (I_{D1}) from the second stage. In the first case, the output voltage v_{op} in Figure 6.33 will fall at the following rate:

$$SR_1 = \left.\left|\frac{dv_{op}}{dt}\right|\right|_{Case1} \cong \frac{I_{D1}}{C_C}.$$
(6.71)

This result follows by approximating the gate of M_{2b} as a virtual ground (the circuit basically operates like a closed-loop integrator). During this slewing transient, the current flowing into M_{2b} is the sum of I_{D1}, plus the bias current I_{D2}, as well as the discharging current required for C_{Ltot}, equal to $SR_1 \cdot C_{Ltot}$. M_{2b} will usually have no issue sinking this current, as its gate voltage can rise to support the required current; this is essentially class-AB operation.

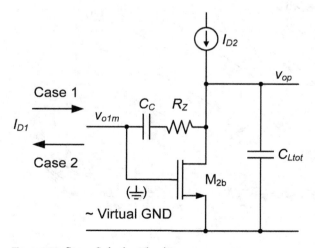

Figure 6.33 Stage 2 during slewing.

The situation is different in Case 2. The maximum rate at which v_{op} can move up is $I_{D2}/(C_C + C_{Ltot})$. The slew rate therefore becomes:

$$SR_2 = \left|\frac{dv_{op}}{dt}\right|_{Case2} \cong min\left\{\frac{I_{D1}}{C_C}, \frac{I_{D2}}{C_C + C_{Ltot}}\right\}. \tag{6.72}$$

In words, the output will slew at a rate that is set by the limiting term in the brackets of (6.72). Therefore, to ensure $SR_1 = SR_2$, the following condition must hold:

$$I_{D2} > I_{D1}\left(1 + \frac{C_{Ltot}}{C_C}\right). \tag{6.73}$$

If this condition is violated, the positive and negative half circuit will slew at different rates, the circuit will become asymmetric and the common mode will drift away. Since the common-mode feedback is typically not as fast as the differential mode, the circuit may not recover by the end of the clock cycle and we may observe slow settling tails and other undesired effects. For robust designs, it is therefore mandatory to obey (6.73).

With this understanding, it is interesting to go back to Example 6.7 and check for this condition. Figure 6.34 shows the same contours as Figure 6.25, but marks the regions where (6.73) is not satisfied with an "X". We see that it is possible to prevent asymmetric slewing by choosing C_{Ltot}/C_C somewhat larger than in the current minimum. As discussed previously, this leads to larger I_{D2}, hence pushing the design in the right direction.

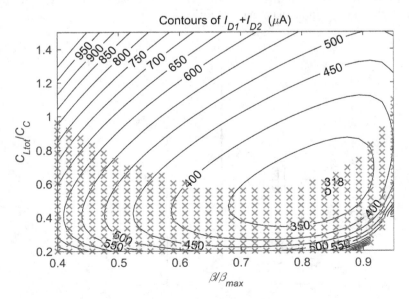

Figure 6.34 Current contours from Figure 6.25, with added markers "X," showing the region where (6.73) does not hold.

With (6.73) satisfied, the slew rate in each half circuit becomes equal to SR_l given in (6.71). The differential slew rate is thus twice as large:

$$SR = \frac{2I_{D1}}{C_C} = \frac{2I_{D1}}{\tau \beta g_{m1}},$$

(6.74)

which is essentially the same result as (6.30), except that C_C acts as the load capacitance. Note that we assumed in this result that $\tau = 1/\omega_{ul} = \beta g_{m1}/C_C$, which is a reasonable first-order approximation to (6.67) that is applied to keep the algebra manageable.

From here, all remaining steps of the analysis of Section 6.1.3 apply, and we arrive at the same expression for the unity gain frequency requirement in presence of slewing:

$$\omega_{ul} = \frac{1}{\tau} = \frac{1}{t_s}\left(X - 1 + \ln\left(\frac{1}{\varepsilon_d} \cdot X\right)\right),$$

(6.75)

where:

$$X = v_{OD,final} \cdot \frac{\beta}{2}\left(\frac{g_m}{I_D}\right)_1.$$

(6.76)

To account for slewing in Example 6.7, we need to find X before we can compute g_{m1} based on the required unity gain frequency in step 3b. However, at this point in the algorithm, only β is known (it is a sweep variable) and g_m/I_D is found only after g_{m1} has been determined. To resolve this issue, we can apply the same solution that was used in the folded-cascode optimization. That is, we compute X and other quantities for a vector of g_m/I_D values and subsequently find the self-consistent point in the feedback factor β to identify physical design points.

6.4 Simplified Design Flows

Much of this chapter was devoted to finding nearly exact minima in current consumption. However, we saw in most cases that these optima are relatively shallow, which means that it should be possible to find reasonably good design point with much less effort, and leveraging the rules of thumb that we have already established though our detailed studies. Shortening the design time is especially important in today's industry environment, where "time to market" is just as important (or sometimes more important) than raw performance.

The first simplification to consider is to fix the feedback factor according to the first-order optimum of $\beta = 0.75 \cdot \beta_{max}$. We have seen in essentially all examples that this leads to outcomes that are close to optimum. In the following two subsections, we will consider additional simplifications for the design flows of the folded-cascode and two-stage OTA. The reader is encouraged to try these and modify/refine as needed for the task at hand.

6.4.1   Folded-Cascode OTA

1. Design the cascode stack as described in Example 6.4.
2. Assume that the circuit will slew for some fraction of the settling time (t_s), for example 30%. Calculate the resulting unity gain frequency target.
3. Assume a reasonable self-loading factor that leaves some margin, for example $r_{self} = 0.4$. This number can be manually adjusted later, but often leads to reasonable outcomes without much iteration.
4. Assume $\beta = 0.75 \cdot \beta_{max}$ for all calculations.
5. Compute the excess noise factor α by assuming that all g_m/I_D in the circuit are the same. The noise in the final result will be somewhat off, but it is straightforward to bring adjust it using the noise scaling approach discussed in Section 6.1.2.
6. Calculate C_{Ltot} using the noise specification. This also fixes C_S, C_F and C_L based on the fan-out, closed-loop gain and self-loading.
7. Decide on a conservative estimate for the current division factor κ, for example 0.7 (which will often leave some margin).
8. Calculate g_{m1} based on the desired unity gain frequency ω_{ul}.
9. Calculate C_{gg1} using β and the feedback capacitor values. Neglect the Miller multiplication of C_{gd1}.
10. You can now compute f_{T1}, $(g_m/I_D)_1$, and thus I_{D1}.
11. Calculate the slewing parameter X and the actual settling time. Adjust your slewing budget in step 2 and repeat all calculations, if needed.
12. Check all other assumptions made in the calculations against the actual outcome. For example, check the actual self-loading. Adjust assumptions as needed and re-compute.
13. Size the circuit and inspect the device widths and inversion levels of all devices. If the input pair is excessively large, it may be worth re-sizing with reduced $(g_m/I_D)_1$, to trade potentially minor increases in current for significant area savings (see discussion at the end of Section 6.1.2).
14. Evaluate the circuit in SPICE. Always begin with small-signal simulations to simplify the debug process.

6.4.2   Two-Stage OTA

1. Assume that the circuit will slew for some fraction of the settling time (t_s), for example 30%. Calculate the resulting unity gain frequency target.
2. Assume reasonable self-loading factors that leave some margin, for example $r_{self1} = r_{self2} = 0.4$. These numbers can be manually adjusted later, but often lead to reasonable outcomes without much iteration.
3. Assume $C_{gs2}/C_C = 1/3$ for all calculations.

4. Assume $\beta = 0.75 \cdot \beta_{max}$ for all calculations.
5. Start by assuming $C_{Ltot} = C_C$ and adjust as you see fit later. This parameter controls which if the two stages will "work harder." The ultimate choice will depend on the circuit configuration (n-channel versus p-channel input, etc.). Adjust this knob if you see vastly imbalanced stage 1/2 bias currents in the subsequent calculations.
6. Calculate C_C based on the noise specification. This fixes all other capacitances in the circuit based on the fan-out, closed-loop gain and self-loading parameters.
7. Calculate g_{m1} using the unity gain frequency f_{u1}. Use estimates for the $g_m R$ products for improved accuracy.
8. Calculate $(g_m/I_D)_1$ given g_{m1} and C_{gg1}. This also fixes I_{D1}.
9. Calculate g_{m2} based on the desired non-dominant pole f_{p2}.
10. Calculate $(g_m/I_D)_2$ given g_{m2} and $C_1 = C_{gs2}(1 + r_{self1})$. This also fixes I_{D2}.
11. Calculate the slewing parameter X and the actual settling time. Adjust your slewing budget in step 1 and repeat all calculations, if needed.
12. Check all other assumptions made in the calculations against the actual outcome. For example, check the actual self-loading. Adjust assumptions as needed and re-compute.
13. Size the circuit and inspect the device widths and inversion levels of all devices. If some devices are excessively large, it may be worth re-sizing with reduced g_m/I_D, to trade potentially minor increases in current for significant area savings (see discussion at the end of Section 6.1.2).
14. Evaluate the circuit in SPICE. Always begin with small-signal simulations to simplify the debug process.

The above-described design flows do not rely on sophisticated sweeps and could in principle be done using basic spreadsheet software. While the outcomes may not be as exact as in the elaborate flows of Sections 6.2.2 and 6.3.2, it is still possible to check all assumptions against the final outcome and track down all inaccuracies. The designer therefore remains in the driver seat, and can use his or her understanding and experience to drive the optimization systematically, and according to known design equations.

6.5   Sizing Switches

Switched-capacitor circuits, such as the one shown in Figure 6.1, rely on a number of switches to time and control their operation. The type of switch that is used depends on the common mode, the signal range and signal type (sampled data or continuous time). Therefore, designers will typically employ a variety of options, including single n-channel or p-channel devices, transmission gates (n- and p-channel in parallel) or a bootstrapped switch [12]. In this section, we will illustrate how one can size such switches based on the lookup tables, and using a transmission gate switch as an example (see Figure 6.35).

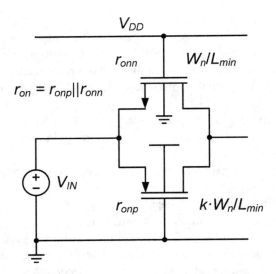

Figure 6.35 Transmission gate switch in the "on" state. Typically, minimum channel length is used ($L = L_{min}$), and the p-channel device is often sized larger than the n-channel device (sizing factor $k > 1$).

The on-resistance (r_{on}) of a MOS device is typically computed for zero drain-source voltage and hence follows from the small-signal drain-source conductance in the triode region, i.e. $r_{on} = 1/g_{ds}$ (for $V_{DS} = 0$). The total on-resistance of the transmission gate switch in Figure 6.35 is thus given by the parallel combination of r_{onn} and r_{onp}. The resistance of the n-channel device (r_{onn}) is lowest when $V_{IN} = 0$, whereas r_{onp} is lowest when $V_{IN} = V_{DD}$. Using the Matlab code below, we can plot the individual and total on-resistances as a function of V_{IN} and observe the typical r_{on} peak near mid-supply (see Figure 6.36).[4]

```
vdd = 1.2;
vinn = nch.VSB;
vinp = vdd - pch.VSB;
gdsn = diag(lookup(nch, 'GDS', 'VGS', vdd-vinn,...
  'VSB', vinn, 'VDS', 0));
gdsp = diag(lookup(pch, 'GDS', 'VGS', vinp,...
  'VSB', vdd-vinp, 'VDS', 0));
ronn = 1./gdsn; ronp = 1./gdsp;
```

Note that the total on-resistance in Figure 6.36 is larger at $V_{IN} = V_{DD}$ than it is at $V_{IN} = 0$. This is because the p-channel in our technology has a lower mobility than the n-channel. Increasing the width of the p-channel will compensate for this and it will also reduce the variation in r_{on} as V_{IN} is swept from 0 to V_{DD}. An important

---

[4] Note that since the input voltage also sets the V_{SB} of the two transistors, we can only step V_{IN} coarsely, according to the number of V_{SB} points available in the lookup table. We obtain a continuous and smooth plot using spline interpolation.

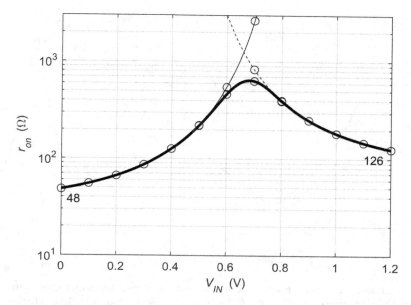

Figure 6.36 Solid thin line: r_{onn}, Dashed line: r_{onp}, Thick line: r_{on}. The marked points correspond to the data stored in the lookup tables. The smooth lines were generated using spline interpolation. The n- and p-channel devices have equal width ($k = 1$) $W_n = W_p = 10$ μm (default lookup table value).

question that comes up in this context is: What is the sizing ratio (k) that minimizes the on-resistance variation (i.e. ratio of maximum and minimum value of r_{on})?

To answer this question with the help of the lookup tables, we can sweep the sizing ratio k in Matlab and plot the on-resistance variation as a function of this parameter (see Figure 6.37(a)). We see that the best sizing ratio is achieved for a k value near 2.6 (which is close to the ratio of n- and p-channel mobility). Figure 6.37(b) shows the on-resistance curve of the transmission gate for this sizing ratio.

With this outcome in mind, the design of a transmission gate switch typically consists of two steps:[5]

1. Size the p-channel to minimize the on-resistance variation across V_{IN}. For the technology used in this book, the proper sizing ratio is $k = 2.6$.
2. Up or downsize size the n-channel and p-channel together (by the same factor) to obtain the desired on-resistance. The designer will often consider the worst-case value near the peak of r_{on} (see Figure 6.37(b)).

We now illustrate this procedure using an example.

---

[5] It should be noted here that another (less common) design option is to make $k = 1$, to achieve (partial) cancellation of the devices' channel charges. It is straightforward to adjust the presented sizing example for this case.

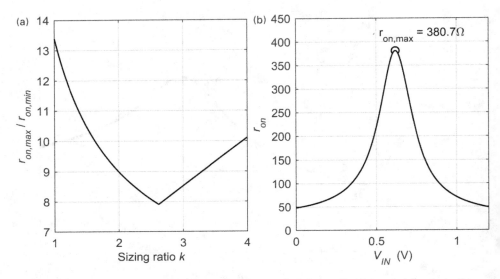

Figure 6.37 (a) On-resistance variation as a function of p-channel/n-channel sizing ratio (k). (b) On-resistance curve for $k = 2.6$. The n-channel device has $W_n = 10$ μm (default lookup value).

Example 6.9 Sizing a Transmission Gate Switch

Consider a transmission gate that is switching a capacitance of $C = 1$ pF. Size the switch such that the circuit is guaranteed to settle to within $\varepsilon_d = 0.1\%$ in one half clock cycle at 100 MHz (f_{clk}).

SOLUTION

We can estimate the required on-resistance based on the first-order settling time (t_s) expression for an RC network:

$$t_s = \frac{1}{2f_{clk}} = r_{on}C \cdot \ln\left(\frac{1}{\varepsilon_d}\right).$$

The Matlab script below uses this equation to compute r_{on}, giving 724 Ω. Using the optimum p/n sizing ratio $k = 2.6$ and the peak on-resistance for the default-width devices from Figure 6.37(b), it then scales the transistors accordingly. The end is result is $W_n = 5.3$ μm and $W_p = 13.7$ μm.

```
% Sizing parameters
k = 2.6;
ron_max = 380.7;
%Design specifications
fclk = 100e6;
epsilon = 0.1e-2;
```

```
C = 1e-12;
% Calculate required ron and device widths
ron = 1/2/fclk/C/log(1/epsilon)
scale = ron_max/ron
Wn = nch.W*scale
Wp = k*Wn
```

The above example showed how to size a transmission gate switch systematically using the lookup tables used throughout this book. It is worth noting that in this context the inversion level (and thus g_m/I_D) is not a meaningful design parameter, since the transistors are not saturated. Nonetheless, the sizing approach based on lookup tables is useful in practice, since it eliminates time-consuming, iterative tweaking in SPICE.

6.6   Summary

This chapter dealt with examples focused on OTA design for switched-capacitor circuits. We first considered a generic SC gain stage to provide the required application context and then looked at implementation options for its amplifier core.

As a first and pedagogically motivated example, we considered the basic two-transistor, single-stage OTA with idealized active loads. While this circuit typically won't meet application requirements in terms of voltage gain, it provides valuable insight on sizing tradeoffs that carry over to more complex topologies. The most significant result of this initial study was that minimizing the current boils down to maximizing the product of the feedback factor squared (β^2) and g_m/I_D, two parameters that trade off against one another via the transistor's transit frequency. Furthermore, we found that for transistors operating in strong inversion, the optimum β is simply 3/4th of the maximum possible value. We saw in all subsequent examples that the true optimum value for the feedback factor always lies relatively close to this value, making it a useful first-order design guideline.

A second goal for our study of the basic OTA was to establish a solid understanding of slewing and its impact on circuit sizing. We found that slewing intertwines the equation set such that the previous analytical optimum is no longer tractable. To overcome this problem, we introduced the concept of running a two-dimensional search involving β and g_m/I_D to find physical design points numerically.

When we moved to the folded-cascode topology, the same two-dimensional search proved again valuable, as there is also no closed-form optimum for the design of this circuit. However, we found that despite the complex equation set, the optimum feedback factor was still in the vicinity of $3/4 \cdot \beta_{max}$. To make the overall design more manageable, we applied a divide-and-conquer flow in which we first designed the output branch based on signal swing and voltage gain considerations, and subsequently optimized the input pair to meet the bandwidth and noise requirements with minimum current.

In the following treatment of the two-stage OTA, we again saw that only a divide-and-conquer approach will let us navigate through the many degrees of freedom in this architecture. We argued that some of these degrees of freedom constitute the main design "knobs," while others can be categorized as secondary. These secondary variables can be set a priori and using reasonable guesses (subject to refinement), while the main knobs will have a much more significant effect on the circuit's performance. The chosen main knobs were the feedback factor β and the ratio of total load capacitance and compensation capacitance. A sweep within this space once again confirmed that the optimum feedback factor lies in the vicinity of $3/4 \cdot \beta_{max}$.

Working through the folded-cascode and two-stage design examples equipped us with a good amount of intuition that we then used to formulate simplified design flows. These simplified flows are based on assuming a fixed feedback factor of $3/4 \cdot \beta_{max}$ and use reasonable, but conservative estimates for several other parameters. This leads to relatively simple and linear design algorithms that the reader can re-use and refine for future designs.

Overall, the examples presented in this chapter reaffirm the fact that the g_m/I_D design methodology can be efficiently applied to circuits that have no closed-form design solution. In addition, we saw that the lookup table based approach allowed us to perform numerical sweeps in a way that reinforces our fundamental understanding of the circuit. This contrasts design approaches that are purely driven by SPICE iterations, without sanity checks against first-order theory. Setting up the scripts used in this chapter can take some time, but subsequent re-use will usually lead to a rapid amortization of the invested effort.

Finally, we showed in Section 6.5 how to size the switches in SC circuits. In this case, since the transistors operate in the triode region, g_m/I_D is not a meaningful design parameter. However, since our lookup tables also contain data for $V_{DS} = 0$ V, they can be conveniently used to compute first-order on-resistances for switch sizing.

6.7   References

[1]   T. Chan Caruosone, D.A. Johns, and K. Martin, *Analog Integrated Circuit Design*, 2nd ed. Wiley, 2011.

[2]   B. Murmann, "Thermal Noise in Track-and-Hold Circuits: Analysis and Simulation Techniques," *IEEE Solid-State Circuits Mag.*, vol. 4, no. 2, pp. 46–54, 2012.

[3]   R. Schreier, J. Silva, J. Steensgaard, and G. C. Temes, "Design-Oriented Estimation of Thermal Noise in Switched-Capacitor Circuits," *IEEE Trans. Circuits Syst. I*, vol. 52, no. 11, pp. 2358–2368, Nov. 2005.

[4]   P. R. Gray, P. Hurst, S. H. Lewis, and R. G. Meyer, *Analysis and Design of Analog Integrated Circuits*, 5th ed. Wiley, 2009.

[5]   D. W. Cline and P. R. Gray, "A Power Optimized 13-b 5 Msamples/s Pipelined Analog-to-Digital Converter in 1.2 µm CMOS," *IEEE J. Solid-State Circuits*, vol. 31, no. 3, pp. 294–303, Mar. 1996.

[6] F. Silveira and D. Flandre, "Operational Amplifier Power Optimization for a Given Total (Slewing Plus Linear) Settling Time," in *Proc. Integrated Circuits and Systems Design*, 2002, pp. 247–253.

[7] H. C. Yang and D. J. Allstot, "Considerations for Fast Settling Operational Amplifiers," *IEEE Trans. Circuits Syst.*, vol. 37, no. 3, pp. 326–334, Mar. 1990.

[8] B. J. Hosticka, "Improvement of the Gain of MOS Amplifiers," *IEEE J. Solid-State Circuits*, vol. 14, no. 6, pp. 1111–1114, Dec. 1979.

[9] K. Bult and G. Geelen, "A Fast-Settling CMOS Op Amp with 90 dB DC-Gain and 116 MHz Unity-Gain Frequency," in *ISSCC Dig. Tech. Papers*, 1990, pp. 108–109.

[10] A. Dastgheib and B. Murmann, "Calculation of Total Integrated Noise in Analog Circuits," *IEEE Trans. Circuits Syst. I Regul. Pap.*, vol. 55, no. 10, pp. 2988–2993, Nov. 2008.

[11] W. M. C. Sansen, *Analog Design Essentials*. Springer, 2006.

[12] A. M. Abo and P. R. Gray, "A 1.5-V, 10-bit, 14.3-MS/s CMOS Pipeline Analog-to-Digital Converter," *IEEE J. Solid-State Circuits*, vol. 34, no. 5, pp. 599–606, May 1999.

Appendix 1   The EKV Parameter Extraction Algorithm

A.1.1   Review of Equations

Throughout this book, we make use of a transistor model that we call the basic EKV model [1]. It is described by:

1. Two equations connecting the normalized drain current i and the pinch-off voltage V_P to the normalized mobile charge density q:

$$i = q^2 + q \qquad\qquad (A.1.1)$$

$$V_P = U_T(2(q-1) + \log(q)). \qquad\qquad (A.1.2)$$

2. Two additional equations that link the gate-to-source voltage V_{GS} and drain current I_D to the normalized variables:

$$V_{GS} = n\, V_P + V_T \qquad\qquad (A.1.3)$$

$$I_D = i\, I_S. \qquad\qquad (A.1.4)$$

While the first two equations define the universal $i(V_P)$ curve depicted in Figure 2.5(a), the second equation pair scales the axes to match any real $I_D(V_{GS})$ characteristic. In a nutshell, V_T and I_S control the horizontal and vertical translations, respectively, whereas n takes care of scaling.

A.1.2   Parameter Extraction Algorithm

The model parameters are extracted from a saturated transistor with constant drain-to-source (V_{DS}) and source-to-bulk (V_{SB}) voltages. The three parameters defining the model are:

- n the subthreshold slope factor;
- V_T the threshold voltage;
- J_S the specific current density (I_S when $W = 1\ \mu\text{m}$).

Figure A.1.1 illustrates the data needed for the parameter extraction:

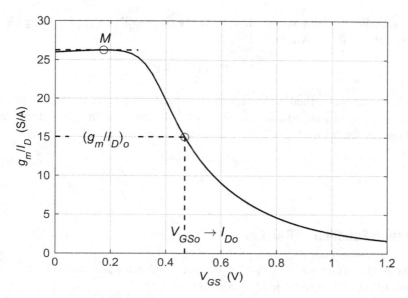

Figure A.1.1 Illustration of the data from which the EKV parameters are extracted.

- the maximum (M) of the g_m/I_D characteristic;
- a reference $(g_m/I_D)_o$ defined via ρ (equal to $(g_m/I_D)_o$ over M) and the corresponding V_{GSo};
- the drain current I_{Do} at $V_{GS} = V_{GSo}$.

With real transistors that show second-order effects, some care is needed in the selection of the reference point. When the gate-to-source voltage gets large, mobility degradation reduces the transconductance efficiency. The reference point should therefore not be chosen in this region. Moderate or weak inversion is preferable. However, the closer we get to the maximum of (g_m/I_D), the worse the accuracy. A good compromise is to position the reference point between 80% and 50% of the maximum of g_m/I_D. The justification is given in Section A.1.4.

The identification proceeds stepwise, starting with the subthreshold slope factor n. Per (2.30), n can be computed from the maximum (M) of g_m/I_D:

$$n = \frac{1}{MU_T}.$$

(A.1.5)

Next, to find V_T, we consider the second point of the g_m/I_D curve. That is, we compute the normalized mobile carrier density q_o that is associated with $(g_m/I_D)_o$ by inverting (2.29):

$$q_o = \frac{M}{\left(\dfrac{g_m}{I_D}\right)_o} - 1 = \frac{1}{\rho} - 1.$$

(A.1.6)

Knowing q_o and n, we can determine the pinch-off voltage V_{Po} using (A.1.2), and compute V_T from (A.1.3):

$$V_T = V_{GSo} - nV_{Po}.$$

(A.1.7)

Knowing q_o, we then evaluate the normalized drain current i_o by means of (A.1.1). The specific current now follows from (A.1.4), since we know the drain current I_{Do} at the reference point:

$$I_S = \frac{I_{Do}}{i_o}.$$

(A.1.8)

A.1.3   Matlab Function XTRACT.m

The Matlab function XTRACT.m extracts the EKV parameters from lookup table data using the steps outlined above. Its syntax is:

```
XTRACT(dev, L, VDS, VSB, rho, TEMP)
```

The function requires four arguments:

- dev structure containing the device lookup data (e.g. nch);
- L the channel length (scalar);
- VDS the drain to source voltage (scalar or column vector);
- VSB the source to bulk voltage (scalar).

The remaining arguments are optional scalars:

- rho the normalized transconductance efficiency. The default value is 0.6 (moderate inversion).
- TEMP the absolute temperature. The default value is 300°K.

Throughout the parameter extraction, the drain and source voltages (with respect to the bulk) remain constant.[1] When the drain voltage is a vector, the extraction algorithm is run for each V_{DS} vector element.

A.1.4   Parameter Extraction Example

Consider an n-channel transistor with $L = 60$ nm, $V_{DS} = 0.60$ V, $V_{SB} = 0$ V and the default values for ρ and TEMP. These data translate into the following syntax:

```
y = XTRACT(nch, 0.06, 0.6, 0)
```

---

[1] This is in strong contrast with the acquisition method proposed in [3], where the extraction algorithm uses measurements performed in the common-gate configuration. Because V_S varies during the extraction, the threshold voltage is averaged over the source voltage excursion. This adversely affects the obtained parameters, especially for short-channel devices.

The output is:

```
y = 0.6000 1.4708 0.4973 0.00000752 ...
-0.0088    -0.0829    0.2175    0.0265    0.0214
```

where: $y(1) = V_{DS}$ (V),

           $y(2) = n$,

           $y(3) = V_T$ (V),

           $y(4) = J_S$ (A/μm) → the specific current density.

The remaining parameters, $y(5)$ to $y(10)$, represent the first and second derivatives of n, V_T and $\log(I_S)$ with respect to V_{DS}. These are useful to study effects related to DIBL (drain-induced barrier lowering) and CLM (channel length modulation).

We now compare the reconstructed basic EKV drain current density J_D and transconductance efficiency g_m/I_D to the lookup table data from which n, V_T and J_S were extracted. The original drain current density J_D and transconductance efficiency g_m/I_D are found using:

```
JD = lookup(nch, 'ID_W', 'VDS', VDS, 'VSB', VSB, 'L', L);
gm_ID = lookup(nch,'GM_ID','VDS',VDS,'VSB',VSB,'L',L);
```

To reconstruct J_D and g_m/I_D, we perform the following steps:

1) Sweep q from weak inversion (e.g. $q = 10^{-3}$) to strong inversion (e.g. $q = 10$):

```
q = logspace(-3,1,20);
```

2) Compute i and V_P using (A.1.2) and (A.1.3).
3) Find J_D and V_{GS} by substituting the parameters computed by the XTRACT function into (A.1.3) and (A.1.4).

The result is shown in Figure A.1.2. It shows that the model fits the original data very well in weak and moderate inversion, but the fit is worse in strong inversion. As shown in Figure A.1.3, the relative difference between the reconstructed and real drain currents does not exceed ±6% for V_{GS} below 0.9 V. This is not bad, considering that the range of currents spans five decades across this interval. Beyond $V_{GS} = 0.8$ V, the model diverges owing to growing effects of mobility degradation.

We mentioned earlier that moderate inversion is the appropriate inversion level for the extraction of the reference point. The default value of ρ (0.6) that we chose is assumed to minimize the difference D (in Figure A.1.3) across weak and moderate inversion.

Figure A.1.4 shows the modifications that V_T and J_S undergo when we change ρ and consider various gate lengths. We see that as long as ρ is between 0.6 and 0.9 (moderate and weak inversion), the choice of the normalized transconductance brings only small modifications of V_T and J_S. The threshold voltage varies by no more than a few millivolts, while the specific current density changes by less than

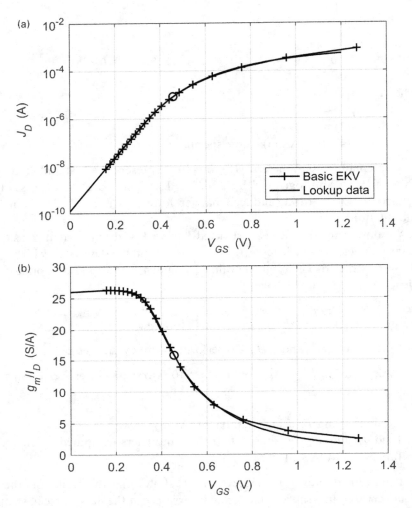

Figure A.1.2 (a) Drain current density of original lookup table data and basic EKV model representation. (b) Corresponding transconductance efficiency. Parameters: $L = 60$ nm, $V_{DS} = 0.6$ V, $V_{SB} = 0$ V. The circles mark the default reference point ($\rho = 0.6$).

10%. This result also holds for the other gate lengths.[2] A similar numerical evaluation shows that the subthreshold slope factor n is nearly independent of ρ. It also varies little with the gate length. For L equal to 100, 200 and 500 nm, n equals 1.2947, 1.2371 and 1.2599, respectively. Below 100 nm, however, n increases rapidly, reaching 1.4741 at 60 nm.

We can re-position ρ to improve the fit at another inversion level, but the price is generally a larger mismatch elsewhere. If ρ lies deep in weak inversion, the

---

[2] Gate length modifications cause shifts that are due to phenomena discussed in Section 2.3.3. These are irrelevant for the present discussion.

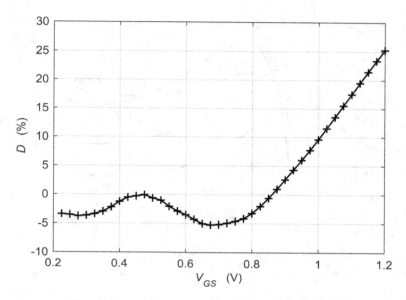

Figure A.1.3 Plot of the percentage difference (*D*) between the reconstructed and the real drain currents shown in Figure A.1.2(a).

EKV parameters essentially adjust to fit the exponential part of the drain current. Making ρ smaller than 0.5 affects the parameters more seriously. Mobility degradation reduces *the* transconductance efficiency, resulting in an erroneous shift of V_{GSo}, which lowers the extracted threshold voltage V_T. The consequences are clearly visible in the plots of Figure A.1.5 when ρ is set to 0.2. The reconstructed data match the original data at and near the reference point (marked by a circle when g_m/I_D equals 5.25 S/A), but cause an overall left-shift of J_D and g_m/I_D due to the reduction in the estimated threshold voltage.

A.1.5   Matlab Function XTRACT2.m

The XTRACT function of Section A.1.3 evaluates EKV parameters knowing the type of transistor, the gate length L, the drain-to-source voltage V_{DS} and the source-to-bulk voltage V_{SB}. The drain current characteristics from which the parameters are evaluated are reconstructed prior to the extraction. However, when the drain currents versus V_{GS} are already available, extraction can be performed using the XTRACT2 Matlab function. Its syntax is:

```
XTRACT2(VGS, ID, rho)
```

The function requires two arguments:

- VGS the gate to source voltage (column vector);

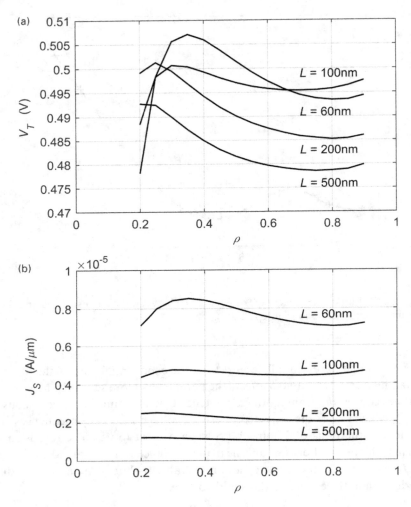

Figure A.1.4 (a) Extracted threshold voltage and (b) Extracted specific current density as a function the normalized transconductance ρ for L ranging from 60 to 500 nm, $V_{DS} = 0.6$ V and $V_{SB} = 0$ V.

- ID the drain current (column vector or matrix).

The third argument is an optional scalar:

- rho the normalized transconductance efficiency that defines the reference point. The default value is 0.6 (moderate inversion).

The temperature is assumed to be 300°K. The function outputs n, V_T and I_S (row vector or matrix).

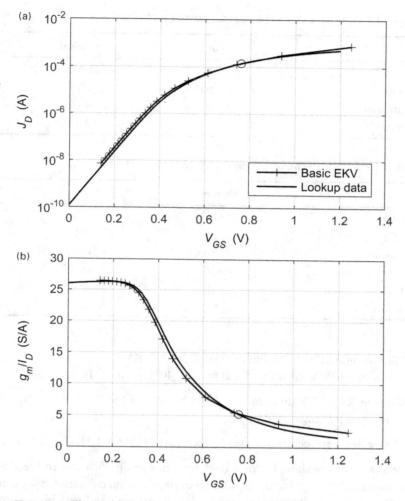

Figure A.1.5 Illustration of discrepancy between real and extracted drain current density (a) and transconductance efficiency (b) when $\rho = 0.2$ yielding $(g_m/I_D)_o = 5.25$ S/A. All other parameters are as in Figure A.1.2.

A.1.6   Corner Parameter Extraction

In this section, we show an example of the possibilities offered by the basic EKV model and the described parameter extraction. We use the XTRACT function to identify the EKV parameters across process corners (slow/nominal/fast) [2]. The numerical data in the lookup tables correctly represents the effects of temperature and process corners on currents, intrinsic gain, transit frequency, etc. However, it is also interesting to get a feel for the changes of n, V_T, and $\beta = I_s/2(nU_T)^2$ (see (2.25) and (2.16)). For our investigation, we extract these parameters for the following conditions:

Table A.1.1 Extracted n-channel parameters, slow/nominal/fast for T = 300°K.

	Slow	Nominal	Fast
n	1.304	1.292	1.180
V_T (V)	0.5299	0.4957	0.4689
J_S (μA/μm)	3.75	4.46	4.92
β (mA/V^2)	2.14	2.57	3.11

Table A.1.2 Extracted n-channel parameters, slow & hot and fast & cold.

	Slow, hot	Nominal	Fast, cold
T (°K)	398	300	233
T (°C)	125	27	−40
n	1.296	1.292	1.318
V_T (V)	0.4975	0.4957	0.4773
J_S (μAμm)	5.11	4.46	4.30
β (mA/V^2)	1.67	2.57	4.03

1. Slow/nominal/fast at room temperature (300°K).
2. Slow at +125 °C (slow/hot) and fast at −40°C (fast/cold).

Using the XTRACT function, we then obtain data for $V_{DS} = 0.6$ V, $W = 1$ μm and $L = 100$ nm, β being $J_S/(2nU_T^2)$:

1. Extracted parameters at room temperature (Table A.1.1).

We see that the slope factor n does not change significantly. In the slow corner, the threshold voltage increases, while the specific drain current density (and also β) decreases, forcing designers to choose larger currents and/or widths compared to nominal conditions. Under fast conditions, the opposite holds true.

2. Extracted parameters under slow/hot and fast/cold conditions (see Table A.1.2). The nominal data from Table A.1.1 is also shown for reference.

We observe that the trend in the specific current reverses, i.e. I_S is smallest in the fast & cold corner. The threshold voltage doesn't change much under slow/hot conditions but reduces for fast/cold.

To check the correctness of the parameters listed in Table A.1.2, we reconstructed J_D and g_m/I_D and compared the outcome to the original drain current density and transconductance efficiency of the lookup table data. As shown in Figure A.1.6, the reconstructed curves agree well with the original data. As discussed above, significant discrepancies occur only in strong inversion due to mobility degradation (see zoomed plots in Figure A.1.6(b) and (d)).

The p-channel counterparts of the tables above for the same device width, V_{DS} and L are listed Tables A.1.3. and A.1.4 for completeness.

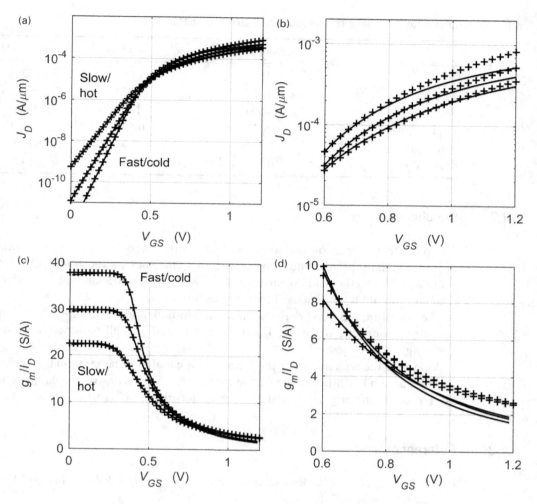

Figure A.1.6 (a) Drain current density of lookup table data and basic EKV model representation across corners. (c) Corresponding transconductance efficiency. (b) and (d) show a zoom into the strong inversion regime, which is affected by mobility degradation. Parameters: $L = 100$ nm, $V_{DS} = 0.6$ V, $V_{SB} = 0$ V.

Table A.1.3 Extracted p-channel parameters, slow/nominal/fast for T = 300°K.

	Slow	Nominal	Fast
n	1.416	1.410	1.292
V_T (V)	0.5838	0.5617	0.5472
J_S (μA/μm)	2.65	3.00	3.25
β (mA/V²)	1.40	1.58	1.87

Table A.1.4. Extracted p-channel parameters, slow & hot and fast & cold.

	Slow, hot	Nominal	Fast, cold
T (°K)	398	300	233
T (°C)	125	27	−40
n	1.4000	1.410	1.454
V_T (V)	0.5371	0.5617	0.5628
J_S (µA/µm)	3.17	3.00	3.09
β (mA/V²)	0.96	1.58	2.63

A.1.7   Conclusion

The parameter extraction method described in this appendix estimates n, V_T and J_S from the drain current and transconductance efficiency characteristics of a real transistor. The method is implemented in the XTRACT.m Matlab function, which reads the lookup table data used throughout this book.

The acquisition method requires the source-to-bulk and the drain-to-bulk voltages to be constant. When V_S, V_D, or both are modified, all parameters require updating.

As long as the reference ρ for the extraction is equal to or larger than 0.6, mobility degradation has little impact on the parameters. The reconstructed characteristics reproduce the original J_D and g_m/I_D characteristics satisfactorily.

A.1.8   References

[1] P. Jespers, *The gm/ID Methodology, a Sizing Tool for Low-Voltage Analog CMOS Circuits*. Springer, 2010.
[2] B. Murmann, *Analysis and Design of Elementary MOS Amplifier Stages*. NTS Press, 2013.
[3] C. C. Enz and E. A. Vittoz, *Charge-Based MOS Transistor Modeling: The EKV Model for Low-Power and RF IC Design*. John Wiley & Sons, 2006.

Appendix 2 Lookup Table Generation and Usage

A.2.1   Lookup Table Generation

As explained in Section 1.2.2, the lookup tables used in this book are created through four-dimensional (L, V_{GS}, V_{DS}, and V_{SB}) DC sweeps and noise analyses in a SPICE-like circuit simulator. Figure A.2.1 illustrates the setup that was used to perform this sweep. The relevant files for this flow (except the 65 nm PSP model file) are available at this book's companion website and also at [1]. We used Cadence Spectre as our circuit simulator, but the approach can be modified to work with

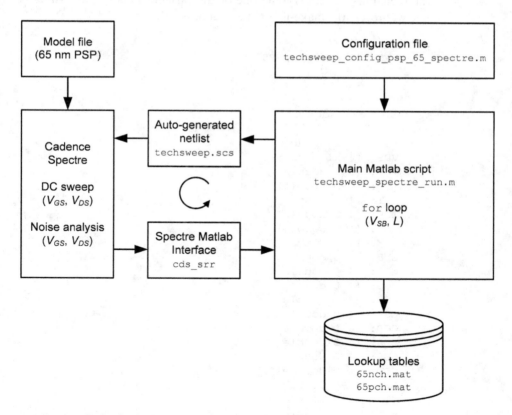

Figure A.2.1 Setup for lookup table generation.

other tools as well (the provided online material includes example files for both Spectre and HSPICE).

Even though most simulators can perform four-dimensional sweeps, we created a flow in which two of the four dimensions are handled as for-loops in the main Matlab routine (techsweep_spectre.m). This prevents excessively large memory usage by the simulator. For each sweep value of L and V_{SB}, Matlab writes a new netlist and calls the simulator via the Linux command line, which then runs an inner 2D sweep (V_{GS} and V_{DS}). For each of these iterations, the results are read back into Matlab via the cds_srr function, provided in Cadence's Spectre Matlab Toolbox [2]. Similar Matlab read-in functionality is available for HSPICE using Perrott's Toolbox [3].

Once all simulation runs are completed, the data set for each transistor is stored in the lookup table output files (65nch.mat and 65pch.mat). To use this data within a Matlab script, one can load the data as shown below:

```
load 65nch.mat
load 65pch.mat
```

This creates two structures in Matlab: nch (for the n-channel device) and pch (for the p-channel device). Typing one of the variable names at the Matlab command line will reveal the make-up of the structure:

```
nch =
      INFO: '65nm CMOS, PSP'
    CORNER: 'NOM'
      TEMP: 300
     NFING: 5
        L: [31x1 double]
        W: 10
      VGS: [49x1 double]
      VDS: [49x1 double]
      VSB: [9x1 double]
       ID: [4-D double]
       VT: [4-D double]
      IGD: [4-D double]
      IGS: [4-D double]
       GM: [4-D double]
      GMB: [4-D double]
      GDS: [4-D double]
      CGG: [4-D double]
      CGS: [4-D double]
      CSG: [4-D double]
      CGD: [4-D double]
      CDG: [4-D double]
      CGB: [4-D double]
      CDD: [4-D double]
      CSS: [4-D double]
      STH: [4-D double]
      SFL: [4-D double]
```

Each structure has a header that contains some identifying information about the data set. The parameter nch.NFING shows the number of device fingers that were used in the simulation; see Appendix 3 for further details. The parameter nch.W contains the total width that was used in the SPICE simulations. Note that throughout this book, we assume units of microns for all geometries. All other quantities are expressed in unscaled SI units, e.g. Amperes, Volts, Ohms, Siemens, etc. The parameters STH and SFL are the thermal and flicker noise power spectral densities of the drain current.

A.2.1.1 Configuration File

The main Matlab script (techsweep_spectre.m) that drives the lookup table generation is kept general and does not contain any user- or process-specific parameters; all of this information is provided through a configuration file (see Figure A.2.1). The configuration file has several sections that define the file path to the simulator, the syntax template for the netlist, the sweep ranges and step sizes, as well as the mapping from simulator output variables to Matlab variables. We will briefly discuss the latter two aspects here.

The code section that defines the sweep ranges is shown below. With this setup, each variable stored in the output file is an array of size 31 x 49 x 49 x 9 (L, V_{GS}, V_{DS}, and V_{SB}). The resulting file size is about 70 MB, which is a reasonable compromise. For all the examples covered in this book, we did not see a need for finer step sizes.

```
c.VGS_step = 25e-3;
c.VDS_step = 25e-3;
c.VSB_step = 0.1;
c.VGS_max = 1.2;
c.VDS_max = 1.2;
c.VSB_max = 0.8;
c.VGS = 0:c.VGS_step:c.VGS_max;
c.VDS = 0:c.VDS_step:c.VDS_max;
c.VSB = 0:c.VSB_step:c.VSB_max;
c.LENGTH = [(0.06:0.01:0.2) (0.25:0.05:1)];
c.WIDTH = 10;
c.NFING = 5;
```

The output parameters that will be stored in the Matlab lookup table are defined in the configuration file through the following array:

```
c.outvars = {'ID','VT','IGD','IGS','GM','GMB','GDS','CGG',...
'CGS','CSG', 'CGD','CDG','CGB','CDD','CSS','STH','SFL'};
```

Most of these parameters are directly computed by Spectre, but some of them must be computed as sums of several Spectre output variables. For example, the total drain capacitance (CDD) consists of an intrinsic component (contributed by the MOSFET), a stray capacitance component, and a junction capacitance component. The mapping from Spectre variables to the final quantity that is stored in the lookup table occurs through a mapping matrix. Below is an example that illustrates the syntax of this matrix, considering only a subset of the variables for clarity (the actual table has additional columns).

```
% Variable mapping
c.outvars =                              {'ID', 'CGG', 'CGS', 'CDD'};
c.n{1}=     {'mn.m1.m1:ids','A',     [1      0      0      0   ]};
c.n{2}=     {'mn.m1.m1:vth','V',     [0      0      0      0   ]};
c.n{3}=     {'mn.m1.m1:igd','A',     [0      0      0      0   ]};
c.n{4}=     {'mn.m1.m1:igs','A',     [0      0      0      0   ]};
c.n{5}=     {'mn.m1.m1:gm','Ohm',    [0      0      0      0   ]};
c.n{6}=     {'mn.m1.m1:gmb','Ohm',   [0      0      0      0   ]};
c.n{7}=     {'mn.m1.m1:gds','Ohm',   [0      0      0      0   ]};
c.n{8}=     {'mn.m1.m1:cgg','F',     [0      1      0      0   ]};
c.n{9}=     {'mn.m1.m1:cgs','F',     [0      0      1      0   ]};
c.n{10}=    {'mn.m1.m1:cgd','F',     [0      0      0      0   ]};
c.n{11}=    {'mn.m1.m1:cgb','F',     [0      0      0      0   ]};
c.n{12}=    {'mn.m1.m1:cdd','F',     [0      0      0      1   ]};
c.n{13}=    {'mn.m1.m1:cdg','F',     [0      0      0      0   ]};
c.n{14}=    {'mn.m1.m1:css','F',     [0      0      0      0   ]};
c.n{15}=    {'mn.m1.m1:csg','F',     [0      0      0      0   ]};
c.n{16}=    {'mn.m1.m1:cgsol','F',   [0      1      1      0   ]};
c.n{17}=    {'mn.m1.m1:cgdol','F',   [0      1      0      1   ]};
c.n{18}=    {'mn.m1.m1:cgbol','F',   [0      1      0      0   ]};
c.n{19}=    {'mn.d1.d1:cj','F',      [0      0      0      1   ]};
c.n{20}=    {'mn.d2.d1:cj','F',      [0      0      0      0   ]};
```

Consider the drain current ID. The single "1" in the column of this parameter means that it is simply computed taking the drain current from Spectre (mn.m1.m1:ids) and no other output variable must be added. Conversely, CGG will be computed by adding up four Spectre output variables:[1]

```
mn.m1.m1:cgg
mn.m1.m1:cgsol
mn.m1.m1:cgdol
mn.m1.m1:cgbol
```

The first one of these is the device's "intrinsic" total gate capacitance, while the other three are fringe (or "overlap") capacitances that must also be considered. Note that the output variables have the prefix "mn.m1.m1" since our model file places the transistors into a sub-circuit for RF modeling (i.e. it includes terminal resistances, etc.). This is not a requirement, but simply due to the composition of our selected device model file.

The unit entry that follows the variable, e.g. "A" for "mn.m1.m1:ids" defines the "type identifier" needed for the cds_srr function; please refer to the Cadence documentation [2] about further details. The attentive reader may have noticed in the above matrix that the type identifier for the output variable "mn.m1.m1:gm" is "Ohm". This appears to be a bug in the Cadence Spectre toolbox that may be fixed in the future.

---

[1] When modifying this matrix, the reader should pay special attention to the sign of the capacitances outputted by the simulator. Some of them may be negative, but may have to be counted as positive in the respective sum. In this case, the matrix entry can be set to "−1".

A.2.1.2   Generating Lookup Tables for a New Technology

The main step for generating lookup tables for a new technology or for additional device types (e.g. additional V_T options) is to create a new configuration file and re-run the main Matlab script (techsweep_spectre.m). Since running this program can take some time (of the order of a few hours), we have also created a simplified version of this script for debug purposes (techsweep_spectre_debug.m). This script executes only a single simulation run and displays the parameters computed via the variable mapping matrix. This makes it possible to detect configuration errors before running the complete lookup table generation.

A.2.2   Matlab Function lookup.m

To simplify the task of reading data from the nch and pch structures, we have created the Matlab function "lookup.m", which provides most of the functionality needed for g_m/I_D-based design. Assuming that the function resides in the current Matlab path, a short description of its usage can be echoed by typing "help lookup". The output is repeated below with some additional details.

The function "lookup" extracts a desired subset from the 4-dimensional simulation data. The function interpolates when the requested points lie off the simulation grid. The function has the following general syntax:

```
output = lookup(data, outvar, varargin)
```

where "data" is a placeholder for the Matlab structure containing the 4-dimensional lookup table data (nch or pch). The mandatory parameter outvar defines which variable should be read from the structure. Lastly, additional parameters can be passed to the function to specify the desired region within the 4-D array (for example, get data for a specific channel length). The exact syntax depends on the function's usage mode. The following functionality is available:

(1) Basic lookup of parameters (e.g. ID, GM, CGG, ...) at some given (L, VGS, VDS, VSB).
(2) Lookup of arbitrary ratios of parameters (e.g. GM_ID, GM_CGG, ...) at some given (L, VGS, VDS, VSB).
(3) Cross-lookup of one ratio against another, for example GM_CGG for some GM_ID.

In usage modes (1) and (2) the input arguments (L, VGS, VDS, VSB) can be listed in any order and default to the following values when not specified:

```
L = min(data.L); (minimum length used in simulation)
VGS = data.VGS; (VGS vector used during simulation)
VDS = max(data.VDS)/2;
VSB = 0;
```

For example, the following code gives the drain current along the VGS sweep (nch. VGS) for $L = 60$ nm, $V_{DS} = 0.6$ V and $V_{SB} = 0$ V:

```
id = lookup(nch, 'ID');
```

An example that gives the drain current at other L, V_{GS}, V_{DS} and V_{SB} values is given below:

```
id = lookup(nch, 'ID', 'VGS', 0.5, 'VDS', 0.8, 'L', 0.1, ...
'VSB', 0.1);
```

In usage mode (3), the cross-lookup between the two ratios is based on evaluating both parameters across the entire range of data.VGS, and finding the intersects at the desired points. In this mode, the output and input parameter ratios must be specified first. The example below outputs GM_CGG at GM_ID values between 5 and 20 S/A:

```
wt = lookup(nch, 'GM_CGG', 'GM_ID', 5:0.1:20);
```

In the above example, L, V_{DS} and V_{SB} are assumed to be at the default values, but they can also be specified explicitly:

```
wt = lookup(nch, 'GM_CGG', 'GM_ID', 5:0.1:20, 'VDS', 0.7);
```

The default interpolation method for the final 1-D interpolation that finds the intersects in mode (3) is "pchip". It can be set to a different method by passing for example 'METHOD', 'linear' to the lookup function. All other multidimensional interpolation operations use 'linear' (fixed), since any other method requires continuous derivatives; this is rarely satisfied across all dimensions, even with the best device models.

When more than one parameter is passed to the function as a vector, the output becomes multidimensional. This behavior is inherited from the Matlab function "interpn", which lies at the core of the lookup function. The following example produces an 11x11 matrix as the output:

```
lookup(nch,'ID', 'VGS', 0:0.1:1, 'VDS', 0:0.1:1)
```

The dimensions of the output array are ordered such that the largest dimension comes first. For example, one-dimensional output data is an (n x 1) column vector. For two dimensions, the output is (m x n) and m > n.

If we want to access only the values of I_D for which $V_{GS} = V_{DS}$ in the previous example, the vector of interest is simply the diagonal of the full matrix:

```
diag(lookup(nch,'ID', 'VGS', 0:0.1:1, 'VDS', 0:0.1:1))
```

The lookup function performs a basic input syntax check and may echo the following error when used inappropriately:

```
Invalid syntax or usage mode! Please type "help lookup".
```

Also, in mode (3), the function echoes a warning when the requested data point does not exist (giving NaN (not a number) values in the output vector). For example, this command will certainly produce the error:

```
wT = lookup(nch, 'GM_CGG', 'GM_ID', 50)
lookup warning: GM_ID input larger than maximum! (output is NaN)
wT =
  NaN
```

The user can turn off the warning using the following syntax:

```
wT = lookup(nch, 'GM_CGG', 'GM_ID', 50, 'WARNING', 'off')
```

Turning off the warning can be desirable if the function is used in a for-loop and repetitive echoing would otherwise cause long delays.

Overall, it is important to note that the provided lookup function is only a simple engineering aid and it is not perfect in terms of syntax checking and other error handling mechanisms that come with more general Matlab functions. The reader is encouraged to make his/her own improvements.

A.2.3   Matlab Function lookupVGS.m

The function "lookupVGS" is a companion function to "lookup." It finds the transistor's V_{GS} for a given inversion level (GM_ID) or current density (ID_W) and given terminal voltages. The function interpolates when the requested points lie off the simulation grid. The function has the following general syntax:

```
output = lookupVGS(data, varargin)
```

where "data" is a placeholder for the Matlab structure containing the 4-dimensional lookup table data (nch or pch). The function supports two usage modes:

(1) Lookup VGS with known voltage at the source terminal.
(2) Lookup VGS with unknown source voltage. This mode is useful, for example when the source of the transistor is the tail node of a differential pair or the intermediate node in a cascode stack.

At most one of the input arguments can be a vector; the others must be scalars. The output is a column vector.

In usage mode (1), the inputs to the function are GM_ID (or ID_W), L, VDS and VSB. Some examples are given below:

```
VGS = lookupVGS(nch, 'GM_ID',10, 'VDS',0.6, 'VSB',0.1, 'L',0.1)
VGS = lookupVGS(nch, 'GM_ID',10:15, 'VDS',0.6, 'VSB',0.1, 'L',0.1)
VGS = lookupVGS(nch, 'ID_W',1e-4, 'VDS',0.6, 'VSB',0.1, 'L', 0.1)
VGS = lookupVGS(nch, 'ID_W',1e-4, 'VDS',0.6, 'VSB',0.1,...
    'L',[0.1:0.1:0.5])
```

When VSB, VDS or L are not specified, their default values are assumed (same behavior as for lookup.m):

```
VSB = 0;
L = min(data.L);  (minimum length)
VDS = max(data.VDS)/2;  (VDD/2)
```

In usage mode (2), VDB and VGB must be supplied to the function, e.g.

```
VGS = lookupVGS(nch, 'GM_ID',10, 'VDB',0.6, 'VGB',1, 'L',0.1)
VGS = lookupVGS(nch, 'ID_W',1e-4, 'VDB',0.6, 'VGB',1, 'L',0.1)
```

The default method for the final 1-D interpolation is "pchip". It can be set to a different method by passing, for example 'METHOD', 'linear', to the function. All other multidimensional interpolation operations use 'linear' (fixed).

The function issues a warning when the output is ill-defined:

```
lookupVGS(nch, 'GM_ID', 50)
lookupVGS: GM_ID input larger than maximum!
ans =
   NaN
```

A.2.4   Lookup of Ratios with Non-monotonic Vectors

All interpolations within lookup.m are handled through the standard Matlab functions interp1 and interpn:

```
Vq = interp1(X,V,Xq)
Vq = interpn(X1,X2,X3,…,V,X1q,X2q,X3q,...)
```

An important requirement for these functions to work is that the input arguments X are monotonic. In usage modes (1) and (2) of the lookup function, this is guaranteed since we are performing interpolations with nch.VGS, nch.VDS, nch.VSB and nch.L as the X-variables. These vectors are linearly swept during the lookup table creation and hence are monotonic by design.

The situation is different in usage mode (3). Here, we are providing the functionality to look up one ratio against another, and this may lead to non-monotonicity issues. To understand why this is the case, let us examine the following lookup command:

```
wT = lookup(pch, 'GM_CGG', 'GM_ID', 28, 'VSB', 0.6, 'L', 0.2)
```

In this example, VDS is not specified and hence defaults to 0.6 V. The VSB and L values are chosen to amplify the non-monotonicity problem that we are about to encounter.

Since neither GM_CGG nor GM_ID are directly stored in the lookup tables, the first step within lookup.m is to compute these two ratios along the VGS sweep vector. For the above example, this essentially corresponds to:[2]

---

[2] The actual implementation in lookup.m is slightly different; we are just highlighting the concept here.

```
gm_ID = lookup(pch, 'GM_ID', 'VGS',pch.VGS, 'VSB',0.6, 'L',0.2);
gm_Cgg = lookup(pch, 'GM_CGG', 'VGS',pch.VGS, 'VSB',0.6, 'L',0.2);
```

The desired final interpolation is then completed using:

```
wT = interp1(gm_ID, gm_Cgg, 28)
```

Unfortunately, this interpolation will fail for the given example, since the gm_ID vector is non-monotonic. Figure A.2.2(a) illustrates this by plotting gm_ID against pch.VGS for the above example. We see that the desired search value of 28 S/A has two intersections with the curve, and hence there is no unique solution for wT at this search value.

We can resolve this issue by noting that the curve intersection at the small pch.VGS value (marked with an X) is caused by unwanted second-order artifacts (see Section 2.3.1). Hence, this is not a design point of interest and can therefore be discarded. Conversely, the second intersect on the right (marked with a circle)

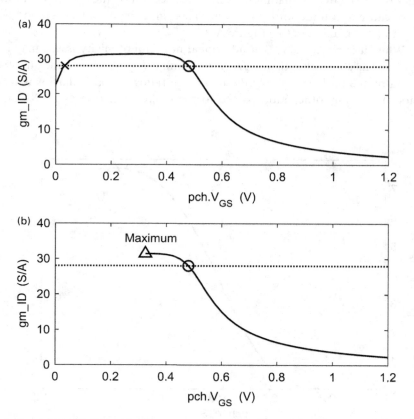

Figure A.2.2 (a) Example of a non-monotonic g_m/I_D versus V_{GS} curve. It has two intersections with the desired interpolation value (27 S/A) a desired one (o) and an undesired one (x). (b) To overcome this problem, the lookup function will consider only points to the right of the maximum gm_ID.

corresponds to a possible operating point in weak inversion. Our implementation of lookup.m makes use of this fact and eliminates the portion of the gm_ID curve that lies to the left of its maximum. This is illustrated in Figure A.2.2(b). The final interpolation can then be executed without any issues.

A similar issue exists when using lookup with the ratio GM_CGG and GM_CGS as the X-variable, for example:

```
gm_gds = lookup(nch, 'GM_GDS', 'GM_CGG', 7.5e11)
```

As before, we can see why this potentially problematic by looking at the GM_CGG data versus nch.VGS (see plot in Figure A.2.3):

```
gm_Cgg = lookup(nch,'GM_CGG', 'VGS', nch.VGS);
```

This command yields the angular transit frequency for a minimum-length n-channel, which suffers from mobility degradation at large V_{GS}, leading to a decrease in $\omega_T = g_m/C_{gg}$ (see Chapter 2). Once again, in mode (3), the lookup function handles this issue by eliminating the unwanted section of the curve that would otherwise cause a non-monotonicity issue. This time, the unwanted section lies to the right of the maximum (see Figure A.2.3)

While it is conceivable that additional non-monotonicity issues may arise with lookup variables other than g_m/I_D, g_m/C_{gg}, and g_m/C_{gs} we limited the automatic handling of the problem to only these variable ratios (whose behavior is well understood). For any other ratio passed to the lookup function as an X-variable, the

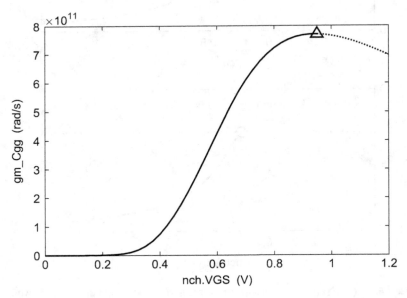

Figure A.2.3 Example of a non-monotonic g_m/C_{gg} versus V_{GS} curve. In usage mode (3), lookup eliminates the doted section of the curve (to the right of the maximum, marked with a triangle) to avoid non-monotonicity errors.

function merely checks for monotonicity and issues an error message when there is a problem:

```
*** lookup: Error! There are multiple curve intersections
*** Try to reduce the search range by specifying the VGS vector
explicitly
Example:   lookup(nch,   'ID_W',   'GM_GDS',   gm_gds,   'VGS',   nch.
VGS(10:end))
```

Consider for example:

```
ID_W = lookup(nch, 'ID_W', 'GM_GDS', 10)
```

Here, lookup will check GM_GDS along the nch.VGS vector and issue the error if non-monotonicity is detected. As hinted in the error message, the problem can then be resolved by manually limiting the VGS search range. This is similar to the way the problem is addressed for the known issues with g_m/I_D, g_m/C_{gg} and g_m/C_{gs}. The only difference is that manual intervention is required to resolve the non-monotonicity problem.

A.2.5 LookupVGS with Non-monotonic g_m/I_D Vector

The issue described in the previous section exists also when g_m/I_D is passed to lookupVGS as the X-variable, for example:

```
VGS = lookupVGS(nch, 'GM_ID', 26.1)
```

The implemented remedy for this problem is the same as described in the previous section. We limit the search vector such that the unwanted part of the interpolation data is eliminated, as shown Figure A.2.2(a).

A.2.6 Passing Design Variables to the Simulator

In design problems that involve a large number of devices, it is convenient to load the Matlab-computed parameters automatically into the circuit simulator. One way to do this is by writing out a text file from Matlab, for example:

```
% Write simulation parameters
fid=fopen('sim_params.txt', 'w');
fprintf(fid,'id1 %d\n', ID1);
fprintf(fid,'w1 %d\n', W1);
fprintf(fid,'l1 %d\n', L1);
fclose(fid);
```

Such a file can be read into the Cadence Analog Design Environment (ADE) using the following script:

```
procedure(ImportDesignVars(filename
   @optional (session asiGetCurrentSession()) "tg")
  let((prt data designVars)
  prt=infile(fileName)
  unless(prt
   error("Cannot read design variable file %s\n" fileName)
   )
  while(gets(data prt)
   data=parseString(data)
   when(listp(data) && length(data)>=2
   designVars=tconc(designVars data)
   )
  )
  close(prt)
  asiSetDesignVarList(session car(designVars))
  )
)
```

This script must be loaded once from the Cadence command window using:

```
load("ImportDesignVars.il")
```

Now the text file generated in Matlab can be imported using:

```
ImportDesignVars("sim_params.txt")
```

More information about such functionality can be found on the World Wide Web.[3]

A.2.7   References

[1] B. Murmann, "gm/ID Starter Kit" [Online]. Available: http://web.stanford.edu/~murmann/gmid.html (accessed May 12, 2017).

[2] Cadence Design Systems, "Spectre/RF MATLAB Toolbox," *Virtuoso Spectre Circuit Simulator RF Theory Man.*

[3] Michael H. Perrott, "Hspice Toolbox for Matlab® and Octave" [Online]. Available: www.cppsim.com/download_hspice_tools.html (accessed May 12, 2017).

---

[3] See e.g. https://community.cadence.com/cadence_technology_forums/f/38/t/20963#sthash.UDavH9n9.dpuf (accessed May 12, 2017).

Appendix 3 Layout Dependence

The g_m/I_D design methodology assumes that some of the electrical transistor parameters scale strictly proportional to the device width. In this appendix, we validate this assumption for the 65-nm CMOS technology used throughout this book.

A.3.1 Introduction to Layout Dependent Effects (LDE)

In modern CMOS technology, the electrical parameters of a transistor show some sensitivity to the way it is placed and laid out in the silicon substrate [1], [2]. For example, the carrier mobility and threshold voltage of a transistor finger depend on its distance from the device edge, due to the strain or stress imposed by shallow trench isolation (STI). According to [2], the variations due to varying distance from the STI edge alone can be up to 15% for mobility and up to 50 mV for the threshold voltage. Typically, such layout dependent effects (LDE) must be considered before a design is sent out for fabrication.

A practical problem with LDE is that the layout is not yet known during circuit sizing. To resolve this issue, the analog designer usually simulates the circuit using parameters that are assumed to be reasonably close to those of the final layout. Specifically, the designer decides on a suitable finger width and number of fingers for each transistor. In the SPICE model, this information is then used to compute the LDE effects to first order. Afterwards, once the physical layout has been created, an extraction tool computes the impact of LDE much more accurately, considering for example the spacing between devices, dummy fingers, well distances, etc. In this flow, the final step is meant to be for verification only, and it should not lead to a re-sizing of the circuit. In other words, the impact of second-order LDE should be small enough so that they can be absorbed into the design margin.

The situation is similar for the normalized design approach within the g_m/I_D design methodology: LDE will shift the device parameters slightly after the exact finger partition and layout have been completed; however, these deviations should not require any sizing adjustments. It is therefore practical to generate the lookup table data used in this book by assuming a fixed reference layout, as described in the next section.

A.3.2   Transistor Finger Partitioning

As discussed above, the designer must decide on a suitable finger partitioning for each device to capture first-order LDE during circuit sizing. This also means that we must decide on a finger partitioning for the creation of the lookup tables for g_m/I_D-based design. For the lookup tables used in this book, the reference transistor layout of Figure A.3.1 is assumed. The device has a total width of $W = 10$ μm and is partitioned into 5 fingers, each being 2 μm wide. The channel length in the drawing is 100 nm, but varies per the intended length sizing.

The chosen finger width is a compromise. As explained in [3], making the fingers too wide leads to poor RF performance due to the increased gate resistance. On the other hand, if the fingers are made too narrow, the relative contribution of parasitic capacitances due to sidewalls and substrate fringe fields increases, which again deteriorates the high-frequency performance. It was shown in [3] that a finger width of 1...3 μm is optimal for the 65 nm and 90 nm CMOS nodes.

When we size circuits using the lookup table data, we typically do not end up with device widths that are multiples of 2 μm. Consequently, since the number of fingers must be an integer, this means that the finger width must vary as we pick

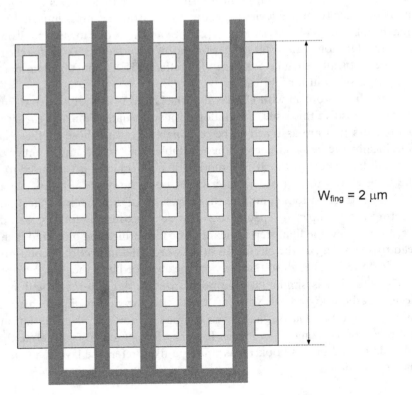

$W_{fing} = 2$ μm

Figure A.3.1 Reference layout assumed for the creation of the lookup tables.

the total width W across some continuous range. This is illustrated in Figure A.3.2, where W is swept from 0.2 μm (minimum width in 65 nm CMOS) to 100 μm. For this plot, the number of fingers (n_f) is computed using:

$$n_f = \max\left(round\left(\frac{W}{2\mu m} \right), 1 \right),$$

(C.3.1)

where "round" refers to rounding to the nearest integer. The plot of W_{fing} crosses 2 μm when W is an exact multiple of 2 μm and it has peaks at odd widths of 3 μm, 5 μm, etc. The largest deviation from $W_{fing} = 2$ μm occurs for minimum width.

A.3.3   Width Dependence of Parameter Ratios

As we have seen above, the finger width and the number of fingers will change as we access the continuous range of total device widths during sizing. We will now quantify the parameter ratio shifts that are caused by this. Specifically, we extract simulation data along the width-sweep of Figure A.3.2, and compare it to the lookup table data that was created using the fixed W and n_f of Figure A.3.1. The setup for these simulations is shown in Figure A.3.3. The device is biased using a current density (J_D) that is computed using the lookup tables, and is set for example by the desired g_m/I_D. The gate voltage of the transistor is set via feedback to force $V_{DS} = 0.6$ V. To quantify width dependencies, we sweep W and compute the number

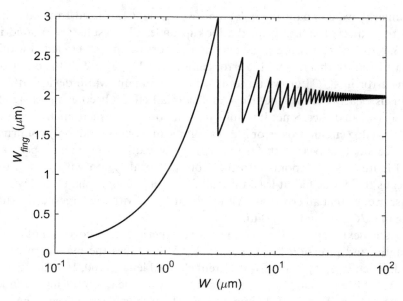

Figure A.3.2 Plot of the finger width $W_{fing} = W/n_f$. W is the total device width and n_f is the number of fingers.

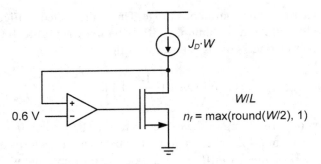

Figure A.3.3 Simulation setup.

of fingers to be used in the simulation per (C.3.1). Ideally, for some fixed current density J_D and channel length L, none of the parameter ratios that we care about (g_m/I_D, g_m/C_{gg}, etc.) should vary as we sweep W. However, due to the variations in finger width and the number of fingers, we observe small variations owing to the LDE effects captured in the device model.

As a first experiment, we consider an example using an n-channel device with $L = 100$ nm and $g_m/I_D = 15$ S/A (moderate inversion). Using the lookup table data, we find $J_D = 9.01$ μA/μm. Now we sweep W in Spectre and compare the expected g_m/I_D with the true value seen in the simulation. We quantify the observed error using:

$$Error(\%) = 100 \cdot \frac{(g_m/I_D)_{Spectre} - (g_m/I_D)_{Lookup}}{(g_m/I_D)_{Spectre}}. \tag{C.3.2}$$

Similarly, we compare the g_m/g_{ds} and g_m/C_{gg} data extracted from the simulations to the expected values from the lookup table. The results are plotted in Figure A.3.4(a). Not surprisingly, the maximum errors occur for very small widths of less than 1 μm. In this region, the errors in g_m/I_D and g_m/g_{ds} are of the order of 3...5%. The error in g_m/C_{gg} grows up to 40% for a minimum width device ($W = 0.2$ μm). This is clearly because the fringe and sidewall effects become more significant for such a small device. Since it is quite uncommon to find such small widths in a practical analog circuit, this error is not a significant concern, but only confirms that the device model appears to be capturing narrow width effects in a proper way.

Figure A.3.4(b) zooms into the more practical region with $W > 2$ μm, which means the device has at least one full finger that is more than 2 μm wide. For this case, we see that all errors are within about 1.5%, with the largest error still appearing in g_m/C_{gg} around $W = 3$ μm.

As a next step, we expand the above experiment to cover additional channel lengths and inversion levels. Figure A.3.5 quantifies the maximum error magnitudes seen in g_m/I_D across five different channel lengths (60, 100, 200, 500 and 1000 nm) and five inversion levels, all for $W > 2$ μm. The first four inversion levels correspond to $g_m/I_D = 5$, 10, 15 and 20 S/A. The fifth value is chosen to lie in the weak inversion plateau, specified by a current density of 1 nA/μm (see Figure 3.11). This

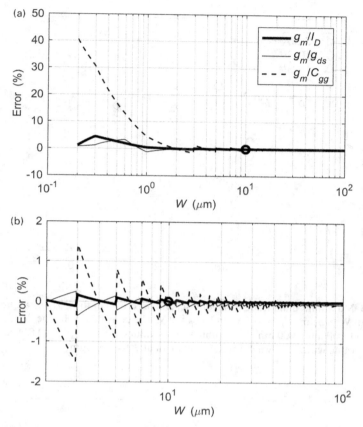

Figure A.3.4 Layout dependent errors in the parameter ratio of an n-channel device with $L = 100$ nm, $g_m/I_D = 15$ S/A, $V_{DS} = 0.6$ V, $V_{SB} = 0$ V. (a) Full sweep range with $W = 0.2...100$ µm. (b) Zoom into the range where $W > 2$ µm. The circle marks the g_m/I_D value at the reference width that was used to generate the lookup tables.

leads to nominal g_m/I_D values of 26.1, 29.8, 31.2, 30.7 and 30.4 S/A for the chosen channel lengths. We observe that the maximum error in g_m/I_D is of the order of 1%. This is acceptable for circuit sizing, especially since such peak errors occur only for relatively small W (see Figure A.3.4)

Figure A.3.6 and Figure A.3.7 quantify the observed errors in g_m/C_{gg} and g_m/g_{gs}. We can see from these plots that the peak errors in the angular transit frequency g_m/C_{gg} are of the order of 2%. This is again acceptable, knowing that the values will be significantly smaller for reasonably large transistors. The peak errors in the intrinsic gain (g_m/g_{ds}) are much larger, reaching up to 10–20% for long channels deep in weak inversion. This indicates that the foundry expects a much larger finger width dependency for this parameter ratio at small channel widths (where the peak errors occur). While the physical reasons for this are not transparent, one can argue that the errors are still insignificant in practice. The intrinsic gain is in any case not a

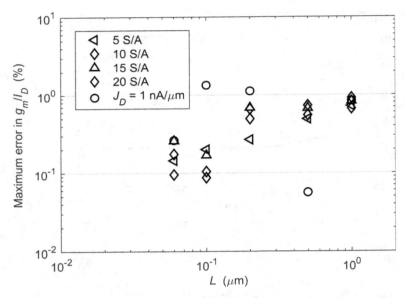

Figure A.3.5 Maximum magnitude of layout dependent errors in the g_m/I_D of an n-channel device with W > 2 μm, $V_{DS} = 0.6$ V, and $V_{SB} = 0$ V. The considered channel lengths are 60, 100, 200, 500 and 1000 nm. The nominal g_m/I_D values are: 5, 10, 15, 20 S/A and an additional value within the weak inversion plateau at $J_D = 1$ nA/μm.

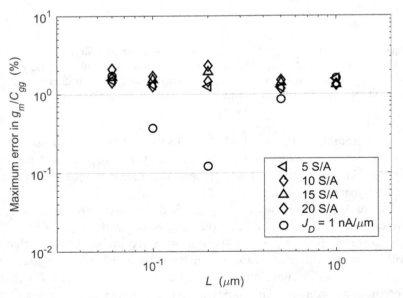

Figure A.3.6 Maximum magnitude of layout dependent errors in the g_m/C_{gg} of an n-channel device with W > 2 μm, $V_{DS} = 0.6$ V, and $V_{SB} = 0$ V. The considered channel lengths are 60, 100, 200, 500 and 1000 nm. The nominal g_m/I_D values are: 5, 10, 15, 20 S/A and an additional value within the weak inversion plateau at $J_D = 1$ nA/μm.

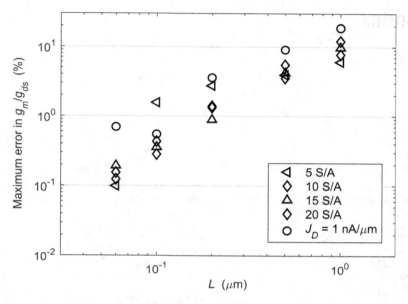

Figure A.3.7 Maximum magnitude of layout dependent errors in the g_m/g_{ds} of an n-channel device with W > 2 μm, $V_{DS} = 0.6$ V, and $V_{SB} = 0$ V. The considered channel lengths are 60, 100, 200, 500 and 1000 nm. The nominal g_m/I_D values are: 5, 10, 15, 20 S/A and an additional value within the weak inversion plateau at $J_D = 1$ nA/μm.

"precision parameter" and it often must be overdesigned to guard against modeling uncertainty. In addition, feedback is typically used to desensitize a circuit against variations in intrinsic device gain.

A.3.4 References

[1] T. Chan Caruosone, D. A. Johns, and K. W. Martin, *Analog Integrated Circuit Design*, 2nd ed. Wiley, 2011.
[2] X.-W. Lin and V. Moroz, "Layout Proximity Effects and Modeling Alternatives for IC Designs," *IEEE Des. Test Comput.*, vol. 27, no. 2, pp. 18–25, Mar. 2010.
[3] E. Morifuji, H. S. Momose, T. Ohguro, T. Yoshitomi, H. Kimijima, F. Matsuoka, M. Kinugawa, Y. Katsumata, and H. Iwai, "Future Perspective and Scaling Down Roadmap for RF CMOS," in *Dig. Symposium on VLSI Circuits*, 1999, pp. 165–166.

Index

Printed in the United States
by Baker & Taylor Publisher Services